Homeowner's

COMPLETE OUTDOOR BUILDING BOOK

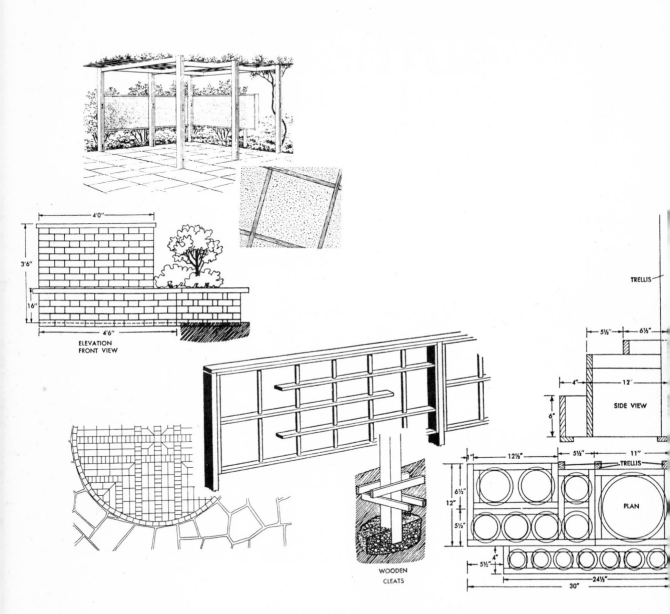

4'0"

3'6"

16"

4'6"

ELEVATION
FRONT VIEW

WOODEN
CLEATS

TRELLIS

5½"

6½"

4"

12'

SIDE VIEW

6"

½"

12½"

5½"

11"

TRELLIS

6½"

12"

5½"

PLAN

5½"

4"

24½"

30"

A Popular Science Book

Homeowner's
COMPLETE OUTDOOR BUILDING BOOK

By JOHN BURTON BRIMER

Illustrations by the Author

Popular Science Publishing Company • Harper & Row
New York • London

*To the memory of my father
and to my mother and brother,
the three original do-it-yourselfers
in my life.*

Contents

Introduction

Part I—WOODWORKING

1. What Are the Most Needed Tools in Woodworking?   10
2. Power Tools   19
3. Choosing Lumber   31
4. Wood Preservatives and How to Use Them   41
5. Little Projects   45
6. The Trellis Story   69
7. Good Fences Make Good Neighbors   91
8. Every Garden Needs a Shelter   152
9. Build Your Own Outdoor Furniture   194

Part II—MASONRY

10. Tools for Masonry   222
11. Materials for Masonry   228
12. How to Lay Masonry   237
13. Choosing the Proper Joint   266
14. How to Make and Use Concrete   273
15. Building Garden Steps   306
16. Planning a Terrace   327
17. Paving Walks and Terraces   348
18. Cast Your Own Paving Blocks   367
19. How to Make Pebble Mosaic Paving   376
20. Driveways for Today and Tomorrow   383
21. Decorative Planters   393
22. The Pool in the Garden   402
23. Versatile Walls   425
    Index   449

Homeowner's

COMPLETE OUTDOOR BUILDING BOOK

Introduction

What will this book do for you? How much will you get out of it? The answers to these questions will depend upon your use of it, of course, but will also be influenced by your need for this book.

If you are already a home craftsman, you will find that the designs in the book will help you to make your home and garden a better place in which to live. There is no question of that. But you needn't follow the designs slavishly, for they have been planned so that you may improvise on them as you wish, in order to find better solutions to your own problems. Those solutions will be original, even though they will be based on this book's sketches and plans; and by their originality they will become even more satisfying to you, as you discover how truly creative you can be once you have made the start.

If you are not yet a home craftsman, you will find that many of the plans and designs in this book will help you to become one by the simplest, easiest, and quickest method possible: by *building*. The projects have been planned, with as much forethought as we could summon, to give the utmost effect with the minimum effort. Wherever we could, we

1

have used stock sizes as our modular basis, thus eliminating the labor of cutting. Only in the more advanced projects will you find any quantity of cutting and fitting required.

But if you are not a home craftsman at all, and don't intend to become one, this book can still be of help to you. Certain men and women may have found that they are not fitted by nature to work efficiently with tools; or perhaps age or physical disability may have ruled out the lifting, hammering, sawing, and other physical effort necessary for many of the projects outlined in this book. These people will have to hire the work done, and for them this book will serve as a source from which to select the designs for the fences, trellises, walls, steps, shelters, and terraces they wish to use for the improvement of their home grounds and for extending their outdoor living pleasures.

Thus, when they are talking to their contractors or builders, they will be assured good, simple designs which will enhance their homes and they won't have to settle for tasteless "carpenter's designs," which often are so commonplace that the owners will wonder, when the job is done, how they ever got talked into accepting the pattern used. These people, too, may be moved to improvise on the designs a bit. If so, there will be a good solid basis from which to work, rather than verbal plans outlined in the air—the usual reason for the final lack of satisfaction in the work produced.

This brings us to the reason we have written this book and spent several years in creating the designs and drawing the practical plans from which you can build:

With the great surge of "do-it-yourself" projects in recent years, hundreds of thousands of people have been fired to build things, finding, as they work, those solid satisfactions which are in manual tasks. This is quite aside from the substantial saving in money which would be spent for labor, which is another satisfaction. Under the skins of most of us lurks a pioneer who either "did-it-himself" or found that it didn't get done at all. In our work today—be it factory work, office work, or the humdrum tasks of the housewife—we are finding fewer of the satisfactions which workers used to encounter before this age of specialization, which has made most of us small cogs in the great machine. There is good therapy in working with one's hands. Particular satisfaction will be found in the act of creation when the object created proves useful and beautiful in one's home.

GIMMICKS OR GOOD TASTE?

But there are all-too-apparent dangers in this great spate of do-it-yourself projects. So many of the designs offered to the home craftsman today for the adornment of his home and garden are in such poor taste that, although his property may be "embellished," it is questionable whether or not it has been "improved." As one disgusted designer recently put it, "These things are so 'gimmicked-up' that we might as well be back in the Victorian Gingerbread Age—might better be, in fact, because those designs were better than most of the junk today and they fitted in with the houses better, too." We couldn't agree more heartily, and that is the main reason we have worked out these designs and plans and written the text for this book. We feel that if people are going to put the time and the effort (not to mention the cost of materials) into projects of this sort, they are entitled to better designs from which to build.

We believe that you will find that most of the projects in this book will be simple and relatively easy to build, that they are in good taste, and that they will have a real function in the average home and garden. We have contributed a few more elaborate projects for the advanced workman so that he can demonstrate his skill, in addition to the aforementioned easy-to-do, simple ones for the beginner. Probably the bulk of the projects will fall between these two poles, so that the craftsman of moderate experience and growing skill will find they are within his grasp—well within it.

ARE THE DESIGNS MODERN OR TRADITIONAL?

We have tried to present a well-rounded group of designs which would fit with a variety of styles of homes of the sort we have observed during our travels about the country in recent years. We know that not all Americans live in the stark, boxlike modern homes at which so many of the designs for do-it-yourselfers would seem to be aimed. Nor do we suspect that everyone lives in Cape Cod cottages, or in homes of elaborate traditional design. Rather, we know that people live in a wide variety of houses built in many styles of architecture—some good, some bad, but all of them *homes*.

Therefore we have tried to give the widest possible choice of different patterns which would enable as many people as possible to find in the book something which would fit in with each particular style of home. Oftentimes, we might point out, a traditional home can be given a fresh look, a new piquancy, by adding a fence, a trellis, or a shelter with a modern flavor to the design. The starkness of modern homes, too, may be tempered by using some such adjunct, which will soften the severity of the architecture, rather than emphasize it. In the custom-designed house of today these two trends are well established, so that the amateur need have no qualms in following these suggestions.

We have tried to avoid the banal, the numerous clichés of design which we find so tiresome in both the modern and the traditional fields. Particularly in the area of traditional design (and occasionally in the modern mode) we find designs so cluttered and so "gimmicky" that they are obviously—at least to the trained observer—out of key with the homes for which they are intended. *Good design is always basically simple and architectural.* We have tried to hew to this rule so that our projects will integrate well with the average, simple, well-designed, small home of today in the cities and suburbs.

"BUT CAN I DO IT ?"

Even though you may never have laid a brick, cut a board, or even built a birdhouse, it is probable that you can develop into a very clever and competent craftsman. To anyone who is uncertain, we offer the assurance that we have seen dozens of office workers, or others who earned their livings by the fruit of their minds rather than by the products of their hands, become very competent craftsmen with a bit of practice. Anyone who has ordinary intelligence, anyone who also possesses a desire to work with his hands and in addition has a degree of patience so that he will allow those hands the time to learn to coördinate with his brain; in short, *anyone at all* can learn to build competently. The important thing is to make a *start* . . .

"WHERE SHALL I BEGIN ?"

If you have never done any woodworking before, we suggest that you choose one of the small, simple projects. Perhaps one of those under the heading of "Little Projects" would be a good starting point. If you are a novice at masonry work, then one of the smaller, simpler projects in this category would be a logical and intelligent choice. Choose one that is not so big that much advance practice will be required to assure its competent rendering. Otherwise, as you construct you may also be building a splendid inferiority complex for yourself which will result in your abandoning the project altogether.

But if you start in a small way, take your time so that you learn as you go, and pick up speed as your mind and muscles learn the rhythm of the work, you may very well develop into the best craftsman on the block.

Women, too, can build many of the projects in this book. More and more women today are learning that building is not solely the man's job. They are finding creative joys in woodworking, in bricklaying, in creating pebble mosaics, and in many other fields of craftsmanship. Wives are helping husbands in these activities, too, and in this coöperation finding deep, rich, and lasting satisfactions in working to make the family home the individual, original place it should be. Not since the pioneer woman assisted her husband as he felled the trees, adzed the logs, and sawed the lumber with which to build their home have American families enjoyed such family effort. And frequently today the children get into the act, too, and prove that they have an interest in the home.

WHAT THIS BOOK IS NOT

It may be that in setting down our aims and intentions we have given the impression that we feel this might be the book to end all books on craftsmanship. Let us hasten to disavow any such intention. We have tried to give as much of basic principle and practice as we could within the space at our command. Good craftsmanship will come to those who go on from here and learn more about their crafts if they wish to become experts. Probably the amateur who wants to engage in two or three of these projects will find that we have given him enough to guide him in building, whatever his choice may be.

To all those who read the entire text we wish to point out that in parts it may be repetitious. This is deliberate, not accidental. Some of our readers may have acquired this book for one specific purpose and will read only that part of the text which is pertinent to the subject in which they wish to engage. There is nothing more frustrating, we feel, than to be forced to stop in the middle of directions to hunt up a cross reference to some other page. Therefore we have kept cross references at a minimum, preferring to repeat where necessary, so that directions might be as complete as possible in each section.

We urge that all tool manuals which come with power and hand tools be preserved. Probably the best way to keep them where they will be available for use is to buy a school notebook of the loose-leaf type, punching holes in the manuals and other material and filing them in the notebook where they can be kept in easy reach of the workbench. Manuals may be removed for use, if desired; replaced when you have finished with them. These manuals should be carefully studied to learn what the manufacturer of your tools suggests for their care and maintenance. In addition, we suggest that the amateur will find in the advertising pages of home magazines many offers of booklets which can be obtained free or for moderate sums from manufacturers of building materials, paints, and so on. These will be valuable as reference material, too, if they are kept on the bookshelf above your workbench and are easily available. They can also be punched and filed in a loose-leaf notebook.

Perhaps the most valuable reference books of all to the amateur will be the mail order catalogs of Montgomery Ward & Co. or Sears Roebuck. They have quite a fine showing of building materials, tools of all kinds, and many other helpful bits of information for the amateur. For those who live in the country, it will be possible to have all the stock of a large hardware and building materials shop available from which to buy merely by writing an order and mailing it. For you who live closer to shops, it will be possible to choose your tools and materials and save yourself many hours of looking, for you will have a ready reference book to study at night at home. You will be fascinated to find large selections of every kind of building material, nearly every tool anyone could want; and with this help you will be able to do a better job of whatever building project you may choose to do.

A FINAL WORD

This book, then, is offered as a good starting point, a foundation on which to build your craftsmanship so that you can improve your home and garden with the designs it offers. To the beginning craftsman or to the already competent craftsman who has tasted the creative joys such work brings, the satisfactions which money cannot buy, it will bring an increase in the monetary value of the property as well as enhance the pleasures of outdoor living.

We trust that you will seriously consider our suggestion that changes may be made in our designs to fit them to your needs. By doing so you will find that your creative faculties will grow and your pleasure will be increased accordingly. Of course, if you find that the designs as they are seen in this book fit your needs, use them as they are, secure in the knowledge that yours is not a commonplace choice, a stereotype which you will find on every other block in your town.

But however you use this book—as a beginner, as an already-started amateur craftsman, or as a source book with which to guide your contractor or carpenter—we sincerely hope that you will find as much enjoyment in its use as we have had in the several years of preparing it for publication.

J.B.B.

WOODWORKING

1

What Are the Most Needed Tools in Woodworking?

The most essential tools for the amateur woodworker are few in number, but they are indispensable. You will want to add others as the need arises and as your budget will permit, but these minimum requirements will be a good base on which to build. Since we, as human beings, are inclined to lose our senses a bit when we are confronted by the tempting displays of the hardware shops, a word of caution may be in order before we launch this discussion. That word is: go slowly in buying tools; be sure you know what you need; and make certain you get your money's worth when buying it.

By that we do not mean to have you seek out only bargains—although no one is averse to taking advantage of a true bargain when it is offered. Instead, what we mean is that *good* tools of honest value—because they are made of good steel and superior quality hardwood and other high quality materials—*good* tools cost a bit more than tools whose makers have skimped on quality in order to produce at a low price. But in the long view, tools of good quality will outlast any others because they *are* made of fine materials, so that in the end they become the most

10

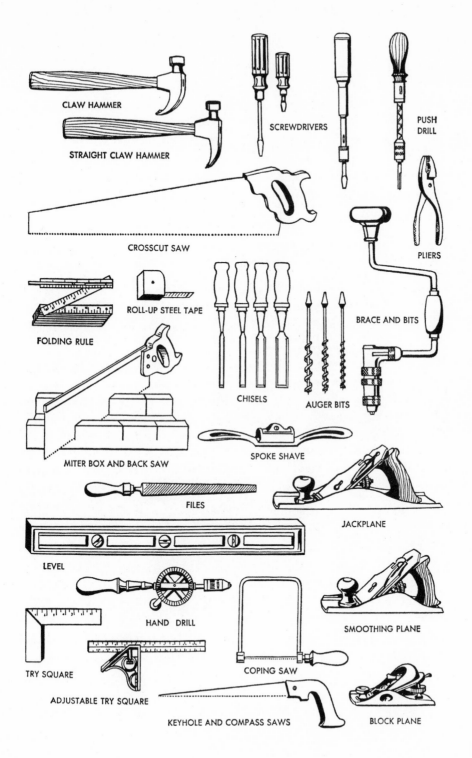

CLAW HAMMER

STRAIGHT CLAW HAMMER

SCREWDRIVERS

PUSH DRILL

CROSSCUT SAW

PLIERS

FOLDING RULE

ROLL-UP STEEL TAPE

CHISELS

AUGER BITS

BRACE AND BITS

MITER BOX AND BACK SAW

SPOKE SHAVE

FILES

JACKPLANE

LEVEL

HAND DRILL

SMOOTHING PLANE

TRY SQUARE

ADJUSTABLE TRY SQUARE

COPING SAW

KEYHOLE AND COMPASS SAWS

BLOCK PLANE

11

economical. Certainly the basic or minimum list of tools should include the very best tools you can afford—or a little better.

HAMMER AND SAW

Perhaps the most natural beginning will be a hammer and a saw, for many of the projects outlined and detailed in this book can be done with only these two indispensable tools. A good hammer is to be prized— one that is well balanced, has a good steel head neither too heavy nor too light, one with a handle which can be easily grasped by the hand of the home craftsman. Try out several in the shop where you trade, if it has a sufficiently large stock from which to choose, selecting the one which fits *your* hand and is right for your purpose. Test the all-steel hammers which have a molded rubber or plastic grip to see if that type pleases you more than the conventional wooden-handled types. If you are going to do a good bit of hammering, however, be warned that they lack the spring of the wooden-handled kinds and are more tiring. On the other hand, there is no danger of breaking the handle or of throwing off the head, which sometimes happens with wooden-handled hammers.

You'll probably find the 10- to 16-ounce curved claw hammer the most useful, the curving claws being well adapted to extracting bent nails. (It is also known as a nailing hammer.) If you need a second hammer, consider the straight-claw type, also called a ripping hammer, which is particularly useful for tearing apart woodwork and boards fastened together. Don't buy a heavy hammer. It will tire you and you'll regret it. On the other hand, one which is too light will use more of your energy in performing the tasks for which you use it, and it, too, will be fatiguing. A good medium-weight hammer is the best answer.

When it comes to saws, the choice may be bewildering if you don't know which saw is used for what operation. Each saw is designed to do a specific job. The crosscut saw (probably the first to buy) is made to cut across the grain of wood for small projects and where fine work is desired. The teeth of a good crosscut saw—and it is by the teeth you will distinguish a crosscut from a ripsaw—are bent alternately outward from the blade, each sharpened on one side only, but with alternates sharpened on opposite sides. Crosscut teeth are usually smaller than those of a ripsaw and are pointed, while those of a ripsaw are squared on the ends like tiny chisels. Crosscut teeth are usually set at a wider angle (i.e., extending outward more from the body of the blade). Probably the most useful size of crosscut saw will be in the 24- to 28-inch range, having 8 to 10 teeth to the inch.

A good saw is ground to taper toward the back from the widest thickness, in which the teeth are cut, making it easier to saw as the blade pushes back and forth in the saw cut. The stainless steel saws are said to be tougher, stay sharp longer than those of ordinary steel, and because they are non-rusting are easier to maintain. Laminated wooden handles are preferred to regular wooden handles because of their additional strength. The new plastic handles seem to be satisfactory. Handles should be firmly riveted or fastened to the saw blade.

Try out the handle of the saw you have selected to make sure that it is comfortable in your hand. If the grasp is too small for the hand, you will find that you'll acquire blisters very quickly. A good saw with a comfortable handle will speed your work and be worth whatever you pay for it. A cheap one will not stay sharp very long and will therefore cost more for upkeep, and it probably won't have a properly designed handle or be balanced well. It will pay to consider well before investing in your saw.

SCREWDRIVERS

Now that the hammer and saw are taken care of, let us suggest that the minimum tool chest will also contain a screwdriver or two, a try-square, a smoothing plane, a chisel (several sizes if you can manage it), a folding rule, a steel tape rule, and a pair of pliers. Probably a brace and bit set of good quality might be added, together with a level 24 to 28 inches in length, and a ripsaw if you plan to do any great amount of sawing *with* the grain of boards.

Screwdrivers come in a great variety of sizes, from about 3 inches long to 18 inches or more, with the width of the blade varying considerably, too. For general use, one about 8 to 10 inches long is best, along with a stubby one about 3 inches long having a medium width blade, which is useful where long-handled regular screwdrivers will not have room to operate. Where many screws must be set, a spiral ratchet screwdriver is a good investment. It will speed the work tremendously. Most of these have a chuck which permits the use of interchangeable blades of various sizes to fit different kinds of screws without injuring the slots and without marring the surrounding wood. Also, it is possible to buy replacements if a blade becomes damaged or is lost. A reversible ratchet makes it possible to remove screws quickly, too, when this is necessary. With this type of screwdriver, once the screw is started, only one hand is needed for setting it.

With the standard type of screwdriver either wood or plastic handles are satisfactory, so long as one doesn't use a hammer to pound them. Handles thus treated will splinter or shatter.

TRY-SQUARES

The try-square has a thin steel blade set into a wooden or metal handle, the blade being marked off in fractions of an inch and usually 6 inches or more in length. Handle and blade are set at right angles (90° exactly) so that the try-square may be used for marking either across lumber or on the edge for true cuts. It may also be used to test and true up cuts and for trueness when boards are planed smooth. An adjustable try-square, although more expensive than the fixed-blade types, will prove a useful tool, for it has a 45° angle as well as a right angle. It has a movable blade which may be adjusted at various lengths for use as a marking gauge, for measuring the depth of a hole, for scaling an odd-shaped piece of wood, and for many other uses in addition to the conventional one. Some varieties of adjustable try-squares have spirit bubbles placed in the stock, making them usable as levels in small quarters.

PLANES

A smoothing plane or a jack plane is also a necessity where any finishing work is to be done. Except in the top-quality woods, most boards will need considerable finishing work to smooth them and true them up so that they will take paint well and not harbor moisture, which will foster decay in the little roughnesses, destroying the wood. Smoothing planes are small, about 6 to 10 inches long, with blades from $1\frac{1}{4}$ to about $2\frac{3}{8}$ inches wide. We recommend the 9-inch size as the most useful. If a larger one is needed for preliminary smoothing, a jack plane—$11\frac{1}{2}$ to 14 inches—or a fore plane 18 inches long will be desirable. Some planes come with the bottom corrugated, rather than smooth as conventional planes are, the theory being that this causes less friction and allows the planes to operate with less effort.

A block plane is the smallest kind of plane, being only about 4 to 8 inches in length, with a blade 1 to 1⅝ inches wide. It is used for smoothing the grain across the ends of boards and for minor smoothing on small pieces of wood, or for small work in general.

All planes should be made of the finest quality of steel, and should have easy adjustment screws and good blades. Handles should be made of hardwood and well finished.

CHISELS

If you can afford only one chisel, buy a ¾-inch size, which is of medium width and adapts to a number of jobs. Other chisels come in a variety of widths, from ⅛ inch up to about 2 inches depending upon their type. Tang chisels (the tang, or top end of the chisel, is driven into the handle) are made for hand use and light tapping with mallets. Socket chisels are of heavier construction, the blade ending in a socket fitted with a replaceable wooden handle which may be hammered with a heavy mallet. A butt chisel has a one-piece blade and shank which extends through the plastic or wooden handle and may be tapped with a steel hammer. Other chisels are made entirely of steel.

Gouges are used for cutting grooves and for finishing edges or paring them down. This kind of chisel has a curved cutting edge to the blade.

For all chisels, forged, heat-treated steel is recommended, vanadium steel being the best, and (naturally) most expensive.

RULES

Because accurate measurements are necessary if the craftsman is to do good work, a 6-foot folding wood rule (plastic or metal may be chosen instead) is of primary importance. Our own preference is for the type of wood rule with a 6-inch brass extension rod or bar in one end which may be extended to take measurements inside shelves and other crannies. The roll-up steel tapes 6 feet in length or longer (tapes up to 12 feet are available) are also useful, the new square-bottomed types allowing for measurements inside nooks by adding the measure of the case. Long steel tapes or white-finished steel tapes of 50 or 100 feet are also useful if fences or other large projects are to be laid out which, unless measured in a straight line, may contain discrepancies. Fabric tapes are also found in these long tapes but are less desirable, the fabric sometimes stretching a little and also wearing out in time, whereas steel tapes are permanent.

PLIERS

A pair of pliers of a size between 5 and 10 inches plus two or three other kinds will always be useful. Those made of forged alloy or carbon steel with a cutter between the adjustable jaws are most versatile, adding wire cutting to their virtues.

BRACE AND BIT

If you are going to make any of a number of projects, such as a dowel fence, a dowel trellis, a bird house, and so on, a good brace and bit will be needed. The brace should have a good center grip of wood or plastic and a hardwood or plastic head, both of which turn freely and with ease. The chuck should open or loosen easily, too, a ball-bearing chuck being preferred. The brace should have a ratchet to permit easy and quick reversal of direction, for the removal of the bit from the hole and also for quarter turns in close, constricted areas. This ratchet probably will be enclosed to protect it from fouling with dirt and rust. The chuck should take all sizes of square-shank bits; good ones will also accept ⅛- to ½-inch round-shank bits, although this is not an absolute necessity for most projects.

LEVELS

Carpenters' levels are made either of wood or of aluminum alloy. Both are good in whatever size you choose. They come in 18- to about 48-inch sizes but the 24- to 28-inch sizes are best for most carpentry work. If you can squeeze out of your budget a 9- to 12-inch torpedo level, you will find it handy for leveling shelves and other narrow or constricted bits of work. The larger sizes made for professionals have as many as six spirit bubbles in glasses set into the body of the level, although the standard ones usually have but three. Sometimes you will find levels with the spirit bubbles adjustable to a 45° angle, which you may find an advantage in some projects. The main advantage in having the maximum number of bubbles is that, whichever way you may pick up the level, it is ready for use without having to be turned over as will be the case with those having only one to three bubbles.

RIPSAW

A ripsaw may be placed very near the top of your priority list instead of at the end, as we have placed it, if you are planning a good deal of sawing *with* the grain of the wood. It is, as its name implies, a saw which rips quickly through wood. It is not recommended for sawing plywood, plasterboard, or hardboard because its chisel-like teeth will tear or splinter them rather than make the good smooth cut possible with a crosscut saw. But for quickly ripping the length of other boards it is unexcelled. Keep it sharp for best results.

OTHER TOOLS FOR ENTHUSIASTS

For those readers planning to make woodworking their main hobby, we commend the theory that the right tool at hand when needed will shorten and make easier any job. For them we present the following suggestions, knowing that some of the tools may become necessities, depending upon the kind of work planned.

A good miter box of metal with a saw carriage or guide rigid enough to guarantee accurate cutting is a good investment. For those not so loaded with ambition—or money—a hardwood miter box slotted with 45°- and 90°-angle cuts will suffice. A good miter box saw (frequently sold as a "back saw") will be necessary for use with either miter box. Keep its fine crosscut teeth sharp for accurate cutting.

A spoke shave is helpful for planing down convex or concave surfaces of wood edges, while a drawknife will rough out and quickly cut to approximate shape all manner of curved and uneven shapes. It is operated by grasping both handles and drawing it toward you across the wood, which should be firmly clamped in the vise. A keyhole saw or a compass saw will cut wood on a curved line, too, the point of the saw being inserted in a hole bored alongside the line of the cut. Hack saws for cutting metal and coping saws for jig saw work are also useful appurtenances in the home workshop. A #2 or #1 half-hatchet is good for certain chopping work. (Hatchets have a nail-driving head, while axes do not.)

A set of files of various sizes and shapes and a wood rasp will prove of value in the home workshop. Certain tools are best sharpened by filing, for power grinding may heat the steel so much in some cases as to destroy its true temper. Filing will prevent such damage to steel. A wood rasp has coarse teeth and is used for rough cutting and finishing of edges and for rough smoothing on other surfaces.

A hand drill, a push drill, or both, will lend themselves to many jobs. The first has a small toothed wheel turned by a handle and has a chuck which accepts small bits and twist drills. The push drill also has a chuck which admits small drills of a type specially made for it, usually coming with the drill and being stored in the handle. It will be useful for making starting holes for screws or for large bits. A countersink bit, which may be used in either a hand drill or a large brace, permits the countersinking of screws to the level of the wood surface or just below it without marring the wood around them.

FINAL WORDS OF ADVICE

May we advise, in general, against the purchase of multi-purpose tools, such as saws with demountable blades, screwdrivers with hammer head and other fittings which can be stored in the handle, and so on. The *best* tool is one which is designed specifically for its job. Multi-purpose tools, however interesting they may be as curiosities, can never do *any one* of the jobs as well as one *made* for the purpose. The best home workshop has no room for gadgets unless they earn their keep.

2

Power Tools

The power tools you may need will depend upon three factors: *what kind of work* you expect to do—both now and later on—*how much work* is to be done; and the final, governing consideration, what *your budget* will permit.

Certainly power tools will provide you with the means of doing with ease a great many jobs which, if hand done, would require many man-hours and back-breaking labor. On the other hand, if you are building only a short run of fence, a small trellis, or a few oddments here and there, and have no plans for future woodworking projects outdoors or indoors, it would be false economy to invest in much power equipment.

Whatever your final decision may be, we urge that you give the most careful consideration to the matter. If you decide to buy power tools, don't rush into their purchase, but do a bit of shopping and looking to be sure that you buy the right kind of tool which will serve you well for years to come. Don't end up with a shopful of gadgets with little use and no real wear in them, merely because you leaped before you looked. *Good* power tools will do many different tasks: sawing, planing, drilling, jointing, making mortises, dovetailing, and other kinds of fitted joints, and in general provide the opportunity of turning out work of professional quality, if they are operated efficiently and properly so as to take full advantage of their many mechanical attributes.

PORTABLE POWER TOOLS

While most power tools are more or less portable in that they may be moved on occasion, the truly portable tools for home use are the small ones which are operated by holding them in the hand, tools which may be carried easily to the location of the job to be done. Precisely because of their light weight, however, it must be remembered that there is a limit to their capacity. They must not be overloaded or overtaxed by your trying to make them do more than they were intended to do. Improper use means *trouble*—burned-out motors or other expensive and annoying inconveniences resulting from lack of consideration for the machine's limitations.

Electric drill.   Because of its versatility, an electric hand drill is probably the first tool you'll want to consider. They range in price from about $15 to $75 or more, coming in various sizes with different types of handles, many accessories, and auxiliary fittings to extend their usefulness. These latter may be purchased separately.

When he begins shopping for a drill, the amateur may be confused by some of the terminology and by the profusion of choices offered. The speed, the power output, and the type of chuck are the three major considerations, but to simplify things let us state that any of the medium- or lower-priced drills made expressly for the home craftsman will perform *well* any of their assigned tasks. The speeds and power outputs which are found in the average ¼-inch drill will be sufficient for any legitimate accessory or auxiliary equipment used with such a drill. There are, however, variable-speed drills now available for home use which are geared for both heavy and light work. This may be a factor to consider if you think you may need a drill for heavy work later on.

As to chucks—the clamping jaws holding the drills—the geared chuck is the best and most trustworthy type, as well as the easiest to use. Other types which require hand tightening or the use of small L-shaped hexagonal steel wrenches to tighten the chuck are satisfactory; but the geared chuck which uses a geared key to tighten the chuck, or to open it to remove or admit drills and other equipment, is the easiest and most efficient kind to use.

Choose your drill by hefting several kinds to see how they fit your hand, how they balance when held in position for use. Possibly you will prefer the type with a pistol grip handle, but try also the D-handled kind to see if your hand may not feel more comfortable in the enclosed handle. Be sure that the drill is *not too heavy* for you to operate easily without unduly fatiguing yourself. On the other hand the very small, very light-weight drills may not be sufficiently sturdy or heavy to perform some of the tasks you may have in mind for them later on. Medium-weight drills, which are comparatively light, would seem to be the best answer for the majority of users.

Drill attachments.   When you are shopping for a drill, consider the many auxiliary kits and appliances which may be purchased separately, or which may even be included in a package price with certain makes of drills. These attachments allow the drill to be used as a sander, polisher, buffer, grinder, circular saw, hedge clipper, jigsaw or saber saw, planer, drill press, and for performing many different tasks. In general, these extra uses will work out satisfactorily if, as noted above, they do not overtax the motor. While it will enable the home craftsman to do these jobs, if he acquires the various attachments, the power hand drill is in no sense a complete substitute for the proper equipment designed specifically to do a *particular* job and no other, nor is it good for prolonged strains during extensive use. Power tools do best the job for which they are designed. In this instance, power drills are made primarily for drilling holes of sizes up to the limit which the chuck will admit (usually ¼ inch). Any other work done by attachments must be considered *extra* functions to be performed as well as their capacities and your skill will permit, and nothing should be demanded beyond the capabilities of such a tool.

Portable sander.   Another great boon to the woodworker is the powered portable sander. Two major types are offered. One is a belt sander having a continuous belt of sandpaper driven horizontally, the upper part of the belt riding within the housing of the tool; the lower just below the bottom of the housing. The other type operates by means of short strokes of a flat piece of sandpaper clamped on a padded bottom plate of the machine. Both have their advantages. The continuous belt type is excellent for quick, rough sanding and for most finishing when an extremely fine grade of sandpaper is used, but some craftsmen find this a more tiring tool to use than the flat-bed type. This latter may also be used with good effect for either heavy duty or fine finishing, depending upon the grade of sandpaper used, the weight of the machine, and also the efficiency of the particular model.

Heavyweight stationary sanders which are attachable to benches or power tool stands are also available, but since they are not needed for the work shown in this book they will not be discussed here.

Circular saw. For cutting plywood and other large sheets of board, sawing 2-inch stock for rafters and joists, as well as for many other uses, a ½- to 1½-horsepower portable saw is most convenient. It can be used to build fences or for other outdoor work right on the spot, preventing the laborious toting back and forth necessary if a stationary bench saw is used, and saving endless muscle strain and man-hours over manually operated handsaws. The circular blades are similar to those used on bench saws (see descriptions of all types of blades under Bench Saws a few pages along in this chapter). In the better type of portable saw, the blade is adjustable to various depths and angles of cutting. A good saw of this kind has a sizable base plate, a built-in adjustable ripping gauge, a sawdust blower, and a blade guard. Both motor and saw arbor are mounted with ball-bearings for easy operation. The saw blade should be set on the right-hand side of the handle, *away* from the operator, with all the adjustments and buttons on the rear of the saw housing so that the operator can see them easily and adjust them quickly. Handles are usually the D-type with a trigger switch enclosed in the handle where the fingers can operate it with ease. A portable circular saw can be converted into a bench saw by the purchase of a steel table made for the purpose, enabling it to do many of the things a bench saw will accomplish. In our opinion, however, it can never supplant the stationary bench saw, although it will be useful.

STATIONARY POWER TOOLS

Most of these tools are large and heavy, coming either with their own tables or ready to be mounted on a bench or support. Although more expensive in general than the portable hand tools, they are capable of astonishing feats, for they are versatile in their accomplishments. For the home craftsman who intends to make woodworking his hobby for many years, a selection of good-quality stationary power tools is a good investment. The amount and type of work intended to be done and the amount of money available for buying the equipment desired will influence the choice here, too.

Bench or table saw. A bench saw (sometimes known as a "table saw" or a "circular bench saw") will cost from about $20 to $25 and up, some of the more elaborate professional models costing in the hundreds. The least expensive kinds are, naturally, less versatile, less durable, and sometimes less accurate than the medium-priced and expensive saws. Also, the cost does not always include the price of motors, belts, stands, and other fittings required to make them ready to perform their tasks.

The higher-priced saws frequently feature built-in motors, and their stands or benches may be a part of the machine. Carefully milled fittings which adjust to a fraction of an inch will assure complete accuracy in the work they do. Many stands are fitted with casters which lock in place when not being used to move the machine into a more convenient position, so that they are stationary in one sense and mobile in another.

Good medium-priced saws have table tops of heavy steel and include a "fence" (a steel bar which can be clamped to the table top parallel to the saw blade, adjusted to the proper width, and thus become a guide for wood as it passes through the saw, so that accurate widths may be cut).

All saws should include a blade guard. Most saws now include a tilting mechanism which enables the saw to cut at any angle up to 45 degrees. Some remain fixed, with only the table top tilting, but this is less convenient and may even be dangerous, in our opinion. The alternative kind has a tilting arbor to angle the saw, with the table top remaining level.

The blades run in diameter from about 6 inches all the way up to 10 inches or more, the most usual sizes in saws for home craftsmen being found in the 7- to 9-inch range. The depth of cut depends upon the size of the blade and, to a certain degree, upon the make of saw, which will govern the angle of the cut with the tilting arbor. All good bench saws can be adjusted for depth of cut, being set for the various thicknesses of wood to be cut, or for the depth of cut when dadoes or other special cuts are being made. The straight depth of cut will vary with individual makes from 2 to about 3½ inches, depending upon the design of the saw, the size of blade, and the power of the motor.

One thing to check on before purchasing is the method of blade removal when replacement or changing to another type of tooth is necessary. This is very important because you may find that you'll change the blade a half-dozen times or more in the course of some projects in order to take advantage of your saw for ripping, crosscutting, dadoing, or whatever cuts are indicated. If the nut holding the blade is located in an open position, enabling you to use a wrench easily and quickly, you will be able to change blades in a trice. If, however, the housing restricts the wrench movement or does not permit insertion of the hand to use the wrench, it will take more time and many curses to complete the change.

Check also on the bore of the saw blade. We recommend a ⅝-inch bore, because replacement blades (unless made by the manufacturer of your saw) frequently come with ⅝-inch bores and small bushings with which to adapt them to a ½-inch shaft. You will find that the bushing pushes out during installation, gets lost in sawdust on the floor, and is generally a nuisance. By purchasing a ⅝-inch bore in the beginning you can avoid all this. If you already have a ½-inch type, wire or scotch tape the bushing to the blade each time you remove it to prevent its being lost.

Kinds of blades.   There are different blades for different kinds of work just as there are in handsaws. Each is useful for its particular job. The crosscut blade is used for cutting *across the grain* of the wood, for mitering, and for finished work. Ripsaw blades are used for quick cutting *with the grain* of the wood. Combination blades which join the good qualities of both rip and crosscut teeth are good all-around blades to use for general work, although they will not cut as quickly as a ripsaw nor as finely as a crosscut blade. Flat-ground blades, usually found in saw blades of lower prices, are perfectly useful and desirable; but hollow-ground blades will give a much finer, smoother cut, although sawing may take a trifle longer.

SPECIAL BLADES: A plywood blade is made for cutting plywood with a minimum of splintering. Nail-cutting blades which rip through old lumber and slice through nails and other metal where encountered are distinctly useful special purpose blades.

Carbide-tipped saw blades are long-lasting and extremely tough, although more expensive than other blades. Usually, however, they more than pay for themselves by lasting better than several ordinary blades. Each tooth is faced with carbide which is set to extend a tiny fraction of an inch above the non-cutting steel part of the blade. These blades will cut through almost anything except masonry, stone, concrete, steel, or other hard metals. They are splendid for cutting plastics, Formica, laminates, plywood, wallboard, asphalt roofing or siding, Transite, asbestos, aluminum, and insulating boards. In addition, they will make any kind of cut on wood—crosscut, rip, bevel, or miter.

One function of the bench saw, which it performs very well indeed, is the groove cut called a "dado." (A groove cut, when the term is used properly, describes a furrow cut *with the grain* of the wood; a dado cut on the other hand is a furrow cut *across the grain* of the wood.) A special cutter, called a "dado head," a "dado assembly," or sometimes just a "dado," is mounted on the saw shaft to replace the usual blade. It can be adjusted to the width of furrow desired, and either a groove cut or a dado cut can be made with it.

There are two types of dado heads. One is a single blade of heavy steel with an adjustable hub which can be set to tilt the blade and make it waggle as it spins. According to the tilt of the blade, the dado will be wider or narrower in the final cut. The other head is an assembly of several chipper blades set between two saw blades, the cut being made wider or narrower by inserting or taking out chippers. Some chippers are $\frac{1}{16}$ inch in width; others are heavier, varying with the individual make of dado head. In addition to making dado cuts across the grain, dado blades also may be used for making rabbet cuts, for grooving, and for certain kinds of joints, tenons, etc.

By using a molding-cutter head, a bench saw can be made to extend its services still further. An auxiliary head holds variously shaped sets of steel cutter bits which may be changed to produce moldings, cut tongues and grooves, beadings, flutings, coves, glue joints, and many other shapes. If you are interested in this feature, be sure to check whether or not the bench saw you are buying will permit the use of this head. It will be most useful in cabinet work and for fine trimming of all sorts, as well as for making the cuts listed above.

Jigsaw. A powered jigsaw is most useful where fretwork, scroll work, and circular cutting are to be done. It may be used as well for making straight cuts, for bevel cutting, for shaping the legs of furniture, etc. It is also useful for cutting tabletops, medallions for decoration, and arcs or double curves.

A good jigsaw has a "throat" size of 18 inches or more (the throat is the distance between the blade and the arm at the rear which supports the saw), although other sizes ranging from 12 to 24 or even 30 inches are made. However, the 18-inch size will permit circles of 36-inch diameter to be cut, which is usually large enough for most purposes.

Good jigsaws usually have a tilting tabletop or a tilting arm to permit bevel cutting, the tabletop being fitted with a clamp to hold the wood firmly against the top while it is being cut. The saw head assembly may also be designed to accept saber saw blades and saber files. The latter is used for finishing interior cuts and the former for making intricate internal or pierced cuts. Such a powered jigsaw will do work as intricate as that accomplished by a hand coping saw, with only a fraction of the effort required for hand sawing.

Power lathes. Useful for turning work such as balusters, table legs, spindles, etc., power lathes are not needed for any of the projects in this book and hence are not discussed.

Jointer. Despite its name, a jointer is not used for making joints but for planing and smoothing the edges of lumber and for cutting rabbets. Bevelling is also done by adjusting the jointer according to the maker's directions, shaping table legs, tapering round wood stock (or squared-off pieces), and doing many other things. It is called a jointer because it smooths wood so precisely that perfectly fitting joints can be made, with roughness of saw cuts smoothed off as well as or better than a jack plane or block plane would do the job by hand. Most of the uses detailed above will be found more in cabinetwork and indoor projects than in the projects shown in this book, but its use on anything which requires smoothing and beautiful, quick finishing will definitely cut down on labor.

Drill press. A drill press, too, is not absolutely essential for use with any of the projects in this book, but it would materially cut down on the labor of drilling holes in those projects which use dowels, such as fences, trellises, and so on. By setting up a jig, or frame, to hold the wood, holes can be drilled at any angle. The depth of drilling can be controlled, too, which will save much time and effort and mental stress. With the proper attachments a drill press is valuable also for sanding, shaping, mortising, and routing. It is very handy for use in metal work.

A portable hand drill may be converted into a drill press by purchasing a frame made for this purpose. It will then serve for performing the lighter tasks of a drill press, but it must be remembered that it should not be expected to do everything that a heavy-weight, stationary drill press can achieve.

Combination tools. There are many versatile tools on the market which make claims of being capable of performing a number of different operations. Particularly in the case of the higher-priced, better-made tools, these claims are justified. It is good common sense to consolidate as many functions as possible into *one* multi-purpose tool with *one* drive mechanism in order to conserve space in the home workshop and also to cut down the basic cost of the tool. A bench saw, lathe, sander, jointer, jigsaw, bandsaw, and so on, can be combined into a master tool with accessories to perform the various operations listed. Some types permit you to buy the basic tool and add to it as your budget can be stretched or the need for accessories arises. For the home craftsman who has limited shop space and the usual budget limits, one of the medium-to-better grade of tools, with whatever combination functions he feels suit his purposes best, is a good investment. But let us not advocate buying gadgets for gadgets' sake—too many craftsmen get carried away and overbuy, only to let the tools rust because there is really little use for such elaborate accessories.

A WORD OF CAUTION

Now that power tools have come into general use, the safety factor tends to be ignored. *Familiarity breeds carelessness,* as the annual statistics of accidents prove. Most accidents would be definitely preventable if only a little *forethought* and *care* were used. All the jokes about the do-it-yourselfer's mishaps are funny enough until they happen to YOU. Then the tragic side sinks in—but too late to save the finger, the thumb, the eye, the limb—perhaps even the life of the unfortunate but careless victim.

Any machine is a splendid assistant as long as the person operating it uses vigilance and intelligence in guiding it through its appointed task. No machine is safer than the limit of the care exercised by its operator.

14 POINTS FOR ACCIDENT PREVENTION

- Wear clothing which fits tightly so that it will not be snagged by moving parts.
- Roll up shirt sleeves or wear short-sleeved shirts. Sleeves of coats or jumpers should button tightly at the wrists.
- Don't wear neckties or other dangling clothing, which may come into contact with shafts and moving machinery parts.
- Use 3-wire extension cords with grounding plugs placed in grounding outlets when operating portable power tools. It is dangerous to use ordinary extension cords or to plug into lamp sockets or some outlets in walls, if wiring is not of sufficient strength to withstand amount of current demanded by the tool. Fires can result, overheated wires in walls are a common cause, or you may become part of the circuit yourself if the tool short-circuits, since you will be the grounding agent. Double-insulated power tools are a fairly recent development and if you are buying a tool, inquire about them. They are much more safe.
- Never turn on the power or put a plug into a socket without first checking the tool to make sure that the operating parts are clear, so that nothing will be snagged from the table and thrown, injuring you or fouling the machine. Always *turn off the power* before pulling out the plug when work is finished.
- Never reach across a *running* power tool. Train yourself to make this a good habit so that you'll avoid trouble before it occurs.

- Always use the safety guards on machines which have them. They are put on the tool to protect you, if you will only let them. (Example: saw guards on bench saws.)

- Never, *never* try to force wood into a bench saw which is taking it with difficulty. When it is necessary to guide wood close to a saw or cutting edge, use as long a scrap of wood as possible; *never use your fingers.*

- Keep your tools clean. Remove pitch from sawblades with turpentine or other solvent; keep sawdust off moving parts and sweep it from the table tops and from crevices. Caution: sawdust causes wear when it gets into working parts of the motor and machinery. Using a vacuum cleaner once in a while and keeping the motor covered when not in use will help to prevent this.

- Keep your tools oiled so that they will remain in good working order. Periodically go over all parts to make sure that bolts, clamps, and so on, are tightened and secure.

- Sweep the floor at the end of each day's work, or more frequently if the job is creating a good deal of débris. You won't track around so much sawdust in other parts of the house, and family relations will remain peaceful.

- A box, keg, or barrel kept near the power tool makes a good repository for scraps of lumber and will also hold sawdust sweepings. Keep scraps out from underfoot—a stumble can cause serious injury.

- Don't try to remove or pick up by hand any wood scraps from a bench saw table *while the saw is running.* Use a long, sturdy pusher stick to shove scraps off the table, or turn off the motor and wait until the saw ceases to turn before removing the scraps.

- Keep children, particularly small ones, out of the shop while you are working. They may get caught in moving machinery while they are exploring, or they may get underfoot and be injured when heavy pieces of material are being moved. Also, even very small children have an amazing adeptness at imitating adults and are quite likely to turn on switches, damaging your work or causing bad accidents to themselves and you. Better to exclude them from the shop than to face this possibility.

- Even in a home with no small children it is a good plan to install a locking switch box which also contains a circuit breaker. Feed all electricity through this to the power tools. The circuit breaker prevents overloading of the lines and short circuiting, with its attendant dangers. It also obviates the need for fuse replacement because as soon as the overloading cause is removed, the flip of a switch will restore the circuit. Throw the switch and lock the box after each use to prevent others from using the tools when you are not present. Another expedient for those who have no such switch box is to pull the plugs, remove extension cords, and put into place an inexpensive plug lock which fits into the wall plugs.

In conclusion, let us sum up the basis for caution in a few words: Your tools need your brain to guide them and to take necessary precautions to make them safe. Tools alone will never be safer than the operator's intelligence—not until someone invents a good $5.98 Electronic Brain to guide them safely. Until then it is up to you.

3

Choosing Lumber

Lumber is not cheap—in fact, it can be very expensive if the wrong sort is chosen for the job you will be doing. Read carefully the suggestions and the listings given herewith, come to know the various kinds and sizes of lumber and their grades, and your job's requirements, so that you may save yourself a good bit of money and in addition have a more satisfactory project. Look into the facts *before* you buy lumber; consider them as carefully as if you were buying a car. Well chosen, carefully cut, and properly installed, lumber will outlast several dozen cars, yet many people (often the same ones who try out every car in the field and buy a car with great care and consideration) will rush out and buy some lumber, *any* lumber, for a woodworking project.

Like cars, lumber has certain definite characteristics. It will perform more or less well in a general way, but usually there is one particular job for which each grade is outstandingly well fitted.

Several methods are used in grading lumber by the various suppliers, but all of them are similar to or approximate the pine grading listed opposite. The thing to remember in ordering lumber is that you need *not* use the *highest* quality or the best grades for *everything* that you build. If you are in doubt as to what kind or grade of lumber to use, talk over your project with your lumber dealer. Tell him what your plans are, how the lumber is to be used, and ask for his advice. Most lumber dealers are reputable and are sympathetic with the home craftsmen. If you have not come to him on a busy Saturday morning when all the other local amateur craftsmen are besieging the yard for materials to use over the weekend, your dealer will be able to advise you and will do so. If you find yourself in doubt as to what he advises, it is always possible to thank him for his advice and say that you will consider the matter. Then go to another dealer to see how the advice checks out, using your common sense to decide which is the better of the two suggestions. It is probable that they will be similar.

After a few projects you will probably be able to choose your own lumber. You will have evaluated the advice given so that you will know which lumber yard to use and which was attempting (possibly) to make a "fast buck" at your expense. The project and where you plan to build it will always govern the type of lumber you must choose. For most outdoor projects you will probably choose one of the common grades of softwood lumber. Properly painted and with the right preservatives applied they will last a long time. They are the cheapest grades of lumber and offer a reasonably wide choice of woods, among which you will surely find one which will fit your needs and give you good service.

SOFTWOODS

Those most frequently available in wide distribution are cedar, cypress, fir (Douglas fir is especially good), hemlock, larch, redwood, spruce, and pine (white and yellow pine are both available in most places).

Softwoods least likely to shrink, swell, warp, or otherwise prove unsatisfactory are white pine, cedar, cypress, redwood, and spruce. For structures needing strength in their frames, use larch, yellow pine, or Douglas fir. All woods, of course, should be properly seasoned or kiln dried.

The kinds most easily worked with in sawing, planing, and general shaping are cedar, spruce, white pine, and redwood. Those most decay resistant are cedar, cypress, redwood, and white pine. If the lumber is cut from the heart of the log where the cellular makeup is particularly dense, cedar, cypress, and redwood will be especially rot-resistant. Contrary to popular belief, not *all* redwood, cedar, or cypress lumber is resistant to decay; boards cut from just under the bark of these trees will decay nearly as soon as some of the woods more usually thought to be susceptible to rot.

PRESERVING WOOD

Modern science has stepped in, coming to the rescue of the home craftsman as well as the professional builder. There are now a number of wood preservatives which can be painted on lumber, or in which the cut ends of posts and other lumber may be soaked for full absorption. Decay-killing elements which are present in these fluids kill or hold at bay the bacteria and moisture organisms which may enter pores and cause the wood to rot. Posts which might otherwise last only five or six years now have their lives doubled and even tripled. Some preservatives have added ingredients which will nearly always make them completely resistant to water and moisture penetration. Since these preservative materials and methods of application are taken up more thoroughly in the following chapter, suffice it to say that they merit your consideration.

GRADES OF PINE LUMBER

#1 and #2 Clear: (In some sections this may be listed as B and Better or First and Second Quality.) These are the best grades of lumber and therefore most expensive. They should not be used for anything but the finest work outdoors and for the best kinds of indoor finished work.

C. Select: This wood may have some small imperfections such as tiny knots or blemishes. Usually one side is practically perfect. This grade may be used for better quality work, but it is usually too good for use outdoors except for furniture.

D. Select: The last of the better grades of wood; it can be used for furniture and other projects where a good finish is required. It contains more blemishes and knots than the C. Grade.

#1 Common: The best of the regular board grades. It usually has small knots which are not likely to dislodge, and a few minor blemishes. It is the best all-around wood for use where these defects are not a factor.

#2 Common: Frequently employed indoors for flooring and for knotty pine panels. It has more knots than #1 Common but should not have knotholes.

#3 Common: Small knotholes may appear in this grade, being open also to possible dislodgement of small knots under rough handling and working. This grade also will check a bit and more pitch may be present in it.

#4 Common: A low-cost grade which is useful for fences and many other outdoor uses. Bear in mind that it will have more knotholes and more knots than the above grades. However, by selecting good pieces from your lot of lumber and using them in the more exposed-to-view sections, and by using the blemished pieces for some of the less visible parts of your structures, you may save money and still have a perfectly satisfactory result.

#5 Common: This grade is so full of knots and imperfections that it should not be used for any project requiring strength, nor (unless you are in love with knots) used where the grain and beauty of wood should show. However, it can be used for many projects in many ways, particularly where a finish of some sort will be applied over it, saving you money and being perfectly satisfactory. Look it over before you buy it, however. You may decide to purchase the better grades instead.

GRADES OF OTHER WOODS

Other woods are graded similarly to the pine wood listing above. Because pine is usually cheapest, readily available everywhere, and because we feel that, properly treated with wood preservative, pine can last well and be perfectly satisfactory, we shall not go into the grading of other woods. Occasionally you may want to consider making a table-top or a bench out of oak wood or some other locally available hardwood. Here, where they can be treated to preserve the natural grain and beauty of the wood so that it will show and be appreciated, the use of better grades of hardwood will be well justified.

HOW TO SAVE MONEY ON LUMBER

Don't feel that you must always use the best grades of wood. Instead, save money and buy one of the lower grades which will be adequate for your project. This is particularly true where wood is to be painted with an opaque color, for paint covers many imperfections and if the knots are properly shellacked and pre-treated they will not bleed through the paint. As we suggested earlier, if you are inexperienced, ask your local lumber dealer to recommend the cheapest grade of wood which you can use satisfactorily on this particular project, showing him the plans and telling him the exposure of the project.

Another way of saving money is to buy a low grade of lumber and, by using your utmost ingenuity, utilize it to good advantage. By careful planning you can often use clear lumber on either side of a major knot and cut your pieces so that it is discarded. Also it is possible to make minor repairs in the wood itself, filling small imperfections with putty or with waterproof outdoor glue and sawdust mixed. Often, warped or bowed boards can be straightened by weighting them on the central parts, where the bowing or warping usually occurs. The method is this: Build up under the ends with cinder blocks, one or two set on top of each other, or use sawhorses for support. Place the board to be straightened on the support and weight the ends in place with a cinder block. In the middle or where the warping occurs, place a stone, a 2- or 4-inch cinder block, a pile of bricks, or any other manageable weight. Don't *over*weight the board, or it may crack. Don't try to rush things: straightening will take time and it will be a mistake to try to speed it up. Permanent results take time and patience. If you have a number of boards of the size of that which is warped, you may pile them on top of the warped one, add a bit of weight, and allow them to do the straightening.

Knots which are a little loose may loosen further as the wood seasons after it has been built into place outdoors, and fall out. To prevent this, examine all knots and press out any which are a little loose, taking care to do it so that the wood is not injured. Coat the inside of the knothole and the edge of the knot with waterproof outdoor glue; then push it back into place, removing any excess glue from the surface of the board before it hardens into place. When the board has been painted, this knot should remain in place as long as the wood itself.

Checking—a narrow crack occurring usually at the ends of boards due to shrinkage of the wood or excess exposure to sun and heat—can be a very serious matter, especially on outdoor projects where water may seep in. Exposure to hot sunlight may cause further shrinkage and extend the checking so that it may eventually split the board. If the checking is severe, it is best to cut off the board as far as is necessary to eliminate the checking. Then treat the ends with wood preservative before building it into your project. If checking is minor, filling the cracks with waterproof glue mixed with sawdust will usually suffice to prevent further damage. Work the mixture into the cracks well with a putty knife. Fence post checks should be thoroughly brushed with a wood preservative and, in addition, the below-ground parts should be well tarred to fill all cracks and prevent decay organisms from entering the wood. On the top of the posts, cap blocks an inch or more larger in all directions will cover the ends and prevent further damage.

MEASURING AND ORDERING LUMBER

Ordinarily, lumber is sold by the board foot. While it is not really necessary to know how many board feet you are ordering, the definition of a board foot may be useful information to possess, because lumber is priced according to the board feet it contains and your dealer will sometimes figure it this way in billing you.

A board foot is a piece of wood 1 inch thick, 12 inches wide, and 12 inches long. Naturally, not all lumber you will be purchasing will be 12 inches wide, but it will be a portion of a board foot and figured and priced accordingly. Let us take an example. Suppose you are using 2 x 6 lumber: each 12 inches of length will be a board foot. On 1 x 6 lumber each 2 feet of length will equal a board foot. Thus, a 2″ x 6″ x 12′ would equal 12 board feet, while a 1″ x 6″ x 12′ would equal only 6 board feet.

ACTUAL VERSUS NOMINAL SIZE

To the novice, perhaps the most puzzling aspect of woodworking and wood ordering is the question of "Actual Size" versus "Nominal Size."

In the accompanying chart you will find a comparison of these two sizes. You will use the Nominal Size in ordering and that is what you will be billed for. This is approximately $\frac{3}{8}$ inch greater than the actual measure of the lumber in all pieces below 8 inches in width, and $\frac{1}{2}$ inch more than the Actual Size of lumber over 8 inches in width. This discrepancy is due in part to the mill work in cutting lumber and in part to shrinkage as the natural sap moisture of the lumber dries out. Milling reduces the thickness of 1-inch wood to $\frac{25}{32}$ of an inch, while 2-inch wood becomes $1\frac{5}{8}$ inches thick after milling and seasoning. If you want lumber to be fully 1 inch, 2 inches, or whatever the measurement may be, you must specify it in your order and pay for it, because it will have to be milled to specifications. Usually it is much simpler to accept the facts of milling and seasoning, and to measure your wood and make your cuts *according to the Actual Size.*

NOMINAL SIZES	ACTUAL SIZES
1″ x 12″	$\frac{25}{32}$″ x $11\frac{1}{2}$″
1″ x 10″	$\frac{25}{32}$″ x $9\frac{1}{2}$″
1″ x 8″	$\frac{25}{32}$″ x $7\frac{1}{2}$″
1″ x 6″	$\frac{25}{32}$″ x $5\frac{5}{8}$″
1″ x 4″	$\frac{25}{32}$″ x $3\frac{5}{8}$″
1″ x 2″	$\frac{25}{32}$″ x $1\frac{5}{8}$″
2″ x 12″	$1\frac{5}{8}$″ x $11\frac{1}{2}$″
2″ x 10″	$1\frac{5}{8}$″ x $9\frac{1}{2}$″
2″ x 8″	$1\frac{5}{8}$″ x $7\frac{1}{2}$″
2″ x 6″	$1\frac{5}{8}$″ x $5\frac{5}{8}$″
2″ x 4″	$1\frac{5}{8}$″ x $3\frac{5}{8}$″
2″ x 3″	$1\frac{5}{8}$″ x $2\frac{5}{8}$″
2″ x 2″	$1\frac{5}{8}$″ x $1\frac{5}{8}$″
4″ x 6″	$3\frac{5}{8}$″ x $5\frac{5}{8}$″
4″ x 4″	$3\frac{5}{8}$″ x $3\frac{5}{8}$″
3″ x 6″	$2\frac{5}{8}$″ x $5\frac{5}{8}$″

NOTE: When you measure, use the try-square to true up, then draw the line. Recheck measures and use the try-square again before cutting, to avoid mistakes.

Lengths of wood usually run as listed, the Actual Length being the same as the Nominal Length. But you cannot always count on the mill's squaring up the ends, particularly in the lower-priced grades of wood, and you may have to square them up and cut them yourself, thus losing a half-inch or so. Therefore it is well to examine and measure your wood first, and then plan how best to utilize it, before rushing the project into production.

An exception to the rule regarding thicknesses of materials as cited above will be found in the "manufactured boards." In making plywood, hardboard, asbestos boards, Transite, insulating boards, and so on, the *listed* sizes become *Actual Sizes*. Thicknesses of ½ inch, ¾ inch, ⅝ inch, or whatever they may be, will measure approximately that. The length and width of the piece will also be accurate—commonly these boards are 4′ x 8′. In certain of the boards other sizes have been developed in modular dimensions which will tie in with the 4′ x 8′ sheets, and these sizes, too, are Actual Size.

PLYWOOD

One of the most versatile and useful of the modern developments in the woodworking field is plywood. Recent years have seen many innovations in this material so that now plywood can be purchased to do practically any job for which you may wish to use wood. Almost invariably when plywood is specified for projects in this book *outdoor* grades are required. Several manufacturers make the claim that their products are so waterproof that they may be boiled without the various layers of wood separating, so they are safe to use outdoors.

Plywood is available in many grades. The best grade should be good on both sides, with no knotholes or other mars in the surface, should be sanded smooth, and should have good, square, unmarred corners. The other grades, in descending order, permit some knots and minor splits in the plies; some are perfect on one side only, the other side having imperfections and being unuseable for finished work but reducing the cost over both-sides-perfect types; still other grades may have circular or diamond-shaped plugs set into the ply to repair knotholes or other blemishes, but are well sanded so that they may be used for painted work, concrete forms, and other jobs requiring good smooth surfaces where perfection of grain is not important. This grade may also possess small, tight knots. Properly shellacked and finished with paint, these will be no problem, either.

Some of the higher-grade interior plywood, in addition to the master panel size of 4′ x 8′, may be ordered in widths of 30, 36 and 42 inches, and in lengths of from 5 to 12 feet in even foot sizes. Special panels made for outdoor use on porches or terraces come in lengths of 14 to 24 feet. It must be pointed out, though, that not all of these may be stocked by your local dealer, and he may be well within his rights in refusing to special order them for you, unless you use at least two-thirds of the minimum "package" his wholesaler may force him to buy. However, if you have some special project in mind, check with your dealer as to which sizes he regularly stocks and the possibility of special orders. If your project is large enough to make it worth his while, he may be able to obtain what you wish, provided that you can wait for the length of time it will take to have the special order filled and delivered. You may find that it will be to your advantage to use the more readily available sizes and adapt them to your project.

USED LUMBER OR SALVAGED WOOD

House-wrecker's storage yards may be profitable places for you to explore, for they can save you money at times. Many fences, garden houses, shelters, and so on, can be built at a considerable reduction in cost by using salvaged wood. Old studs, joists, planks, and boards *which are in good condition* may be found in the wrecker's yards "filed" according to size. The thing to remember here is that key phrase "in good condition," for it is important to you. You must carefully look over the lumber where the dealer has stored it, usually having arranged it by its size and length. Most often, dealers will permit you to examine and select what you want. You may either take it away yourself, if you have a station wagon or if the pieces are small enough to be accommodated in your car, or have the dealer send them to you in his truck and pay for delivery. You will have to pull out remaining nails and be careful when sawing to avoid any which may have broken off and been left in the wood. It would be well to use the wood in its rough condition rather than to try planing it, because of the hazard of nails damaging the plane blade. If time is less of a factor than money, you can certainly save money this way.

Another way to save would be to purchase the wood in some building that is being wrecked and pull it down yourself, doing your own house-wrecking and nail-pulling. This is a strenuous occupation but it need not be dangerous if intelligent care is exercised; climbing, prying, and pulling down structures always demand the caution which common sense dictates. The author has pulled down a large shed with the help of his wife, and some assistance from two other men when large beams and other pieces had to be brought down. It was a large task to remove all the nails, but the wood was so well seasoned and in such good condition that is was easily re-used to make a fine, sturdy structure. Examine old buildings and wood to be sure that neither dry rot, termites, nor other insects are present, or the lumber will be a poor investment of time and money.

Another source of lumber usually overlooked is old crates and boxes. These can be utilized in numerous ways, depending upon the quality of the wood and its size. Retailers usually will give away the crates or sell them for a nominal sum. If they are carefully pulled apart the wood can be used in many different projects. When the wood is rough it can be planed or sanded smooth with a power sander, but sometimes the very roughness will be an asset in adding an attractive texture to a fence, even though it will not add to the ease of painting.

These foregoing facts about lumber do not exhaust the subject, but they will give the amateur a good groundwork on which to build further knowledge about woods, and we believe this is sufficient for a beginner.

4

Wood Preservatives
and How to Use Them

One of the most important aspects of working with wood—and one all too frequently neglected—is proper finishing. This is not a matter to be lightly brushed aside, for, unless properly finished, the wood and your handiwork will not last very long. Not only will you have the labor of doing the job over again, but you must pay a second time for the materials—and that is no small consideration in these days of increasing prices.

Painting will help to preserve the surface and therefore the interior cells of the wood you use, particularly if it is done properly, but paint alone will not achieve your goal. There are two main causes of deterioration of wood: decay and insect damage. Your problem is to outwit both of them, perhaps in one operation, as cheaply and easily as possible. Many woods, such as redwood, cypress, locust, and cedar, have natural qualities of resistance to bacteria and even to insects, but most woods of the less expensive grades (and some of the more expensive as well) are susceptible to these twin causes of disintegration. The naturally resistant woods do not break down so soon and permit rot to enter; they

41

resist penetration by termites and other insects, or are immune to them even when in contact with the soil or placed in areas of constant dampness. Other woods left untreated, however, will soon succumb in such situations.

Formerly, creosote was the only effective means of preserving wood. It was used extensively on fence posts, trellis posts, and on other wood which was to be in constant contact with the soil. It is still widely used, and you can buy posts already treated with creosote under pressure, the treatment forcing the preservative deeply into the pores of the wood. However, it is not easy for the amateur to deal with creosote. Brushed on, it is not very effective; and soaking achieves penetration best only by using hot creosote and immersing the wood in pressurized vats, something which only large industrial plants are equipped to do.

There are some disadvantages, too, in its use on wood structures in the garden by the amateur. The fumes and gases of creosote permeate the soil around the post and, unless the post is set in concrete to the level of the treatment, may injure or kill the roots of shrubbery or vines planted close to the post. Add to this the fact that it is nearly impossible to paint over creosote—unless it has dried into the wood for a long period—and you will probably turn, as have most other amateur craftsmen today, to one of the fine new chemical wood preservatives which you can apply easily. They are relatively cheap and have the additional virtue of permitting the wood to be painted as soon as they have thoroughly dried into it.

SOME MODERN PRESERVATIVES

Pentachlorophenol, often called Penta in trade language, is the name of the main ingredient of a widely used and very effective low-cost wood preservative. It comes ready to use, diluted in oil for easy spreading and quick penetration, whether it is brushed on the wood surface or used in a pail or large can for soaking the cut ends of posts or boards. It can be painted over with most outdoor paints when it has thoroughly dried.

Zinc and copper naphthenates are chemicals which are not only odorless, but they, too, can be painted over when dry. *Copper* naphthenate leaves a pale green stain in wood, something to consider if you are not using an opaque-finish paint. *Zinc* naphthenate preserves the natural color of the wood and may be used under clear varnishes, lacquers, shellac, and other colorless finishes. It is a little more expensive than copper naphthenate. Both preservatives may require a sealing coat of shellac or an opaque primer-sealer before final paint coats are applied to the wood.

Copper sulfate crystals dissolved in water can be used effectively for treating wood, particularly green, unseasoned kinds. Use 1½ pounds of the copper sulfate to 1 gallon of water, dissolving it in a watertight wooden barrel. Fill the barrel to a little above the depth to which the wood is to be buried and let the wood posts soak in it for a half-day or so to permit deep penetratration; then brush in the liquid well on the parts of the wood which are not being buried. It will leave a bluish stain, be warned, so that you must count on painting the wood rather than finishing it in its natural color.

All of the modern preservatives can be either brush-applied or used to soak the wood. Because the efficiency of the preservative is measured by the depth to which it penetrates, soaking is obviously the better method. You will, therefore, find that soaking cut ends (particularly those of fence and trellis posts) is advocated in the text of this book.

If you have a project such as a large lot of fence posts or a big structure to be treated, look into the possibility of buying your preservatives in quantity and treating the wood on a large scale.

A WORD OF CAUTION

Creosoted posts, particularly freshly treated ones, may poison the soil with gases and make the plants around them sick or even kill them. Modern preservatives are also poisonous to a degree, although quite safe for plant life once they have dried into the wood and been exposed to the air. They should not be stored where children or animals can reach them or where they may be upset on plants. Remember that they *are poisonous to human beings*. Be sure to wash your hands well after using them and to wear clothing which can be laundered. Occasionally one of the preservatives may affect sensitive skins. You may buy a pair of cotton gloves to wear while painting, and if your hands or skin feel irritated, wash immediately with warm water and a mild soap.

ARE PRESERVATIVES REALLY WORTH WHAT THEY COST?

Don't be misled by the dollar sign on the cost of the preservative. If you have a sizable job to do—the cost may mount up to a startling sum, it is true, but don't just discard wood preservatives because of their price. Think through the whole problem, and don't salve your conscience with the thought that you have also got rid of one more operation in building. Instead, figure up the cost of replacing the wood and estimate the hours of labor you expect to put into the building. Balance these costs against the added years of life the expenditure of the cost of wood preservative would assure you, and you will agree that it is worth while, we are sure.

According to reliable reports, even green wood posts when treated with copper sulfate will last up to ten years, with seasoned wood lasting from fifteen to twenty years. Untreated posts will last between half and two-thirds of this time, according to wood and to climatic conditions. Both softwood and hardwood can be treated with preservatives, the operation being the same. Although no reliable figures are available to us at this time, claims are made that some preservatives will double or triple the life of posts. Pressure-treated creosoted posts last from twenty-five to thirty-five years, again depending upon conditions of soil, etc. All treated posts last longer when set in concrete. Therefore we feel the case is good for using wood preservatives, particularly on posts.

WHY NOT JUST PAINT?

Although paint alone is better than *no* treatment with preservatives, paint films can easily be broken, will deteriorate and crack, allowing moisture to penetrate. Decay organisms can follow and work undetected beneath the paint for some time. Therefore paint alone is not an effective protection. It is imperative to use wood preservatives on *all* wood which is to come into contact with dampness, whether it is to be buried or not. Use it on all wood surfaces so that its chemical action will destroy decay bacteria. It will also protect the wood from insects by making it unpalatable or poisonous to them for years to come.

5

Little Projects

We present a number of small projects which will not take much time to build, or very much in the way of material, either. They will serve, we hope, as a good introduction to craftsmanship for beginners. More advanced workers will also be interested in them because they are all useful and beautiful in their way—objects which one can make to use for one's own garden or build as gifts for friends who love their gardens.

The simplest of these projects are the bird feeders, birdhouses, and shelters. They may be built of scrap lumber if you wish, requiring but little initial skill in construction in order to complete them. But if the novice perseveres and does the work carefully, step by step, he will find that many of the principles and practices he uses will train him for more ambitious projects.

The plant-stands and outdoor shelves for summering houseplants on the terrace are practical solutions to an everpresent problem of the indoor gardener. Whether they are the very simple demountable ones or more permanent types, they will assist the gardener to keep his plants in good order during their summer vacation outdoors.

The plant shield shown can probably be adapted to many more uses than those shown here. Every gardener recognizes the need for shielding tender plants from the winter sun and cold blasts of wind. By using a decorative shield (something more pleasant to look at out the window than the torn and ragged strips of sagging burlap sacking one usually sees), the sad winter landscape can be made more appealing. Such shields can be painted any color you wish, making them distinct assets rather than bitter necessities when they are installed in your garden.

The cold frame is another project which may be built with very little tool experience. We recommend buying the glass sash rather than making it—sash is cheap and making it is rather an intricate job—and then constructing the cold frame to fit the sash. Advanced workers may wish to go a step further and build the sash. There are many changes which can be rung on the construction of the frame. Permanent sides of brick, concrete, or other masonry may be used, or a wooden frame made to fit over the top to hold the sash. For using such a frame as a hot bed, permanent masonry sides are recommended because of high temperatures and humidity which may rot the wooden sides of the type we show.

Our "Ever-Blooming Plant Box" can be adapted to whatever sizes of pots you wish to use in it, and may be built with or without the trellis shown. It will make a bright spot of color beside your front door, on the terrace, or wherever you finally decide to use it. By fitting it with wheels, you'll be able to move it about from spot to spot, changing it about each day or each week, or whenever you give a party.

The "Ugly Duckling Doors" also require but little skill and only a bit of patience for their rebirth and refurbishment. Old cracked panels and odd, mismatched doors can be made into smart, fresh ones and will last many years when they are covered and brought up to date this way.

These, then, are our Little Projects, which we hope will inspire you to find fun in building and will give you many years of pleasure and profit from their use when they are completed.

May all your Little Projects become big ones as your skill and confidence grow.

BIRDHOUSES

With the encroachment of building developments around cities, more and more wooded areas are being decimated, sometimes being destroyed altogether. Birds are having a rather thin time of it these days when it comes to finding suitable quarters for building nests and rearing families. Every person who has the faintest interest in outdoor life and in conservation will want to help to remedy this deplorable situation, not only for the amusement afforded by watching the birds as they go about the business of rearing their young, but for the very sound, practical reason that they do so much good in ridding gardens and trees of pernicious pests.

What can you do? Build birdhouses and shelters and make feeding stations, so that during the winter you can supplement their insect and seed rations, particularly during the severe days when natural foods may be iced over and unavailable.

A great many birds will adapt themselves to dwelling in man-made houses and also are tolerant of living near human habitations. If you wish to lure birds into your garden, the main consideration will be to construct a house which will fit the requirements of the kinds of birds you wish to attract. For instance, robins, phoebes, wrens, and bluebirds, all fairly common in a good many areas, are amenable to living in man-made shelters. Purple martins and flickers, song sparrows, titmice, nuthatches, barn swallows, house finches, woodpeckers, and even some of the owls, are known to be open-minded about setting up housekeeping in civilized houses. But each kind of bird prefers a certain kind of house, usually liking it placed at a certain convenient level and in an environment suited to its needs or attractive to the species.

Anyone who can hold a saw, hammer a nail, use a screwdriver, or bore holes with a brace and bit can make birdhouses and feeders which will not only be suitable, but will be received with open wings by the feathered populace. It needn't cost much, either, to do your part in relieving the housing shortage in the bird world. You can either buy the materials or use remnants of boards left over from other projects, packing-case boards, plywood scraps, or seasoned wood from old pieces of furniture which have been discarded. The only limit is the ingenuity of the builder, and he may well be someone who is embarking upon the hobby of becoming a craftsman.

Materials chosen particularly for the purpose may look better to the human eye, but the birds won't mind if you use oddments. Cedar, cypress, redwood, and hardwoods will last longer; but poplar, white pine, and practically any kind of wood which is easily worked will do if it is of suitable size and thickness. Wet and green lumber should be avoided. When the house is finished, it should be stained or painted. It should then be taken down yearly, cleaned, and given another coat of stain or paint to make it last for as long as possible. Make any necessary repairs at that time, too.

You can build a birdhouse as good as or better than the best offered commercially. Some professionally-made birdhouses (and many of the home-made kind, too, regrettably) are "cute little numbers" painted in garish colors and made to simulate human dwellings. They look as if they belonged in the plastic department of the local five-and-ten-cent store. The birds don't like them very much, usually, so you may be wasting your time if you go off on this tack.

The most successful birdhouses are engineered for the birds' use, not to tickle the eyes of humankind. They strive to emulate as far as possible what is known about the natural preferences of the birds. While some may nest in brightly-painted houses, most birds seem to prefer a nest-house which is less conspicuous. This doesn't mean that you have to go to the length of making everything brown and green, as some bird lovers advocate. Neutral gray, soft blue-green like that of spruce needles, medium blue, terra cotta, or any of the pleasant colors which are not too brilliant will not offend the birds, and will charm the human eye, too.

An exception might be made of the phoebe shelter, which is best placed under the eaves of the house or in a protected spot around human habitations. It can be painted to conform with the walls or trim color of the house so that it assumes protective coloration by not standing out against its background. Don't paint houses too often, for birds seem to like houses which have weathered a bit.

Feeders may also be painted, unless you happen to subscribe to the nature school, which insists that everything out of doors must be stained a dingy brown. Actually, we believe that it is the oil smell which birds dislike and not the color. We have always painted our feeders and houses the same soft but not dull blue-spruce color we used on the shutters of our home, and we have never lacked either tenants or free-lunchers. However, we always try to get the houses and feeders painted several weeks ahead of the need for them outdoors. If the structure is built early in winter and painted by late winter, any paint odor will have dissipated by house-hunting and nesting time. Put out the house somewhat earlier than you think it may be needed. This will allow it to weather a bit and will also have it there ready for early arrivals, who may very well stake out their claim and start building ahead of schedule. You can check with your local Audubon Society group as to the approximate times of arrival and nesting in your climate and region, if you want to be really certain. Don't be alarmed if birds don't nest the first year; you may have missed their arrival time, or they may have taken out their season's lease where they were last year. Next year you may be their landlord.

WHERE TO PLACE THE BIRDHOUSE

A suitable site for the house is of primary importance, naturally, for if the location is not to the liking and needs of its potential tenants it will remain vacant. There should be safety from marauding squirrels and cats; there must be sun (or shade, according to species); and the house must be placed at a suitable height.

Be sure to face the house away from prevailing winds, so that rain will not be driven into the entrance hole. Never tilt the front of the house upward if you are affixing it to a tree trunk, or rain will enter. Either place it vertically or slant it downwards. Some houses may be hung from two wires below a horizontal limb, the wires preventing it from spinning in the breezes. Select a large enough limb so that the house is not constantly jouncing about.

Except in isolated cases, the minimum height at which to set bird houses is 4 to 5 feet, or even at the height of a fence post. Although in nature bluebirds nest much higher, they will sometimes adapt themselves to the low placement. Most birds seem to prefer a height of 8 to 12 feet or more from the ground for their houses. The table at the end of this text will give you recommendations for placement, as well as other information on building houses for specific species. If you are troubled with squirrels—they often steal eggs and rip up the nests— or with neighborhood cats, the answer may be to place the house in the open where it will be beyond leaping distance from trees or buildings. Neither cats nor squirrels can climb a pipe, and their jumping capacities are limited, too.

BUILDING THE HOUSE

If you are buying new wood, the best kinds are cypress, cedar, and redwood. White pine and poplar may also be used. Purchase the 1-inch thickness (note that the *finished* size is less than that, when you are making your measurements and preparing to cut the wood) except if plywood is to be used, when ½-inch or ¾-inch may be bought. (Thickness measurements in plywood are reasonably exact.) All woods will last longer if they are treated with preservatives, and if aluminum or brass screws and non-rusting nails are used. Similarly, if hinges are required, try to purchase non-rusting types. Countersinking screws and puttying all holes will also make a better-looking and longer-lasting birdhouse or shelter.

Two coats of stain or paint should be given initially, and repainting of the houses should be done every three or four years, oftener if necessary, to keep the structures in good condition. Yearly cleaning of the houses to remove nesting materials will help to keep them attractive to birds. Although many bird lovers take their houses down and store them indoors in winter, it is recommended that they be left out as winter shelters, particularly in cold regions. On cold nights many birds will enter houses, often piling up on top of each other to keep warm, even if the house is not the type they would use for nesting.

Several types of cleanouts are shown in the accompanying sketches. They may be adapted to any of the houses shown in the design pages. When you have cleaned the house, it may be well to take it down and paint it so that it will withstand winter moisture better, and also to be certain that the paint smell will have dissipated long before spring nesting time.

Don't fail to drill the holes for drainage in the floors, or the vent holes around the tops where recommended. The bottom holes will permit any moisture to drain out should a hurricane force it into the house in quantity, and both floor and top holes will help to ventilate the house on hot days. The thickness of the wood in the box-type houses will help to insulate the houses somewhat.

Observe the sizes given for the entrance holes and *use them*. While birds may nest in houses with larger holes, competitors or enemies may also be able to enter. For example, if you make the hole in a bluebird house 1¾ to 2 inches in diameter, starlings will enter and displace or fight with the bluebirds or even usurp the nest. If it is 1½ inches, these slum-children of the bird-world will be excluded. Similar considerations have brought about the choice of sizes for the entrance holes for all birds.

Some species have more than one nesting each season, and each nesting requires a new nest. If all the old material is cleaned out after the fledglings of the first brood have departed, it is quite possible that the house will be used for the second nesting. The cleanouts spoken of previously will facilitate cleaning. If you have woodpeckers, they will appreciate 2 inches of sawdust placed loosely in the bottom of their box nests. They seem to require it for their nest, probably because they are used to having wood chips or rotting wood in the natural nesting cavities in dead trees in the woods.

As indicated in our plans, entrance holes are placed high so that the mother bird is concealed when she is sitting on the nest. Holes must not be placed lower or higher (this latter will make it difficult for the bird to enter), but at the position indicated. A natural twig or a dowel placed in a hole on the front of the house will give birds an alighting place from which they can hop up into the entrance hole. Some bird observers claim that robins need a good-sized branch extending out and below the shelter, stating that robins like to alight and walk up to the nest. This theory would seem to be borne out by the fact that robins like to nest on flat limbs in orchards.

Much more might be written about attracting birds and building homes for them (much more has been, in fact), but that is not the function of this book. This section serves as an introduction, so we give as much space as we can spare to material which we hope will help you to get houses built properly and birds coming around regularly each year to nest and to be fed. For those wishing further information, two publications on sale by the Superintendent of Documents, U. S. Government Printing Office, Washington, D.C. 20025, will probably prove helpful with more comprehensive information. They are

> *Homes for Birds,* Conservation Bulletin #14 of the U. S. Department
> of the Interior.
> *Bird Houses and How to Make Them,* Farmer's Bulletin #609, U. S.
> Department of Agriculture.

In addition, the Boy Scouts of America have a pamphlet, one of their Merit Badge Series, on *Bird Study.* Younger builders will be interested in this publication and older ones should not scorn it.

For those desiring a really comprehensive study of what the requirements of various birds are, what to plant for natural food for them, and other aspects of attracting and housing birds, *Songbirds In Your Garden* by John K. Terres of the National Audubon Society, published by Thomas Y. Crowell, New York, is to be highly recommended. If you wish to join the National Audubon Society or one of its branch groups, write for information to The National Audubon Society, 1130 Fifth Avenue, New York, N.Y. 10028.

But whether you make a full-time hobby of birdhousing and birdwatching, or just have a house or two for the pleasure it gives you, you will enjoy the company of birds in your garden.

SAME HOUSE DIFFERENT SIZES FOR WRENS OR BLUEBIRDS

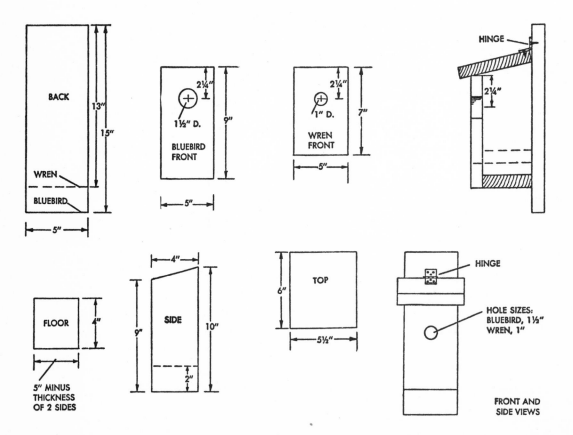

BIRDHOUSE CLEANOUTS

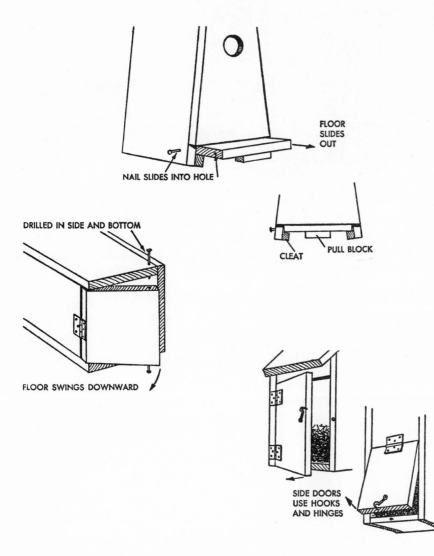

FLOOR
SLIDES
OUT

NAIL SLIDES INTO HOLE

DRILLED IN SIDE AND BOTTOM

CLEAT   PULL BLOCK

FLOOR SWINGS DOWNWARD

SIDE DOORS
USE HOOKS
AND HINGES

After the brood has flown, the house should be cleaned out to allow it to be used again for a second nesting if the species of bird desires to do so. Birds don't use the same nest twice and are more likely to renest in a clean house than in one they must clean themselves. The house needn't be taken down for cleaning if easy cleanouts such as those we show here are constructed —hinged bottom, slide bottom, and hinged sides. Note that all of them provide for securing the cleanout during the nesting periods.

53

LITTLE PROJECTS

OPEN HOUSE FOR "FRESH-AIR BIRDS"

Certain birds will nest only in shelters with open fronts. By placing the shelter under the eaves of a building with the open front protected from prevailing winds, you will attract phoebes and barn swallows to nest in it. Robins sometimes nest near houses, but more frequently prefer the shelter to be placed fairly high in a tree on the trunk or on a major limb. Here, too, wind protection is a major essential for attracting the birds.

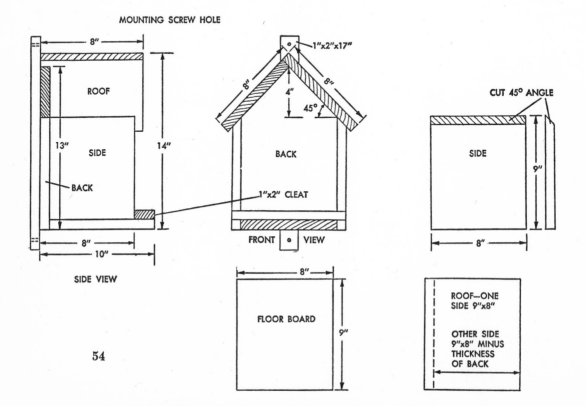

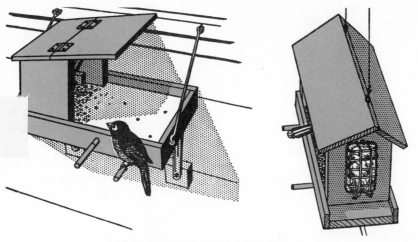

TWO WAYS TO FEED THE BIRDS

Conveniently placed below a window so that it is easily seen by inmates of the adjacent room, this feeder features a hinged roof, allowing the glass-fronted seed hopper to be easily refilled on inclement days without your leaving the house. Brackets may be used to attach it to the wall, or merely to stabilize it if long hooks screwed firmly into the windowsill are used. This feeder can be swung below a tree limb on copper wires, the hopper letting down more food as it is needed. The wire soap dish holds suet and is kept in place by three cup hooks. The angle hooks inside will hold paper cups of peanut butter or other food. Note the short dowels run through screw eyes and nailed in place to provide safe, comfortable landing places for small birds.

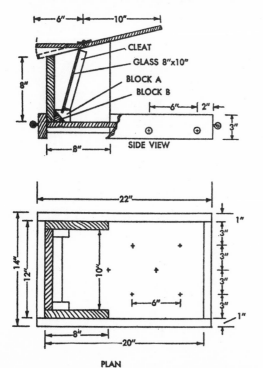

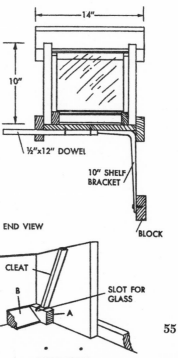

55

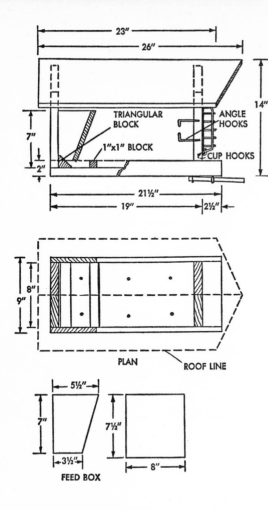

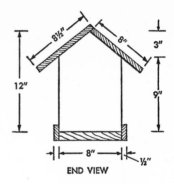

TRIANGULAR BLOCK

1"x1" BLOCK

ANGLE HOOKS

CUP HOOKS

END VIEW

PLAN

ROOF LINE

FEED BOX DETAIL

FEED BOX

OTHER WAYS TO FEED BIRDS

There are, of course, many other ways of feeding birds which are approved by bird lovers. One very simple and inexpensive method is seen at the right: A short length of tree limb with a few shortened twigs has had some holes 1″ to 2″ in diameter bored an inch or so deep. These are filled periodically with peanut butter, with suet, or with wild bird seed mixed with peanut butter. Where there are no twigs, use short dowel sticks to provide alighting spots for the birds, inserting them in holes bored just below the food holes. Hang by a copper wire through a screw eye in the top. Using paint may be frowned on by some purists, but the birds do not seem to mind—they'll nest in houses and eat from feeders long before the paint smell dissipates. Also, do not feel that you must stick to stains of woodsy brown or leaf green for all your birdhouses and feeders.

ALL YEAR BIRD SHELTER-FEEDER

An open shelter, comprised of a shelf, roof, and half sides is preferred by phoebes, barn swallows and robins. But this shelter can also be used as a feeder when it is not being used for nesting. Its roof is hinged, allowing it to lift so that the glass which holds the seed can be removed or inserted, and for replenishing the seed when necessary. Aluminum screws are used to attach it to the house or to a tree for nesting, making it easy to remove it for placement adjacent to a window of the house for use as a feeding station.

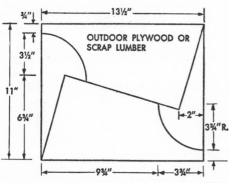

OUTDOOR PLYWOOD OR SCRAP LUMBER

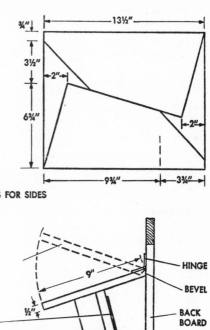

ALTERNATE CUTS FOR SIDES

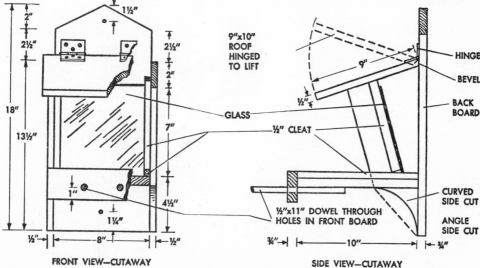

FRONT VIEW—CUTAWAY

SIDE VIEW—CUTAWAY

A TABLE OF DIMENSIONS*
Nesting Shelves For Birds (One or more sides open)

Kind of Bird	Min. Floor Size	Depth of Box	Preferred Height Above Ground
Robin	6" x 8"	8"	6'-15'
Barn Swallow	6" x 6"	6"	8'-12'
Song Sparrow	6" x 6"	6"	1'-3'
Phoebe	6" x 6"	6"	8'-12'

Birdhouse Dimensions

Kind of Bird	Floor Size	Box Depth	Ht. of Entrance Above Floor	Diameter of Entrance	Height Above Ground
Bluebird	5" x 5"	8"	6"	1½"	5'-10'
Chickadee	4" x 4"	8"-10"	6"- 8"	1⅛"	6'-15'
Titmouse	4" x 4"	8"-10"	6"- 8"	1¼"	6'-15'
Nuthatch	4" x 4"	8"-10"	6"- 8"	1¼"	12'-20'
Wrens: House and Bewick's	4" x 4"	6"- 8"	4"- 6"	1"-1¼"	6'-10'
Carolina	4" x 4"	6"- 8"	4"- 6"	1¼"	6'-10'
Violet Greenswallow and Tree Swallow	5" x 5"	6"	1"- 5"	1½"	10'-15'
Purple Martin †	6" x 6"†	6"†	1"†	2½"	15'-20'
House Finch	6" x 6"	6"	4"	2"	8'-12'
Starling	6" x 6"	16"-18"	14"-16"	2"	10'-25'
Crested Flycatcher	6" x 6"	8"-10"	6"- 8"	2"	8'-20'
Flicker	7" x 7"	16"-18"	14"-16"	2½	6'-20'
Woodpeckers: Golden Fronted, Redheaded	6" x 6"	12"-15"	9"-12"	2"	12'-20'
Downy	4" x 4"	8"-10"	6"- 8"	1¼"	6'-20'
Hairy	6" x 6"	12"-15"	9"-12"	1½"	12'-20'
Owls: Screech	8" x 8"	12"-15"	9"-12"	3"	10'-30'
Saw-whet	6" x 6"	10"-12"	8"-10"	2½"	12'-20'
Barn	10" x 18"	15"-18"	4"	6"	12'-18'
Sparrow hawk	8" x 8"	12"-15"	9"-12"	3"	10'-30'
Wood duck	10" x 18"	10"-24"	12"-16"	4"	10'-20'

*These dimension tables are taken by permission from *Homes for Birds*, Conservation Bulletin #14, U. S. Dept. of the Interior. The publication may be purchased from the Supt. of Documents, U. S. Government Printing Office, Washington, D. C. 20025.

†Dimensions for one compartment of a large martin house. Usually the martin houses are built in sections comprised of eight compartments each.

Foods Preferred By Garden Songbirds
Or Non-Game Birds

Blackbirds Cardinals Towhees	Sunflower seeds, corn, shelled and broken peanuts, scratchfood.
Juncoes Finches Native sparrows	Scratchfood, millet, wheat, screenings, small seed mixtures, bread crumbs.
Robins, Catbirds, Mockingbirds, Thrashers Hermit thrushes	Cut apples, oranges, currants, raisins, bread crumbs.
Chickadees Titmice Nuthatches	Suet, cracked nuts, shelled and broken peanuts, sunflower seeds, bread crumbs.
Woodpeckers Jays	Suet, cracked nuts, corn, peanuts, sunflower seeds.

NOTE: Food cakes attract many birds. They are prepared, by scalding or partly cooking cereals such as cornmeal or oatmeal. Next, add bread crumbs, peanuts, raisins, and an egg. Mix well, stir in melted suet, and bake.

MORE LITTLE PROJECTS ▶

LITTLE PROJECTS

A GARDEN SHRINE

Light enough to be hung on a fence, yet of sturdy enough construction to stand by itself, this little shrine will add interest to the garden. Shown here is a reproduction of a plaque of St. Fiacre, the patron saint of gardeners; the box at his feet will hold either flowers or green vines At left below and at right are shown two versions of the shrine. At left is a pot pocket with holes bored in the bottom for drainage. At right is the version shown in the top illustration with a hole cut in the bottom to admit a pot of flowers. Version at left has pocket fastened to back and has short side pieces screwed to pocket. Plaque hangs on hook.

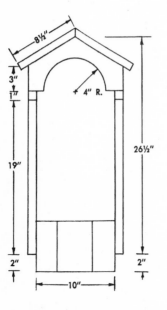

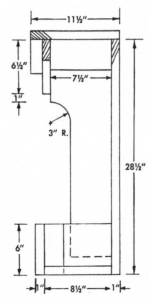

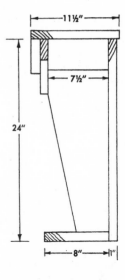

ALTERNATE DESIGN

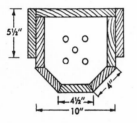

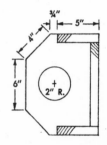

60

SHOW-OFF FOR POTTED PLANTS ON THE TERRACE

Either houseplants or potted plants prepared for terrace use outdoors will find graduated shelves like these the best possible way to make their contribution. Note the alternate methods of making the shelves, with wooden gratings to allow circulation of air and good drainage. Use either bolts or dowels to hold shelves in place, decorate with trellis slats.

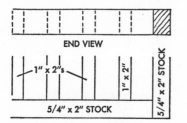

END VIEW

1" x 2"s   1" x 2"

5/4" x 2" STOCK   5/4" x 2" STOCK

SHELF DETAIL — SQUARE TYPE

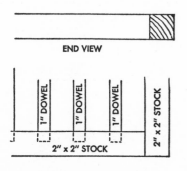

END VIEW

1" DOWEL   1" DOWEL   1" DOWEL   2" x 2" STOCK

2" x 2" STOCK

SHELF DETAIL — DOWEL TYPE

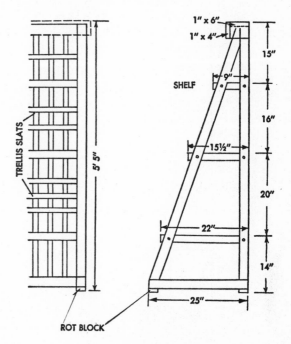

TRELLIS SLATS

5' 5"

ROT BLOCK

1" x 6"
1" x 4"
15"
SHELF
9"
16"
15½"
20"
22"
14"
25"

FRONT AND SIDE VIEWS

61

VACATION SPOT FOR HOUSE PLANTS

If you have a few steps down from a north door of your house, you have an ideal spot for summering your houseplants. Build steps of outdoor plywood to match stairs, topping them with wooden gratings so that plants can drain easily. Plant steps are easily transportable for winter storage because all of the gratings are removable. Use a wood preservative and give all wood three coats of paint. Note "rot blocks" which are replaceable.

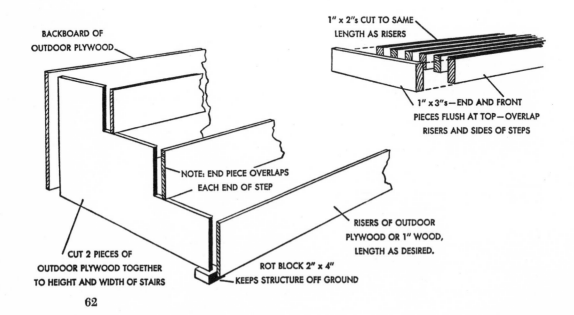

BACKBOARD OF
OUTDOOR PLYWOOD

1" x 2"s CUT TO SAME
LENGTH AS RISERS

1" x 3"s—END AND FRONT
PIECES FLUSH AT TOP—OVERLAP
RISERS AND SIDES OF STEPS

NOTE: END PIECE OVERLAPS
EACH END OF STEP

RISERS OF OUTDOOR
PLYWOOD OR 1" WOOD,
LENGTH AS DESIRED.

CUT 2 PIECES OF
OUTDOOR PLYWOOD TOGETHER
TO HEIGHT AND WIDTH OF STAIRS

ROT BLOCK 2" x 4"
KEEPS STRUCTURE OFF GROUND

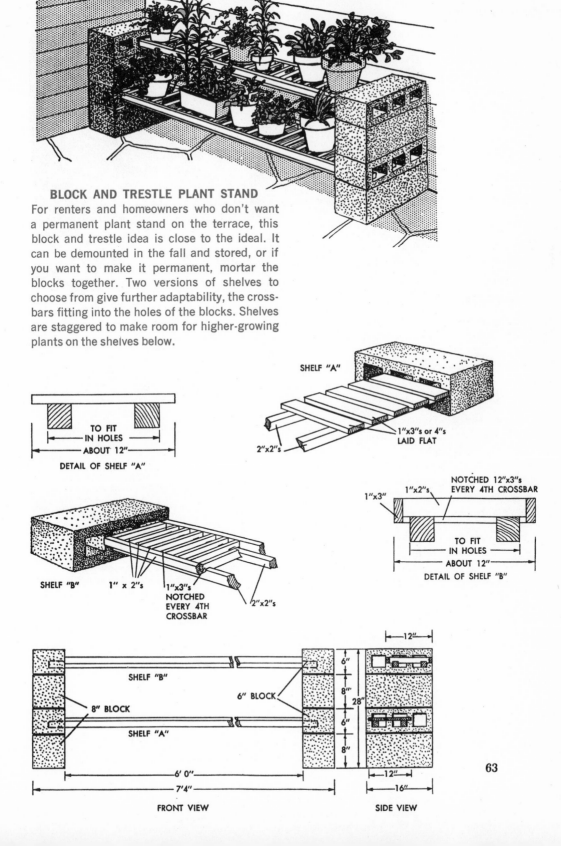

BLOCK AND TRESTLE PLANT STAND

For renters and homeowners who don't want a permanent plant stand on the terrace, this block and trestle idea is close to the ideal. It can be demounted in the fall and stored, or if you want to make it permanent, mortar the blocks together. Two versions of shelves to choose from give further adaptability, the crossbars fitting into the holes of the blocks. Shelves are staggered to make room for higher-growing plants on the shelves below.

SHELF "A"

1"x3"s or 4"s LAID FLAT

2"x2"s

TO FIT IN HOLES

ABOUT 12"

DETAIL OF SHELF "A"

NOTCHED 12"x3"s EVERY 4TH CROSSBAR

1"x2"s

1"x3"

SHELF "B"   1" x 2"s   1"x3"s NOTCHED EVERY 4TH CROSSBAR   2"x2"s

TO FIT IN HOLES

ABOUT 12"

DETAIL OF SHELF "B"

SHELF "B"

6" BLOCK

8" BLOCK

SHELF "A"

6'0"

7'4"

FRONT VIEW

12"

6"

8"

28"

6"

8"

12"

16"

SIDE VIEW

63

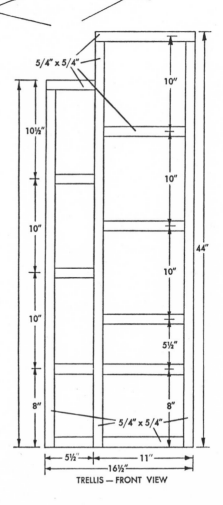

EVER-BLOOMING PLANT BOX

One way of dressing up the front door is to place beside it a plant box, which will always greet the visitor with a burst of blossom. Let the plants be "demountable" at a moment's notice so that they may be taken out and replaced easily as soon as they begin to fade, thus assuring yourself of all-summer beauty at your doorstep. Inexpensive to build—scrap lumber may be employed for most of it—the plant box features a series of tiers which allow the shoulders of the pots to rest on the box edges, giving good drainage and also support for the pots. The trellis shown may be used or not, made a bit taller or shorter, according to the plants you wish to use. Geraniums can be trained up on the trellis, or vining houseplants such as ivy or philodendron (if there is shade) or a pot of morning glories can be used. Paint the wood or stain it, according to what you prefer in the way of finish, but always use a good wood preservative over the entire surface of the box first, particularly on cut edges, then paint or stain it.

TRELLIS — FRONT VIEW

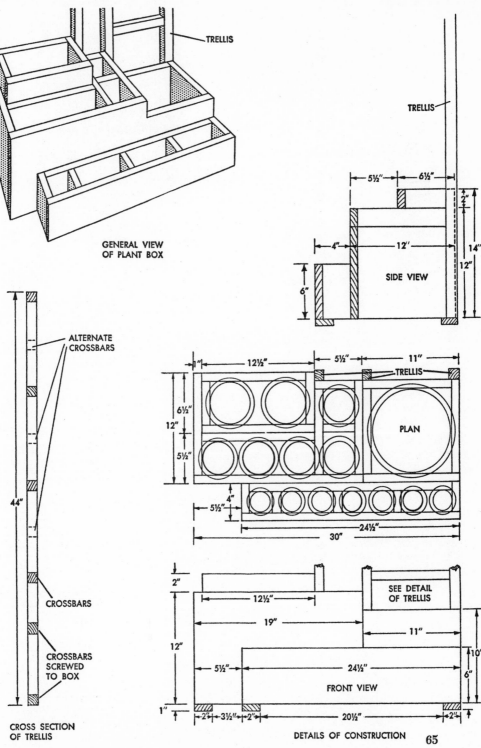

GENERAL VIEW
OF PLANT BOX

TRELLIS

TRELLIS

SIDE VIEW

5½"   6½"

2"

4"   12'

14"

12"

6"

ALTERNATE
CROSSBARS

44"

CROSSBARS

CROSSBARS
SCREWED
TO BOX

CROSS SECTION
OF TRELLIS

TRELLIS

1"   12½"   5½"   11"

6½"

12"

5½"

4"

5½"

PLAN

24½"

30"

2"

12½"

19"

12"

5½"

SEE DETAIL
OF TRELLIS

11"

24½"

10"

6"

FRONT VIEW

1"   2" 3½" 2"   20½"   2"

DETAILS OF CONSTRUCTION   65

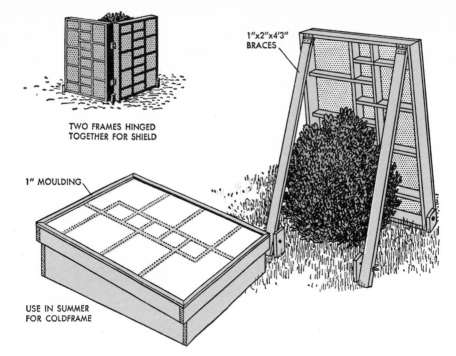

TWO FRAMES HINGED
TOGETHER FOR SHIELD

1"x2"x4'3"
BRACES

1" MOULDING

USE IN SUMMER
FOR COLDFRAME

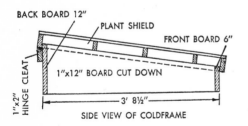

BACK BOARD 12"

PLANT SHIELD

FRONT BOARD 6"

1"x2"
HINGE CLEAT

1"x12" BOARD CUT DOWN

3' 8½"

SIDE VIEW OF COLDFRAME

VERSATILE PLANT SHIELD

Wrapping plants with burlap or setting up a screen of old sacks makes a garden very unsightly. Make a wind-and-sun screen like this and anchor it with back braces and four stakes, using bolts and wing nuts to secure it, or hinge two frames together and anchor. Boxwood or evergreens will be protected and your garden will look neat all winter. In summer, use the shield as a coldframe cover, making the cold-frame to fit inside the shield, and use the same hinges to attach to frame. Cover with plastic wire-screen cloth, which lasts for years.

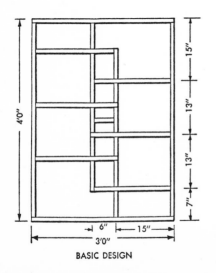

4'0"

15"

13"

13"

7"

6"

15"

3'0"

BASIC DESIGN

SCREEN IN PLASTIC

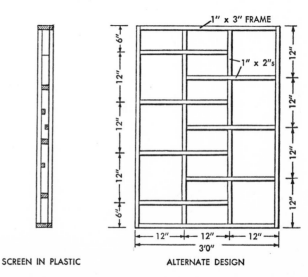

1" x 3" FRAME

1" x 2"s

6"

12"

12"

6"

12"

12"

12"

12"

12"

12"

3'0"

ALTERNATE DESIGN

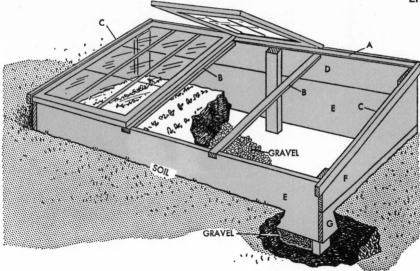

EVERY GARDEN CAN USE A COLDFRAME

For starting seeds early, for wintering-over tender plants, and for starting perennials in late summer, nothing can replace a coldframe. We recommend buying cypress sash already made up because they are relatively inexpensive; then build the walls to fit the sash. Sizes are usually: 2' x 4', 3' x 4', and 3' x 6', and come glazed or unglazed. For walls, cypress is preferred because it is long lasting; but other hardwoods may be used equally well if they are treated with wood preservative, replacing them in a few years if necessary; or walls may be built of concrete or brick with wood frame topping to hold the sash. Make a set of sash to fit frame, covering them with laths, to use when the weather warms enough to make shade desirable and glazed sash unnecessary. Propping block holds sash up at various levels for ventilation.

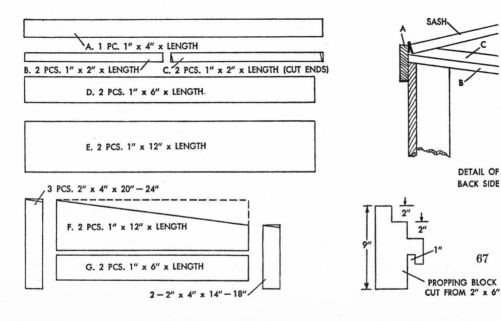

A. 1 PC. 1" x 4" x LENGTH

B. 2 PCS. 1" x 2" x LENGTH   C. 2 PCS. 1" x 2" x LENGTH (CUT ENDS)

D. 2 PCS. 1" x 6" x LENGTH.

E. 2 PCS. 1" x 12" x LENGTH

3 PCS. 2" x 4" x 20" – 24"

F. 2 PCS. 1" x 12" x LENGTH

G. 2 PCS. 1" x 6" x LENGTH

2 – 2" x 4" x 14" – 18"

SASH

DETAIL OF BACK SIDE

PROPPING BLOCK CUT FROM 2" x 6"

67

MAKE OVER UGLY DUCKLING DOORS

Old doors in bad condition or those which have panels of an ugly shape can be brought beautifully up to date. Cover them with a sheet of ½″ outdoor plywood with moldings framing shutter-effect panels made from overlapping trellis slats. Use a base of 1½″ doorstop molding, secured to plywood with waterproof glue and brads, a second molding on top overlaps slats ½″. Either high-crowned or flatter moldings may be used for the second one, depending on the effect desired. Bore holes of lock, knob, and cut slots for hinges if necessary. Use wood preservative before painting, fasten plywood to door with waterproof glue, or screws and nails of rustproof metal.

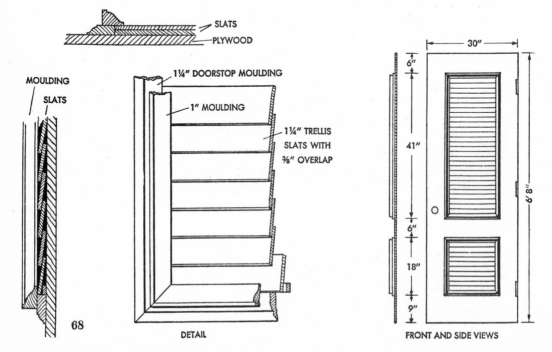

SLATS
PLYWOOD

MOULDING
SLATS

1¼″ DOORSTOP MOULDING
1″ MOULDING
1¼″ TRELLIS SLATS WITH ⅜″ OVERLAP

30″
6″
41″
6″
18″
9″
6′ 8″

68

DETAIL

FRONT AND SIDE VIEWS

The Trellis Story

Strictly speaking, a trellis is a frame of light wood strips crossing each other, with open spaces between the strips, particularly such a frame as is used for supporting vines. However, we shall show in the succeeding pages a number of trellises which are built of heavier wood. Further, we shall advise you to take any design and improve on it as you will, building it of heavier or lighter wood, or repeating it several times to make a wide screening-fence trellis on which growing vines will complete the cover-up during the outdoor season. Perhaps you will want to make two units and set them at an angle to give a three-dimensional quality to the trellis, lending an individual air to your front door or wherever you will be using it. But as you improvise be sure that you make the trellis fulfill its primary function, that of supporting vines.

CHOOSING A TRELLIS

In choosing a trellis, you will naturally start with the site where it is to be placed. This will govern the size, the design, and what you will plant to grow on it. In turn, what you are going to plant will influence the choice, because heavy vines require a heavy trellis, twining vines or vines with tendrils need a different kind of support, and so on. All these factors are tied in together, and all must be considered and worked out before the trellis is built.

Annual vines, as anyone who has grown morning glories can attest, may become quite heavy with the quick rush of growth in a single season. They often break the fragile strings which hopeful gardeners use as support, up which the vines twine so gaily when they are slender young plants. Perennial and woody vines also tend to become quite heavy with age and exert great pressures on any structure on which they climb, sometimes pulling roofs apart and even causing weak structures to collapse. You can see, therefore, that there is much more to choosing, planning, and building a trellis than would appear at first sight. On the other hand, there is nothing more calculated to give any house that final look of "home" than a bit of greenery twining around the door, on a trellis which is as good to look at in winter as it was in summer.

Does your front door satisfy you? Is your back door merely a way of getting out of the kitchen to go to the trash can? Is there a starkness about the front of your garage which a trellis filled with roses would relieve? Perhaps the answer in all cases is a trellis and a vine. Your front door can be made different from its neighbors, your kitchen door a charming entrance which it is a pleasure to use each day.

Is your home modern, traditional, French provincial, of Spanish or Mediterranean inspiration; built of brick or frame; one or two stories in height? All of these factors will be pertinent in your planning.

If your home is modern, perhaps it could stand a little restrained design: something slightly oriental in feeling, or perhaps just geometric without being dull and rigid. That is the trend in the newest modern homes today. If your house is traditional—Cape Cod or Georgian—you can give it a lift by using a trellis which is somewhat modern, but is restrained enough not to clash with the pure lines of the traditional house.

Possibly you will want to follow the 18th century precedent of using one of the modified Chinese patterns such as were employed for balustrades or porch railings, arbors, and fences in those early homes. Because they were simple in design, they somehow fitted very well with that classic inspiration from Italy which brought forth the English Pal-

ladian houses in the 18th century from which our own colonial architecture descended. Therefore, if you feel you need a precedent, you
have an excellent one for choosing a Chinese fretwork design for your
trellis.

If you are fencing the back garden or building a privacy fence
around your terrace, as so many of us are doing today, use a trellis as a
gateway or as a fence topper, or to tie the fence in with the house. This
will add a friendly touch and keep it from looking too coldly architectural
by softening it with living greenery all through the growing season. And
when the leaves have fallen, it will give the fence more interest during
the bleakness of winter.

Multi-level decks of varying sizes are unified by the trellis of beams that are
supported with doubled 2″ x 6″s and 2″ x 4″s while privacy is given by 1″ x
1″s that form screening fences. All wood is redwood, even the floors which
are laid both crosswise and lengthwise on decks.

If you don't find exactly the trellis you have in mind in these pages, look through the other sections of this book and see if you may possibly find a design among the fences or other structures which you like. Adapt it. Be creative and use the design you like, but make it of the weight of material needed for your purpose and adapt its proportions as best you can to the space your trellis must occupy. Only one caution must be observed: keep the design open and simple, making sure that it will integrate well with the lines of your house. And, of course, never forget that a trellis is primarily a vehicle for the support of vining plants, and only secondarily is it an entertainment for the eye.

THE VINE FOR YOUR TRELLIS

Most vines, like most children, are charming and inoffensive when they are small and youthful, but within a few years they may become veritable Frankensteins of monster proportions. But also like children, most vines, even rampant ones like wistaria or the trumpet vine, can be kept within bounds and relatively civilized by exercising care and discipline. Pruning shears in a firm hand will keep them at bay, and will even benefit wistaria, forcing it to bloom well. Trumpet vines may grow sulky and push up more shoots from the roots, making you think the "Frankenstein" story has changed into the "Legend of the Dragon's Teeth." In any case, pruning is a chore under which you will chafe in years to come, so it is well to outwit it right at the beginning by choosing some lightweight, less rampant, and less demanding vine, possibly an annual one.

Roses, of course, leap to the mind when a trellis is mentioned. It is a delight to see the wayward curves of a climbing rose threading its way through the crisp architectural lines of a good-looking trellis. If a pillar rose is planted, rather than one of the rampant climbers which may go up as high as 20 feet or more, the rose will not get out of bounds but will beautifully clothe your trellis with its blossoms, with a minimum of pruning on your part. Ever-blooming roses, such as Dream Girl, American Pillar, Summer Snow, or any of the other excellent modern roses, are obtainable from all good nurserymen. Yearly pruning, as recommended by your nurseryman when you buy them, will keep them strong and blossoming as you want them to do.

Clematis is another lovely vine, airy, light in weight, and with a wealth of fine flowers once it is established. It offers a good choice of color, time of bloom, and size of blossom. It blooms best when its roots are shaded, so a planting of low shrubs or perennial flowers to perform this function will protect it during the heat of summer. Perhaps you would like to plant two clematis of different colors on the same trellis to bloom at the same time, or to follow one another so that blooming time is extended. Consult your nurseryman for his recommendations for your particular area. A lightweight trellis is adequate for clematis in most cases, for it is a lightweight vine.

Akebia is another good perennial vine, the five-leaved variety being preferred by many who like its chocolate-colored flowers, which come in spring.

The Bittersweet vine, *Celastrus scandens* for the North and *Celastrus angulatus* or *C. hypoleucus* in the South, can be kept within bounds easily and will give good results. The Northern Bittersweet makes a most spectacular presentation each autumn of masses of orange-colored pods. By doing your pruning each autumn, you will be able to use the cut pods indoors for bouquets each winter.

The Silver Lace-vine, sometimes called the Chinese Fleece-vine, is a fast-growing climber. It will clothe the trellis each year with its largish leaves and a wealth of greenish-white flowers in clusters which do not belie either the lace or the fleece of its names. Cut back sharply each autumn or each spring, it will come back strongly if fed well.

For the northern sides of buildings or shady spots elsewhere, ivy can be trained over a trellis if a bit of patience is exercised. Tie it to the frame of the trellis until its rootlets attach, or until it is strong enough to support itself. Rough wood trellises will allow these rootlets to attach more readily than will well-planed wood.

Climbing Honeysuckle is another possible vine to use, but check first with local plantsmen to see how it performs in your area. Some varieties in certain places become rampant climbers, which makes them unsuitable for small trellises, charming though the vine is.

A vine which is suitable for a trellis, whether placed on a fence, a roofed-over arbor, or even a pergola kind of trellis, is the Grape. There is something rather pleasant about having one's food crop climbing a trellis beside one's outdoor room. It gives beauty of leaf all summer and makes a thick roof if allowed to climb on a shelter from the trellis, its ripening grapes giving off a perfume from each cluster. I remember a luncheon *al fresco* a few years ago in Italy beneath a pergola burdened with grapes, with the table laid beside a fence on which grapes also grew. It was very convenient to pick one's dessert without moving from one's seat. But in America we must give a bit of thought to pests for which we must spray in many areas. This will discolor a painted trellis and may contaminate chairs and tables if it contains poisonous substances. Be sure to consider this before planting grapes.

The Climbing Euonymus is an evergreen plant which can be kept within bounds by a bit of judicious pruning now and then. Its handsome dark-green leaves are very pleasant on white-painted trellises. It needs a good sturdy support, however, because it is a really permanent plant.

We must not forget the Perennial Sweet Pea. This is a climber whose shoots die down annually, leaving the trellis clear for painting and other maintenance work and allowing its pattern to show and give distinction to its surroundings during the dreary days of winter. Pink, red, or white flowers are produced over a long season during the summer, the vine needing a minimum of care. They may be used as cut flowers, although they do not have the scent of their little annual sisters.

ANNUAL VINES MAY BE THE BEST CHOICE

Most trellis builders will agree that the annual vine is the best choice. There is a wide range to select from: lacy, finely-cut leaves, bright green or dark green, heart-shaped or with multiples of various shapes on one stem—and in addition many colors and shapes of flowers from which to choose. Annual vines, of course, die down each autumn like the Perennial Sweet Pea, permitting the growth to be removed and the trellis laid bare for winter maintenance and for displaying its beauty of design. Annuals grow from seeds and in most cases the growth is very rapid. In northerly regions they may be started indoors; then taken from their pots and placed in their summer location when the weather is sufficiently warm.

Possibly the favorite annual vine in America is the Heavenly Blue Morning Glory, but Pearly Gates (white) and Blue Star (blue with a white star in the throat) are also beautiful. Scarlett O'Hara or the new wine red Darling varieties may be used if red is desired.

Moonflower vines are nocturnal bloomers, and then grow quickly to cover large areas in a manner similar to the Morning Glory. Their blossoms are white, opening in early evening to give forth a delicious scent, a factor to remember when you are planning a trellis beside a terrace used in the evening. Planted on a trellis fence, alternating with blue Morning Glories, they give a day and night combination which will keep your trellises in bloom around the clock. In September, Moonflowers stay open in the morning and Morning Glories remain open well into the afternoon, so that they bloom together in many areas at that time. Moonflowers are sometimes perennial in warm climates, and are tender and definitely annual only in the Northern part of the country.

Perhaps you will want to grow Sweet Peas at the base of your trellis or on a trellis fence. This is a delightful idea, but in places with very hot summers Sweet Peas do not last beyond June, if that long. If you want to have them for the early summer only, all well and good, but be sure to have something to take over later on.

Cobaea, or Cathedral Bells, is another annual climber which does well in many areas, its lavender-purple blossoms and rather open growth making it a good foil for a decorative trellis. It needs a long season in which to develop.

Cypress Vine has very finely cut leaves and sports scarlet or white trumpet-shaped flowers. Lacy and delicate, it has the added attraction of staying below ten feet, usually, as does its relative, the Cardinal Climber. This is sometimes called Hearts-and-Honey because its red trumpet flowers have a showy yellow center. These latter two vines are for decoration only, never growing thick enough or tall enough to make a good screen or to cast much shade.

Another dainty climber is the Canary-Bird Flower, which has fringed, nasturtium-like flowers of rich canary yellow. It will grow in semi-shade—even prefers it—and likes a moist soil. Given these requirements it will be a good choice for a trellis.

The Balloon Vine has large, deeply-cut leaves, and is more noted for its inflated seed pods than for its small white flowers.

Also more famous for its fruits than for its handsome, showy leaves is the Gourd Vine. It produces its fruits in a bewildering array of shapes and colors, beautiful to see on the vine and equally beautiful and also useful for indoor decoration during the winter.

Mock-cucumber is a quick-growing native vine which will cover a trellis in jig-time. Although it covers thickly, it is actually a rather open plant in appearance, with lacy light-green leaves and tiny white flowers in profusion. The fruit is melon shaped, rather than long like a cucumber, and covered with soft spines all over its pale green surfaces.

Another plant which is decorative in flower is the Scarlet Runner Bean. It has spikes of bright red flowers and leaves somewhat similar to garden bean, though darker in color. Its relative, the lavender-flowered Hyacinth Bean, is similar, producing lavender spikes of great beauty. Both will grow on strings which can be tied to the trellis, or they can be twined through the openings of a trellis, preferably one of light wood.

The Japanese Hop is a speedy, dense-growing vine which laughs at dry weather and heat. It will also grow in shade. A variegated form is popular in some sections and grows as well as the regular form, covering large areas quickly with decorative leaves. The fruits, or hops, are papery, resembling spruce cones, and are also very decorative.

MAKING YOUR TRELLIS

Once you have decided on the design and have considered carefully the site of your trellis so that you know what sort of vine you want to plant, you will come to the point of working out the kind and quantity of lumber necessary to build the trellis. Even though lumber is rather expensive these days, we recommend that you make the trellis heavier than you may think it needs to be. It is better to err on the side of strength than to find, in a year or two, that the vine which was so tiny and innocuous has become a colossal "Frankenstein" of a vine.

In our text we have shown the suggested dimensions and the sizes of lumber to use. These are useful as guides and you may want to follow them just as they are, but you may increase or decrease the dimensions as you wish, bearing in mind the probable weight of the vine you will be using. We submit, also, that for trellises only first grades of lumber should be used. They should be free from structural weaknesses such as knots and flaws, which may weaken the trellis and cause it to break just as your vine is getting to the point where it will be a thing of beauty. Then you would have to rebuild the trellis and spend double the time and money, in addition to waiting for the vine to recover.

Along this same train of thought, we might urge you not to rush on this work, either, but to take your time and cut the lumber carefully, fit it well, and put it together solidly. A job well done is done once. A job badly done must be redone, within a number of weeks, months, or years. Save on the cost of replacement and the labor involved by doing it right the first time.

Another means of saving labor and replacement costs is to use one of the good wood preservatives (see Chapter 3) to prevent decay bacteria and moisture from entering and destroying your handiwork. Pay particular attention to those parts underground and to all cut edges, soaking the posts to a depth of an inch or two more than the depth they will be buried, even though they may be set in concrete. Do not use creosote, by the way, or any other preservative which gives off noxious gases in the soil and may injure or kill roots of vines planted nearby. Posts set in concrete are less likely to release toxic gases and cause trouble, however.

Speaking of concrete, we strongly recommend that posts always be set in a good solid base of concrete, not only to preserve the wood, but also to give stability to the trellis when the vine becomes heavy or when winter winds whip the structure about.

FINISHING THE TRELLIS AND AFTER-CARE

After the trellis is completed and the wood preservative applied (some craftsmen prefer to give each individual piece its coating of wood preservative and even paint it before gluing or nailing or screwing it finally in position) it is time to finish it with paint or stain. Follow the recommendations of the paint manufacturer as detailed on the can, or in folders put out by the better paint companies. You will find that at least two, but more frequently three, coats of good outdoor paint should be used. The first coat fills the grain, the second seals it, with the third coat completely sealing the first two and preserving them so that they will not need attention for two years. Every winter, when the vine has shed its leaves and you can see the trellis well, inspect it carefully to see if there are any sprung joints, blistering paint spots, or cracks in the paint coat where decay organisms may enter, particularly around the base.

If you find the joints sprung, try to force the wood back together, after filling the joint with waterproof glue mixed with sawdust. If the joint is a narrow one, fill it with either putty or white lead and let it dry for a few days before repainting it. If blisters occur, wire brush them and sand down the edges of the solid paint; then use wood preservative and let it dry for a half day or more before repainting, giving the spots two coats and repainting the entire trellis with the third coat. Where cracks occur, treat in the same manner before giving the final coat.

Each time you repaint, wait for a warm and sunny day, or at least a warm one. Cold weather slows down the absorption of oils into wood cells and may not give as complete adherence as is needed. By the time the weather has again warmed sufficiently to permit the oils to soak in, the surface oils will have evaporated and dried, decreasing appreciably the penetrating capacity of the paint and consequently its lasting qualities. Never paint unless the weather will be above the freezing mark during the day on which the painting is to be done. Early fall or late spring are the preferred times. Many home craftsmen feel that the best time for painting is in the early days of spring, when the weather is more likely to become warmer than colder, rather than in the autumn. The paint will completely dry within a few warm days, and they claim that because the fresh paint job is spared the ravages of winter it will look smarter and better through the growing season. We believe that it is immaterial *when* the job is done as long as it is done whenever the trellis *needs* painting. If the old coat is sufficiently intact to carry it through the winter, all well and good, provided that you are sure spring garden tasks will not interfere with the painting in the spring.

A WORD ABOUT COLOR

When you are considering the paint you will use and choosing the color, take note of the color of the house and its trim, of the fence, and of adjacent buildings. Consider also the color of the blossoms or the leaves of the vine you will be using. White is always a good, safe paint color, but not always the most imaginative. Some trellises, particularly those which must endure winter bareness because of the departure of annual vines, might be painted either to match the trim of the house, if the walls of the house are white, or keyed to the color of the blossoms of the annual vines. The color should harmonize with the leaves and flowers of any flowering shrubs or plants in the vicinity, too. If you have bright yellow- or blue-green-leaved plants, be sure to take this into account in choosing your color.

Occasionally a trellis may be painted with two colors, picking up the basic wall color of the house to tie it into the general effect; then using the trim color for the inner frame, for crossbars, or in some other way echoing the second color of the house without ever allowing it to become too obtrusive. Never use color for color's sake or you may find that even *you* don't like it when it is finished, let alone how it affects the neighbors! But if color is used well, you may find yourself top man on the street, even if nobody tells you so. You'll find them copying the principle, and you'll know.

WHAT A TRELLIS IS NOT

You may well want to consider the facts of life before constructing a trellis. You will have to face the fact, sooner or later, that children live not only in the big world but in your immediate vicinity. Even though your own little angels may not be inclined to emulate Tarzan and climb up the trellis, others will. It is part of the child's nature. If this is a problem for you, then be sure to build the trellis strong enough so that it not only will support a child but can become an informal Jungle Gym if necessary. Maybe the answer will be to build some sort of climbing apparatus elsewhere for the children; and then to make doubly sure of your trellis, if you are a demon gardener as well as a parent, plant a good thorny climbing rose on it, standing guard with shotgun or club until the rose is large enough to fend for itself.

A trellis is not a ladder, nor is it a trapeze. We might add that it should not offend the eye of the beholder, either. This is our main objection to the usual trellises offered for sale. They were never good designs and they haven't changed in fifty years or so; or if they have, it has only been in the direction of fanciness. Bad design is boring, and a million trellises of bad designs are something more than a million times as boring. You will take pride in your own trellis, on the other hand, because it is unique and was chosen to fit your house, and yours alone. It will be worth any effort you may wish to put into its building, if only for that reason.

Most of the commercially made trellises are cheap because they are badly made of too-light materials. They are too flimsy to support the weight of vines, and that is why, after a season or two of use, they sag and look so sad. Moreover, because they are made to get the most out of the wood rather than to be of pleasing proportions, they have little relation to the lines of any particular house. This is one of the penalties of buying goods made for the mass market. While they may be "cute," they will never make anyone proud to have them attached to his house, because they have no distinction. Therefore we submit that the answer is to build your own trellis, choosing it for its suitability with the architecture of your house, and thereby achieving distinction.

Metal trellises, a recent innovation, are usually to be avoided, in our opinion. Aside from the un-beautiful curves and curlicues of most of them, there is the problem of heat. They are likely to get very hot in the broiling sun of summer, and certainly such heat will do the vines no good. However, some of the metal mesh or hardware cloths can be used on trellises because they are light in weight and would not hold heat for long, being made of such thin wires. They are very well suited for use with certain types of vines which like to cling with tendrils. The metal meshes which can be detached and rolled up when the house is to be painted, or when the vines have gone their way with the season, also may be worthy of consideration, but our concern is not with them.

Pipes and iron rods are not usually conducive to easy handling nor to true beauty, unless elaborate jigs and various kinds of tools are obtained in order to bend them into shape. That is why we have chosen wood as our material in designing these trellises. It is the most easily worked, least expensive and most long-lasting material at our disposal today. The home craftsman can do a good job with it, using a minimum of time and labor and only a few simple tools.

And now we release your attention so that you may leaf through the designs for trellises in the following pages. We trust that you will find at least one which you feel will fit into your home conditions and integrate well with your house, or which may be adapted without much trouble. Good luck with your trellis and long live your vines.

To Aid You In Selecting Annual Vines

Name of Vine	Usual Height	Cultural Notes	Flower Color
BALLOON VINE	10'-15'	Full sun, good soil.	White
CANARY BIRD VINE	10'-18'	Shady, moist soil.	Canary yellow
CARDINAL CLIMBER	10'-15'	Sunny, good soil.	Red
COBAEA SCANDENS Cathedral Bells	10'-30'	Sow seed edgewise. Germinate quickly.	Lavender- purple
CYPRESS VINE	10'-15'	Soak seeds in warm water 4 hours, then plant.	Red and white
GOURD VINES	8'-20'	Full sun, good soil, not particular.	Fruit various
HEARTS-AND-HONEY	10'-15'	Sunny, good soil.	Red with honey-yellow throat
MOCK-CUCUMBER	8'-15'	Sun, any soil.	White, green fruit
MORNING GLORY Heavenly Blue Pearly Gates Blue Star Scarlett O'Hara Darling	10'-15'	Any good soil, full sun, soak seeds 4 hours before planting outdoors.	Blue White Pale Blue Red Wine Red
MOONFLOWER	10'-20'	Same as for Morning Glory.	White
SWEETPEAS Cuthbertson varieties	4'- 8'	Early planting, cool climate best. Sub-irrigation may take them through to July.	Various

TRELLISES

FOUR-SQUARE TRELLIS

A trellis for a vine which will clothe all four sides of the central post, which will be interesting in winter because of its unique pattern and yet which will afford summer security for heavy vines, roses or other climbing plants, is hard to find. This one will look well all year and will also be transportable—a factor for renters—if you should decide to move. If desired, of course, the central post may be set in concrete. Note that the post is tapered at the top to the width of the $\frac{5}{4}''$ x $\frac{5}{4}''$ stock used for cross pieces. They are overlapped as explained in the detail drawing and toenailed to the post where dowels are not used at base and top.

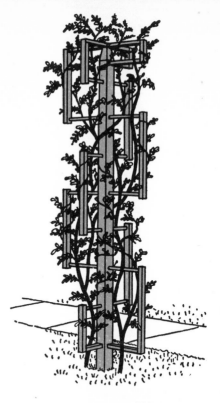

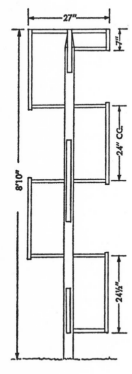

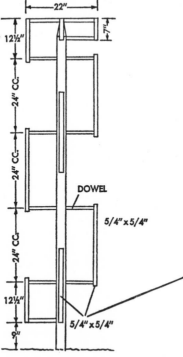

DOWEL

5/4" x 5/4"

5/4" x 5/4"

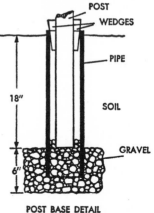

POST
WEDGES
PIPE
18"
SOIL
6"
GRAVEL

POST BASE DETAIL

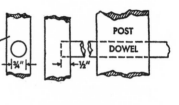

POST
DOWEL
3/4"
1/2"

DOWEL DETAIL

82

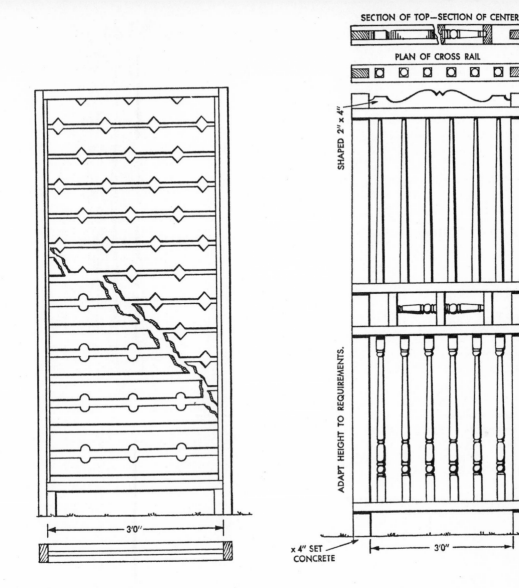

SECTION OF TOP—SECTION OF CENTER

PLAN OF CROSS RAIL

SHAPED 2″ x 4″

ADAPT HEIGHT TO REQUIREMENTS.

3′0″

x 4″ SET CONCRETE

3′0″

FROM THE SPANISH

A trellis adapted from grillework or gates seen in Spanish architecture is scaled down to fit small homes. Shown here: top half, plain stair balusters; bottom half, turned balusters (buy them at lumberyard, housewreckers). Center grille uses parts cut from center balusters. Use one version or the other, but same center grille for both. Cross-rails are cut out to fit around uprights (see details) holes being bored to admit the balusters or penetrate rail with long screw to hold all square balusters in place.

SQUARES ON CIRCLES

A trellis that is also a view-stopper is a useful structure beside a door or window. Cut triangles from 1″ x 6″ boards, then set boards 1″ apart, alternating squares to form diagonal patterns. In the bottom half is shown another version; 1″ x 8″ boards have holes carefully bored at 12″ intervals, and then board is sawed down the middle, edges smoothed and boards set $1\frac{1}{2}$″ apart with circles in regular vertical patterns. Boards are held in place on 2″ x 4″ frame by 1′ moldings. Train vines through holes.

83

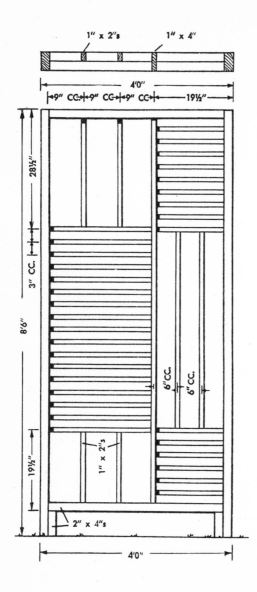

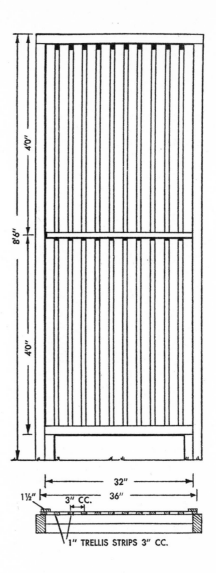

STRIPES AND RECTANGLES

This trellis combines both strong horizontals and verticals, making a good contrast to the informal, natural curves of a vine, and will give many good strong shadow lines as well as a sturdy framework to which can be fastened climbing or pillar roses, or some other vine which needs a strong support. The framework is made of 2″ x 4″s while the interior portion is made up of rectangles of 1″ x 2″s in various combinations, fastened in place with rust-resistant nails or screws.

STRICTLY VERTICAL

For twining annual vines a light frame of this sort is sufficient. Morning glories, moon-flower and runner beans will twine about the vertical trellis strips nailed or screwed to the back side of the 2″ x 4″ frame with a 1″ x 4″ center bar. Frame the back side with a wider trellis strip on all four sides. Note that in both of these trellises the frame 2″ x 4″s are doubled at base, bolted together and set in concrete for protection against high winds.

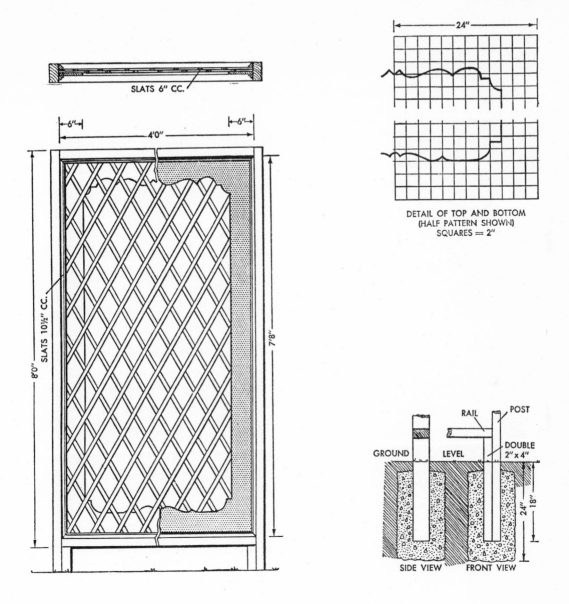

SLATS 6" CC.

6" 4'0" 6"

SLATS 10½" CC.

8'0"

7'8"

24"

DETAIL OF TOP AND BOTTOM
(HALF PATTERN SHOWN)
SQUARES = 2"

RAIL   POST

GROUND   LEVEL   DOUBLE 2" x 4"

18"   24"

SIDE VIEW   FRONT VIEW

FRENCH PROVINCIAL TRELLIS

So many houses, both old and new, have taken the French Provincial style for
their motif of design that a trellis might well follow out the general scheme.
Outdoor plywood jigsawed out to the graceful lines shown in the scale detail
at right, makes the "window" which frames the lattice and vines. The frame
of 2" x 4"s is securely set in concrete, the lattice and plywood held in place
by mouldings mitred at corners. A 4' x 8' sheet of ¼" plywood is used in our
version but design may be adapted as desired. Plywood center may be used
for another project if care is used in cutting. Use wood preservative on exposed edges.

85

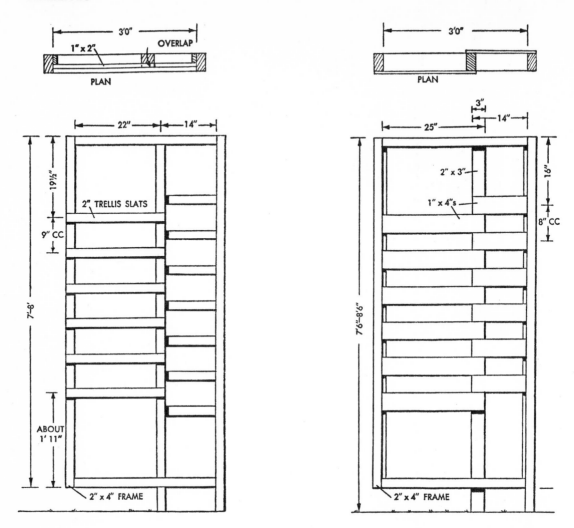

SIMPLE MODERN DOORSIDE TRELLIS

Simplicity is the keynote of good modern design, and the crisp architectural lines of this simple trellis will complement and contrast well with the flowing natural lines of the vines that will grow on it. The two versions shown are related but give completely different effects. On the left is a design that has the slats all on the front, set into the frame and set flush with each other, while that on the right has narrow slats attached to the outside of the frame, and one set on the front, the other on the back, alternating in placement. If annual vines are used, so that trellis is bare in winter, perhaps painting slats in contrasting colors would give a gay, all-year effect, providing the colors were chosen to harmonize with summer blossom colors. Use wood preservative on exposed edges.

TRELLIS-TOPPED VIEW-BREAKER FENCE
A simple fence to embellish terraces while protecting them from prying eyes, hoists the fence part to screen the eye level, and leaves the ground level open for ventilation. Shown here for covering is wire screen set in plastic, but various other materials will be adaptable: 1" x 2" mesh welded wire fencing; thin, wide basketwoven boards; bamboo roll-up shades; or snowfencing as shown.

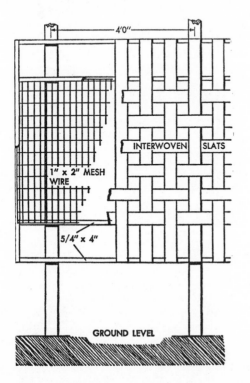

4'0"

INTERWOVEN SLATS

1" x 2" MESH WIRE

5/4" x 4"

GROUND LEVEL

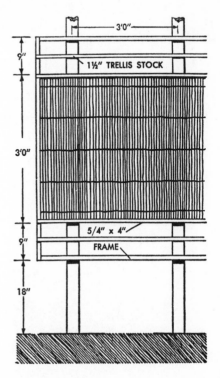

3'0"

9"

1½" TRELLIS STOCK

3'0"

5/4" x 4"

9"

FRAME

18"

TRELLISES

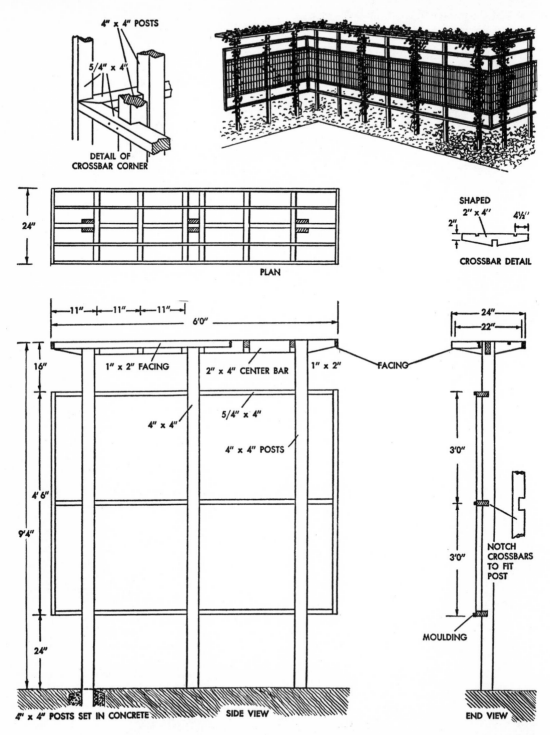

4" x 4" POSTS

5/4" x 4"

DETAIL OF CROSSBAR CORNER

24"

PLAN

SHAPED

2" x 4"

2"

4½"

CROSSBAR DETAIL

11" **11"** **11"**

6'0"

24"

22"

16"

1" x 2" FACING

2" x 4" CENTER BAR

1" x 2"

FACING

4" x 4"

5/4" x 4"

4" x 4" POSTS

4' 6"

9'4"

3'0"

3'0"

NOTCH CROSSBARS TO FIT POST

24"

MOULDING

4" x 4" POSTS SET IN CONCRETE

SIDE VIEW

END VIEW

Vary the measurements of the viewbreaker fence to suit the material you choose. Bend fence around a corner, set it at an angle, adapt it in any way to fit your own needs.

THE DOORSIDE TRELLIS

Many a door is perfectly good, but dull. The addition of a trellis can make it a distinguished feature of the house. Use a design which will go well with the style of the house, and choose a vine which will stay within reasonable bounds and add its picturesque quality to the design of the trellis. All three designs are built on one basic framework of 2″ x 4″s with the horizontals and verticals inside built of 1″ x 1″s or ⅝″ x ⅝″s. Trellises can be attached to the house as shown on left, or be freestanding like those below, with the two major vertical members sunk into concrete after being carefully treated with wood preservative. The vertical at right angles to these members should have a ½″ to 1″ space left between the end and ground level to prevent moisture, rot.

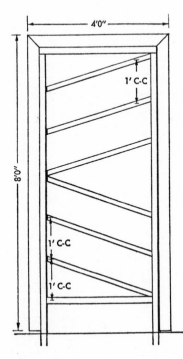

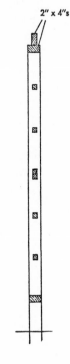

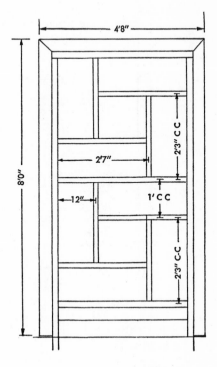

TRELLIS FENCE ENDERS

To relieve the starkness of a fence beside a bare wall, end the fence with a trellis such as these on this page. Construction is similar to that on the previous page, the top members merely being extended to meet the house and bridge the walk. The fence shown is built of grooved plywood but might be built of 1" x 4"s or 6"s nailed on the 2" x 4" framing, posts sunk in concrete. Both designs below would complement modern houses.

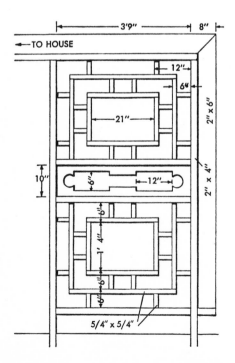

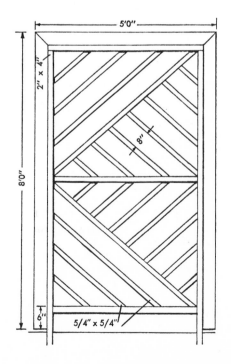

7

Good Fences Make Good Neighbors

Sometime in the dim prehistoric days beyond the memory of man, an ancestor of the race began to cultivate the land and grow food plants. He soon found that the animals of the forest also enjoyed this food, so that protection became necessary; and in order that other humans might know that these superior food products were the result of private enterprise and not the bounty of nature, some boundary line was needed to mark the limits of his endeavors. Therefore the Fence came into being.

Since that long-ago time, of course, the fence has gone through many developments. It has been an important means of adding to the beauty of a garden; it has frequently provided its owner with a means of conspicuous display of wealth, as it embellished his property; and, of late years (as well as earlier), it has been an important way of providing visual privacy. Sometimes the fence has served merely as a definition of the property lines; or as a kind of openwork background for a shrub border or flower bed; again, it has at times been a kind of horizontal trellis for vines, full of beauty when the flowers and leaves of the vine clothe it, and continuing through the drab days of winter to offer its own charm of architectural pattern to brighten those dismal, barren months of the year.

FENCES OF YESTERYEAR

During the Victorian era fences had developed into amazing structures, serving a function far beyond that of excluding unwanted animals or marking the boundary between the public street and the private areas of the home property. They became as elaborate, as loaded with gingerbread design, as the whatnots inside the houses. Sometimes they were built of iron, with imposing stone gateposts and with stone posts supporting the weight of the elaborate fence. More frequently, jigsawed wood attempted to imitate the whorls and arabesques of the iron fences. Even the circular flower beds in the lawns were fenced with low iron or wooden fences, and low iron fences surmounted the ridgepoles of house roofs. Eventually, of course, people tired of this nonsense and sought something new.

In America the fence began to disappear, and front yards all along the street were incorporated, visually, into the public area of the street. The resulting open, park-like look of our streets has often excited comment from visiting Europeans. They do not like the absence of fences, particularly in back gardens where fences have been omitted or wire netting used to clothe the fence frame; and it must be admitted that their remarks on the American lack of privacy are completely valid. That is why there is a return of the fence today: primarily to give us privacy from trespassers and from intrusive glances of passersby.

The fence today fulfills several definite functions. It encloses the yard to keep children in and animals out; it offers, many times, an architectural finish to the home composition, enhancing the lines of the house, lengthening or shortening the property and buildings visually; and, above all, the modern fence produces the privacy that America is now mature enough to want. We are discarding the old pioneer idea that we live in a wilderness, and we are realizing that communal life in civilized areas entails a loss of that privacy which is necessary to the mature individual.

Let us take up these points and examine them further. Our small children must play out-of-doors to get healthful sunshine and air and to get them out from underfoot in our small homes of today. All gardeners will recognize the need for keeping animals (except one's own) out of gardens, where they may trample flowers and ruin garden effects in other ways. Therefore, the back garden at least must be fenced. A good fence will provide a beautiful backdrop for the scene we set for our summer activities. Adults, too, are taking the air more frequently out-of-doors, and often we find that the children's play yard is enclosed beside the terrace, where it can be overseen from the house as well as from outside.

As we drive around the subdivisions which have sprung up about every city and many towns, we see how well fences can fulfill their function of extending the lines of the house, of providing an architectural finish to the property. Many a little row house in a subdivision, indistinguishable from its neighbors, suddenly achieves an identity when a good fence is added which harmonizes or contrasts with the house. The use of materials which lend pattern or texture, or both, is one of the most stimulating and interesting of the many new facets of the trend in home building. Look through the designs for fences which follow this section of text and see how the long lines are emphasized in some; how fences are kept low to give long, wide lines to a property which may be quite narrow. If the house is one of the long, low modern homes, this will harmonize with it and produce a visual width quite at variance with the actual measurements.

PRIVACY FENCES

As plots of ground become increasingly smaller with the mushrooming expansion of suburban areas around our towns, Americans have felt a need which is met by the "privacy fence." This is not to be confused with what our grandmothers called a "spite fence"—a high fence built to show a neighbor the degree of contempt felt for him and all he represented. A "spite fence" was high, ugly, and as undecorative as possible on the side facing the scorned neighbor. It bears little resemblance to the handsome, tall, good-neighbor fences which today insure privacy on both sides and are usually erected with mutual consent; sometimes with mutual funds. Such a well-planned fence, good to look at on both its sides, removes a cause for friction immediately, and good neighbors are more likely to remain good neighbors because of it.

Panels alternate on both sides of a 2" x 4" framework on 4" x 4" posts, overlapped 3 inches to maintain privacy, yet provide for air circulation. Masonite grooved siding is moisture-resistant, yet can be painted. Note the matching gate in the fence by the house.

An attractive garden, viewed from an indoor bathroom, is surrounded by privacy fences of two kinds: redwood framed fence, right, is faced with solid light-colored boards and redwood staves with shaped tops echo the line effect (left) of the platform outside windows.

Outdoor privacy alone is not the entire problem. Today so many homes are being built with large areas of glass or with "picture windows" that interior privacy is often destroyed or made negligible. Since not everyone is an exhibitionist, it is disconcerting, to say the least, to look out of a picture window and see passersby or neighbors finding a picture from their side of the window, too. The answer is a privacy fence, one which is a part of the over-all architectural scheme and not too obviously an afterthought.

The first means of obtaining privacy is to build a high fence, tightly constructed around the entire area to be screened. But this is not always necessary. Sometimes the height of the fence need be maintained only to the far edge of the angle of vision from which the fenced area is to be screened. The rest of the fence can step down or even be dispensed with entirely. Materials, design remain the same; only height changes.

Fairly frequently it is better not to have a fence which must be solid, for in that case it may cut off air circulation. Privacy can be gained without creating problems of ventilation in the living areas. Similarly, a fence must not create a wind trap where drafts make it unpleasant for outdoor living. Perhaps a baffle fence or a louvred fence such as those shown in these pages will offer a good solution, blocking the public gaze, but allowing free circulation of air; breaking the force of the wind where that is a problem, but letting through enough air for good ventilation. Baffle fences also have the advantage of controlling traffic through the living areas, directing it away from flower beds and terrace areas but permitting a good easy flow of air, thus adding to the pleasure of living outdoors.

Short sections of a fence may be attached to the house wall at a right angle, or at whatever angle is proper, or even curved outward, assuring privacy for bedrooms which front on the garden and terrace, and at the same time providing an excellent background for terrace-side planting. These short baffle fences may be elevated to cut the line of view into the windows not only from the terrace but also from neighboring gardens or the street. Occasionally a tiny terrace or garden will be fenced with a high privacy fence to provide a pleasant view from the house and to permit sun-bathing in privacy.

The kind of fence chosen will, of course, be limited by the degree of privacy desired and by the factors of wind control and ventilation. Tight

Oversized basketweave fence is easy to build, allows ventilation through openings. Inexpensively made of ¼″ exterior plywood, cut to any width that divides equally into 48″ plywood width, it is railed to alternating 4″ x 4″ posts and 2″ x 4″ divider posts.

board fences or those made of plywood, asbestos board (Transite), outdoor hardboard, and other complete coverings will give the maximum privacy and protection. Wire fencing, post-and-rail, picket, and lattice fences naturally give the least privacy and wind control. Between these two extremes will be found various other fence designs, each of them with a definite function as well as a degree of beauty all its own: the louvred fences, both horizontal and vertical; closely- or more widely-spaced pickets of various kinds; woven lattice fences very closely spaced or more open in construction; and boards and trellis slats used either vertically or horizontally, in even or uneven spacing, and sometimes utilizing a variety of widths as part of the charm of their design.

Sometimes the answer will best be found in the use of a sturdy framework clothed with some more or less opaque but translucent material, such as wire screen in plastic, corrugated sheets of fiberglass or plastic, or, where there is a good view that must not be shut out, clear glass. In cost, glass will be the most expensive, of course, with plastic and fiberglass costing somewhat less, and the wire screen in plastic being cheapest of all. This last is light in weight and most attractive. It is not so durable as many other materials but, if plants are placed in beds close behind it, the sun will make very beautiful translucent shadow-pictures on the fence all day long.

Where privacy is needed only during the outdoor-living months, roll-up bamboo porch shades or snow-fencing may provide an answer to the question of what to use on the fence frame. Annual vines can augment and produce a living privacy screen; welded wire fencing in 1″ x 2″ or 2″ x 4″-mesh provides good clinging facilities for them. In winter, when the vines have been removed, the attractive oblong-patterned mesh permits sun, air, and light to enter, which may be most desirable at that time, particularly in the modern home. Gardeners, too, will welcome such a trellis, for the vines will strip off easily once they have died back in the autumn.

FENCES FOR PROTECTION

A fence protects property in two ways. It keeps *out* unwanted animals and people. It keeps *in* one's own children and pets. Not all human intruders can be excluded, of course, but a fence acts as a definite deterrent to the casual trespasser. It must be remembered that post-and-rail fencing, widely spaced louvres, and fences with the lower rail set fairly high off the ground will, of course, admit small animals and permit one's own pets to stray.

Cooperative fencing is often seen today, with several congenial neighbors planning their gardens so that the openness is maintained and each small plot, visually augmented by its neighbors, given an effect of space. Except for a perimeter fence, the area may be kept open, or low wire fences may be used, with posts painted deep green or black so that they disappear into the shrubbery or are inconspicuous. In this way the view is almost unobstructed but the object of the fence is achieved: pets and children are kept in the home yard.

FACTS ABOUT FROST

It will be well to consider the possibility of frost and its action in the northern parts of the country. Frosts flow down a hillside like water, collecting in low spots, being diverted around obstacles even as water flows around a log or stone in a stream bed. Realizing this, you can build your fence to protect your plants. If your plot slopes, don't dam up the path of frosty air currents, but place your fences and plantings so as to keep the currents in motion down the hillside. Solid fences placed across the lower end of a hillside plot will create a frost "pool" and help to kill or injure any plants affected by frost. Instead, use an open fence which will permit free flow of air through it. If you have a solid fence creating such a gardening problem, opening up a gate in it or providing an open section in one or more places will help to overcome the condition.

FENCES FOR WIND DEFLECTION

Frequently the logical place for a terrace or outdoor sitting-room would be alongside the house except for the fact that the prevailing winds are on that facade. It is not pleasant to sit outside and be lambasted by winds; and to the gardener this may be a distinct deterrent to growing tender or fragile plants. A baffle fence or louvred fence, even a solid one with a few holes to permit ventilation of the area, can provide the answer here. Curve the fence, angle it, step it up or down to fit the contours of the location, but make it high enough to do the job.

Fences can also act as temporary or stand-in structures which will permit small inexpensive plants to grow into thick and tall windbreaks. Unless prohibitively expensive nursery stock were used, it would be many years before plants would grow large enough to become a wind barrier. Here, a fence will not only give the plant material protection while it is growing and provide proper conditions for the terrace and garden, but it will also provide a sightly background for the plants until they have reached such a size that the fence can be removed.

It is not necessary to block wind off absolutely. It can be slowed down, diverted, and guided according to the need felt. Sometimes it may be *necessary* to block the wind in a place where there is a view which one wishes to keep unobstructed. In this case window or plate glass should be used. Remember that they can often be purchased for very reasonable sums from wrecking yards or from dealers who take in broken plate in part trade for new glass. Set in sturdy wooden frames, with moldings holding it firmly in place, a glass fence can be a thing of beauty. A word of caution may be in order, however. Where winds are very strong, extra-large-sized panes of glass may buckle and break under wind stress. Also be sure that no dead limbs are on trees nearby where they may be dislodged and hurled into the glass fence. Aside from these drawbacks, however, a glass-enclosed terrace will permit the householder to enjoy his terrace and garden in many problem-locations.

Because of many factors, the prevailing winds may have capricious courses. Large buildings, heavy plantings of trees—particularly ever-greens—the slope of the land, and the placement of buildings: all may account for a wind problem. Therefore you may have to keep a watch on the winds for several months, or even an entire year, to observe their action before knowing just how and where to place your fence to outwit nature.

Drive 2-foot stakes at intervals around the garden and the terrace and attach strips of white or bright cloth to the tops so that you can see them easily. Taller stakes may also need to be used in extreme cases to see whether the winds vary in direction close to the ground or higher above it. Note down your findings, and, when you feel you have seen enough to evaluate the problem, figure what the pattern of the winds' courses may be and plan your fence accordingly. However, this procedure is advised only for extreme cases. In most places the winds will not be strong or capricious enough to warrant going to these lengths.

CONCEALMENT FENCES

Unlovely but necessary appurtenances of modern living can be hidden from view by a fence. Garbage and trash cans, incinerators, work areas where cold frames, toolsheds, and compost piles are placed—any of these strictly necessary but unsightly objects may be beautifully obscured by a fence.

Either try to make such a concealment fence as integrated and as unobtrusive a part of the general scheme as possible, or else be bold and make it into a definite *feature* on the garden side by using some contrasting material or well-planned design, thus turning the concealment of some necessity into a triumph. If possible, tie the fence in with a garage or house, or with another fence, so that it seems to have a reason for being rather than standing on its own, calling attention to itself.

CHILDREN'S PLAY YARDS

The problem of keeping small children away from the hazards of traffic and from the danger of straying animals can be solved by constructing a play yard. Locate it near the kitchen or the family room, where the mother can keep her eye on the children as she works. In later years this may become a little private garden for herb-growing or the pursuit of some other hobby. It can be made a sun-trap where sun-bathing can be indulged in privately, early and late in the season when less protected areas are not warm enough. The fence should be high, of course—high enough to prevent active young children from scaling it and escaping—and the gate should have some means of fastening it high enough that they cannot reach it, and preferably placed on the outside of the gate, so that children may be "forcibly detained" when that is desirable.

BOUNDARY DEFINER AND DOORYARD FENCES

All over the country we see a growing tendency to return to the dooryard garden, a charming area surrounded by a low fence. Where boundaries are to be defined, but where no particular purpose would be served by the use of a high fence, the low picket, wire, or board fence may be used to protect the areas. Low fences contain the plantings, giving to the outer aspect neatness and definition, which may be lacking when there is merely an edge of a lawn for a boundary. Also, a fence provides an architectural contrast for the flowers and plants used in the dooryard garden or along other boundaries, counterpointing textures with crisp lines.

ENCLOSING THE "OUTDOOR ROOM"

In a sense, a terrace and even an entire garden may be an outdoor room. A fence, therefore, becomes the "walls" of that outdoor room. But unlike indoor walls, a fence need not be box-like and conventional, a squared-off structure. Fences can be built at an angle, can curve, be serpentine or zigzag, can even be freestanding structures. As with indoor walls, fences segregate living areas from service areas, giving more room by limiting each activity to its proper sphere. By clever planning a fence can make a house look longer, higher or lower, more important or less aggressive, and even help a house to settle into its site better. When materials are used which are compatible with those of the house, the whole landscape composition will be more pleasant and attractive, because all of the architectural elements "belong." A fence saves wear and tear on the garden, too, by guiding traffic into the proper channels, keeping errant footsteps from wearing out the lawn and trampling flowers.

Among the pages of designs in this section you will find many devoted to "walls" for the outdoor room which we hope will serve as an inspiration and a guide rather than as something for you to copy. Improvise, adapt and change as you wish, using materials of your own choice; or combine some elements from one design with those of another so that they will suit your location and help to solve your own particular problem.

CHECK THE REGULATIONS

There are a few preliminary matters to be settled before you dig the post holes, saw the lumber, or even settle finally on the design so that lumber can be ordered. Before going ahead with your fence, check with local authorities to see if there are any ordinances or regulations concerning the height, location, or materials of fences. Check, too, to be sure there are no new zoning laws which may have a bearing on fences. You may inquire about this at the bank which holds your mortgage. They can tell you where to make the proper inquiries of local officials. Then, when you know the limitations, plan your fence accordingly.

It is wise, too, to have the place surveyed if you are not absolutely certain of your property lines. If you build your fence so that it encroaches on your neighbor's side of the line, it becomes his property and he can remove it, paint it, do anything he likes with it. Even if it is correctly situated on your property line, it will be to your advantage to have an agreement with the adjoining owner as to the placement, the materials used, and, if it is centered on the property line and he pays half the cost, whether or not he will stand half the maintenance. Usually neighbors reach such agreements amicably (which is the best way), but occasionally there is lack of agreement. In this case you still have a perfect right to build your fence and to maintain it, so long as it is wholly on your own property and does not violate any state, county, or local law or ordinance, and so long as there are no restrictions in your property deed concerning fences.

If your neighbor proves hostile to the idea of a fence, you may place it from one to six inches inside your property line, after you have made sure that it conforms with regulations. *Be sure* to check and double-check your property lines so as to avoid unpleasantness in the future. If your neighbor is amenable, or even wants to join in paying for the fence, have a lawyer draw up an agreement in writing and have it recorded locally so that it will be legally binding on future owners of the property, should either of you sell his home or should there be a change of heart later on.

If it is at all possible to reach an amicable agreement with your neighbor, do so. He will probably be living next door to you for many years to come. Whatever your *legal* rights may be, many headaches and worries can be avoided if the relationship is pleasant and the fence is mutually satisfactory. If, however, your neighbor is adamant in his opposition to the suggestion, you may very well have other trouble with him. A fence is likely to keep that trouble at bay, and you will be able to enjoy the privacy which is the legal right of every property holder.

Sometimes in local regulations or deed restrictions you will find that it is obligatory that "party-line" fences be paid for and maintained equally by the property owners. Or it may be that, if the adjacent lot is not built on or improved as yet, there is no obligation on the part of the landowner to contribute. Then you may have to bear all of the cost yourself and, if you do, you may want to build the fence entirely on your own land. In any case, the wisdom of careful exploration of the regulations on fences has been demonstrated. Look before you build.

HOW TO PLAN YOUR FENCE IN ADVANCE

When you have selected the design for your fence and established its location, stake out the area with relation to the property lines, as indicated above. Run a cord along the stakes so that you can see exactly what your problems may be. By measuring the distance of the run of the fence you can determine pretty exactly how much lumber will be required, how many posts to buy, etc. It is always a good idea to draw a plan of the layout of the fence, though it needn't be absolutely accurate. It will assist you materially in determining what you need and where to put the fence. It is always easier to make revisions on a piece of paper than on the ground. Perhaps you will even find that in the plan you can figure out a better way to place the fence for greater privacy or for better traffic flow than when you are working outdoors, where you may be more confused by details.

In designing the fences shown in these pages we have tried to use stock lengths of lumber in the construction so as to eliminate as much cutting as possible. The maximum span recommended is 8 feet—in many cases 6 or 7 feet (a 14-foot-length cut will secure this) is better where the "skin" of the fence is heavy or there is some other need for shortening the span. Look into the possibility, when figuring costs, of buying the double length and cutting it, if you have a portable power saw. It may well save you money.

Check the lumber when it is delivered to see how accurately it has been cut. Sometimes the ends are cut at a rough square, but not accurately enough for use in butting two pieces together. In this case you must square them and cut them to remove the discrepancy. Where a number of boards need trimming you can save labor by cutting them all at once, using a large C-clamp at both ends to hold them together accurately, thus assuring that all boards will be cut to exactly the same dimensions.

FENCE POSTS

Usually posts can be ordered already cut to the length you wish to use. For the average fence the length is from 18 to 30 inches longer than the aboveground height of the fence when finished. For heavy fences

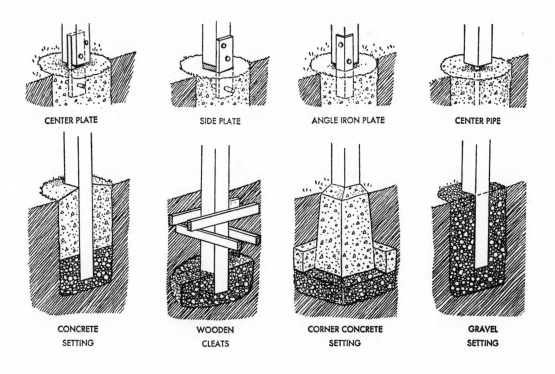

CENTER PLATE   SIDE PLATE   ANGLE IRON PLATE   CENTER PIPE

CONCRETE   WOODEN   CORNER CONCRETE   GRAVEL
SETTING   CLEATS   SETTING   SETTING

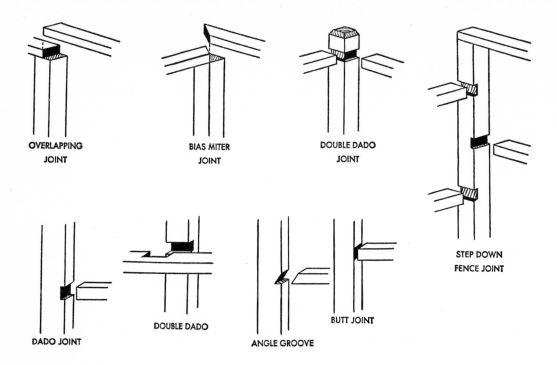

OVERLAPPING JOINT

BIAS MITER JOINT

DOUBLE DADO JOINT

STEP DOWN FENCE JOINT

DADO JOINT

DOUBLE DADO

ANGLE GROOVE

BUTT JOINT

made of solid boards or other weighty materials, longer and heavier posts will be needed. For taller fences, too, where there will be more strain from wind and weather, longer posts are indicated. A 6'0" tall fence will require posts 8' to 9' long; these should usually be 4" x 4"s if the covering is heavy. Fences 48 inches tall need posts 6'0" long, and 6 to 12 inches longer than that if the material is heavy. For fences up to 3'0" in height posts may be made of 2" x 4"s for average or light covering materials, if doubled below ground. Other post sizes which may be used are 3" x 3"s, 4" x 4"s, 6" x 6"s, or larger sizes if your budget and the job warrant the expense. For dooryard or other low fences 2" x 3"s will usually serve.

Corner and gate posts are usually made heavier (4" x 4" where a 2" x 4" is the intermediate post size; 6" x 6" where the intermediate posts are 4" x 4"s) and also should be 6 to 12 inches longer than intermediate posts. If they are well set in concrete (see the Post Base Setting sketch) this extra length will not be necessary unless there is an unusual amount of wind or other strain to be withstood.

FENCE RAILS

Top and bottom crossrails should generally be made of 2″ x 4″s for most types of fences. Occasionally, as in the low Dooryard Fences or for taller but light-weight kinds of fences, 2″ x 3″s, 2″ x 2″s, and even 1″ x 4″s may be used. Even if there is no weight with which to contend; there may be other strains to consider—children climbing on the fence, the weight of vines if the fence is to be used for trellis work, and wind and weather stresses. You may want to stick to the classic basic frame: 4″ x 4″ posts for gates and corners, 2″ x 4″s doubled below ground for intermediate posts, and 2″ x 4″s for top and bottom and center rails.

The span of the top rail may be double that of the lower rails because it will be supported by the posts on which it rests, and this will save a good deal of time and labor in cutting. Lower rails are supported by being dadoed into the post in various ways (see Fence Post Joints sketch) or by blocks nailed to the post under the rail. This latter is less satisfactory and less sightly than dadoing, however.

FACING BOARDS AND OTHER "SKINS"

Almost anything can be used for surfacing the fence these days, from boards of random widths to grooved plywood, from corrugated asbestos or flat plastic to wire screen in plastic, which comes in rolls.

When you are deciding on the length and height of each span of fence, consider the stock lengths and widths of the materials and save yourself much labor by using fractions or multiples of these stock sizes. For instance, plywood comes in 4′ x 8′ sheets (other stock sizes are made but are not always readily available), and so do such materials as hardboard, Transite or asbestos board, and so on. If you can adjust your fence units to this size you can build your fence with little trouble. It is possible to have the lumber yard cut the materials on their power machinery to whatever sizes you need, if the job is large enough to warrant the added charge they will make. This will save much wear and tear on your shoulder muscles. It will also eliminate the need for having another person help you to wrestle with the large sheets (it is not always possible for one person to handle them alone).

Boards, where used for the "skin", are economical, particularly when random widths may be employed. Instead of using the more expensive grooved plywood, it is possible to use grooved siding, shiplap, or some other stock lumber and save a bit on it. But the amateur will probably feel that the extra cost is justified when he compares the time consumed in cutting a large number of boards with the convenience and saving of time effected when a piece of plywood is used. Consider the surface interest and textural beauty achieved when plywood is used with various slats or 1″ x 1″s in patterns applied to it. This is a field which is new, so you may want to develop your own patterns.

Where strength is not mandatory, or where the fence is to be used as a kind of horizontal trellis for light annual vines, lighter weights of wood may be used. Because of their lighter weight they will cost less, but they will not stand up as well as more sturdily built fences. They must be used with discretion, therefore, but with the knowledge that in many places nothing will ever give the same effect as the lacy, trellis-like fence. On the other hand, a good, sturdy, blocky fence has its uses, too. It will complement certain houses as no other fence can, lending a solidity which is most attractive and correct for a particular kind of house and location. And, of course, for privacy this type of fence is ideal.

PICKET FENCES

The picket fence, traditional as it is, has a rightness and a very pleasant quality with traditional homes. Certain types are adaptable to more modern structures, too. It is possible to purchase pickets already cut in various decorative traditional shapes of suitable lengths in many parts of the country and from the large national mail order houses. However, for those who wish to make their own pickets, we recommend using stock from 1¼ to 4 inches in width and from ½ to 1 inch in thickness, the length of the picket depending upon the height of the fence. Note, in the pages of designs which follow, the details of how to cut pickets, and also the various designs given which can be adapted as you may wish for your own fence.

Square pickets may also be used, 1″ x 1″, ⁵⁄₄″ x ⁵⁄₄″, 1½″ x 1½″, or even 2″ x 2″ if you wish a very heavy fence. Round pickets may be made of dowels of ½ inch in diameter upwards, but ¾ or 1 inch is preferred. Larger sizes of dowels run into considerable expense, but if you wish to use clothes pole stock, which is about 1¼ inch in diameter and comes in various lengths, you may be able to cut down the cost considerably.

UNUSUAL MATERIALS GIVE UNUSUAL EFFECTS

In some sections of the country various unusual materials are available locally. For instance, on the West Coast, grape stakes may be purchased. They are rustic, rugged and good-looking, and may be driven into the ground, mounted on or between rails, and used in many other ways. By driving them into the ground to an even height, placing them in serpentine curves, semicircles, squares, or in any pattern you wish, you can make a very unusual low fence.

A material which is available nationally from mail order houses is snow-fencing: rough-sawn pickets or boards ½″ x 1½″ x 4′ spaced 2 inches apart and held by several double strands of twisted wire. They are cheapest when bought in 50-foot rolls from Sears Roebuck or Montgomery Ward, but sometimes in local supply houses shorter lengths may be purchased. They are used by highway departments for protecting the roads from drifting snow, but when mounted on a fence they are very good-looking. Usually they come stained a pleasant soft red, but they may be painted any color.

Picket fencing, with pointed pickets ½″ x 1″ and of various lengths— 36, 42, and 48 inches—mounted in twisted wires similar to snow-fencing is also available from the mail order houses. Matching gates can be supplied. Posts may be either round or square, with light rails or no rails at all being used for the fence, since the wires will support it reasonably well. This would be an ideal fence for a renter, since it could be rolled up and taken with him when he moved.

Where privacy is needed without cutting off all light, translucent fences may be made of corrugated or flat plastic, fiberglass, or some other translucent material mounted on a wooden frame. Steel posts may be driven into the ground and strips of flat plastic woven between them basket-fashion, making an absolutely rot-proof fence, and a good-looking one. A less permanent, but equally good-looking, fence for view-blocking without loss of light can be made of plastic-encased wire screening, using steel or aluminum wire. This tough, durable plastic will last for a number of years in most climates and, being fairly inexpensive, will cost but little to replace. Where high winds are encountered it is not advocated, because it is likely to whip loose under the stress. It can be cut with tin snips or strong shears, and it generally comes in 36-inch widths. However, mail order houses list additional widths of 28 and 48 inches in 4-mesh, 9-mesh, and 14-mesh screen, the price increasing with the density of the screen. Mouldings or wood strips are used to hold the material in place on the fence frame. Similar materials are sold under various trade names, such as Cel-O-Glass, Sun-Ray Wire, and so on.

Bamboo roll-up blinds or porch shades also make interesting textures on fences and may be used for many other purposes, too. Being woven together with cotton string, they are not eternally durable, but they are cheap enough so that they can be replaced every few years. It is wise to watch end-of-season clearance sales, when stores are disposing of summer stocks, but sometimes you can find slightly damaged odd lots which you can buy and cut to fit your spaces. They are usually 6 feet long and come in widths of from 2 to 12 feet. Their pleasant color and texture, even when weathered, make them desirable in many locations.

ROLL-WIRE FENCING

Wire comes in many interesting patterns. It is ideal for use where annual vines will clamber on the fence to clothe it for the summer months. Welded wire fabric, usually made from rust-resistant copper-bearing steel, comes in various square meshes, ¼" x ¼", or ½" x ½", up to 1" x 1"; and in oblong meshes 1" x 2" or 2" x 4", the oblongs running with the length of the roll. This pattern is particularly good with modern, simple homes.

Although considerably cheaper, chicken and rabbit wire is not so attractive. It might, however, be adaptable in some places, particularly if it backed up an attractive wooden frame.

A FENCE CAN BE A WALL

Not all fences, baffles, and enclosures are made of wood, of course. In the past most boundary fences were built solidly and securely of stone or other masonry. There is a section in Part II, Masonry, on walls and enclosures, retaining walls and view-blockers which may also be of interest to the reader. Often masonry can be combined with wood construction to produce the desired effect, adding a difference in texture that will be most welcome and attractive. For instance, where a bank has been cut away to level land on which to build a terrace, a permanent retaining wall to hold the cut soil may be necessary.

At the same time it may be desirable to have a fence to block the view, to protect the terrace from wind, or to provide a background for plantings placed atop the wall, at least in some portions. Brick, concrete, block, or stone masonry may be employed to give the permanent holding quality, the face of the masonry contributing either a formal, smooth appearance, if that is desired, or it may be textured for a rustic natural surface if a more informal or rugged look is indicated. The fence, likewise, may be formal and architectural, light or heavy, finely finished and painted, or rugged and rustic in appearance, to make the total effect exactly what the builder wishes or feels is best with the surroundings.

NOW IT IS UP TO YOU

In the following pages you will find 75 designs for fences from which to choose. You will find low fences, high fences, open fences, solid fences, privacy fences, and fences we call "Bikini" because they define the property without interfering with the view. Among them we hope you will find the fence that is right for you. We wish to emphasize that you should feel free to improvise, to adapt and revamp any of them to fit your location and your taste, bearing in mind that a *good* fence is a *simple* fence; that only occasionally is there a place for an ornate fence. Bizarre, gimcracky effects are to be avoided at all costs because they grow tiresome sooner or later. Materials cost too much in these days of inflation to be used for anything which you will regret and, in a few years, want to replace. The good, simple, architectural pattern, on the other hand, is always tasteful and always pleasant.

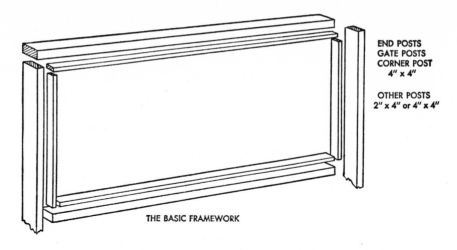

END POSTS
GATE POSTS
CORNER POST
4" x 4"

OTHER POSTS
2" x 4" or 4" x 4"

THE BASIC FRAMEWORK

BASIC FRAME TO USE WITH DIFFERENT COVERINGS

In contrasting fences, the same basic framework can often be used in a great variety of ways, the different aspects coming about from the kind of "skin" one chooses to apply on the basic "bones". Note that in all these fences the central section may be prefabricated in the shop or on the ground, then secured to the posts with nails or screws, the top railing finally bracing and holding the whole together.

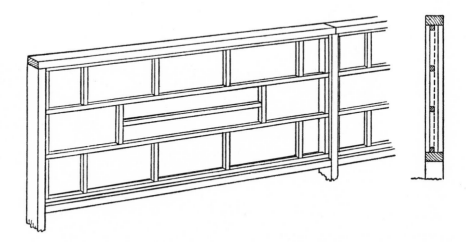

Use 1" x 1"s, $\frac{5}{4}$" x $\frac{5}{4}$"s or other stock for the crossbars and intermediate rails set into 1" x 2"s or 1" x 3"s used for the framework to be fastened to posts. Use 2" x 4"s for top, bottom rails and posts.

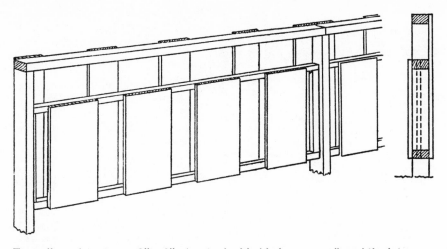

Top rails and posts are 2″ x 4″s (posts doubled below ground) and the intermediate rails either 2″ x 2″s as shown or 2″ x 4″s if you desire. Twelve-inch boards on fence front alternate with 6″ boards on back.

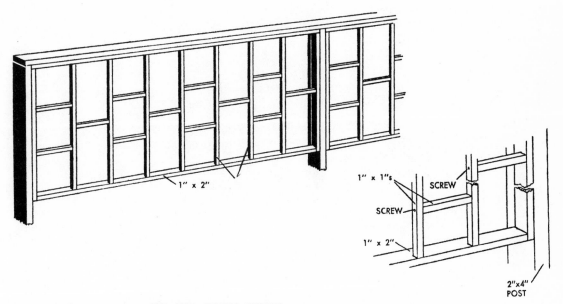

UP AND DOWN FENCE

Alternating square and oblong patterns are interesting even without vines twining through them. One-by-one verticals and horizontals are framed by 1″ x 2″ stock, but 2″ x 2″s or 2″ x 4″ might be used if a stronger fence should be desired. Use 4″ x 4″ corner and gate posts, 2″ x 4″s for intermediate posts set firmly in concrete.

FENCES

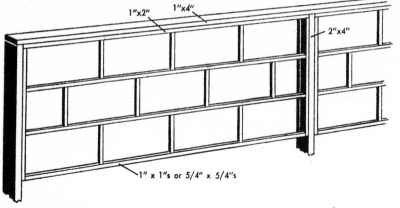

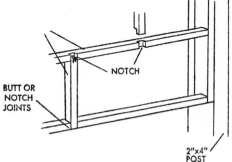

ALTERNATING OBLONGS

By centering the alternate rows of vertical pieces on the oblong above, an interesting and stable pattern results. Probably this is the simplest of the Bikini fences to construct because the alternating placement allows nails or screws to be driven with the greatest of ease. Heavier materials may be adapted to this design as indicated above if you wish.

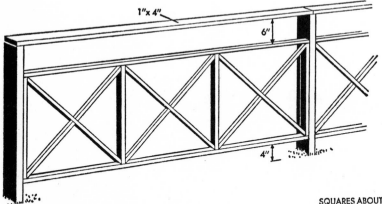

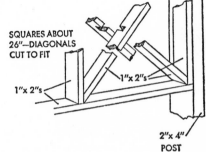

SQUARES WITH DIAGONALS

The lines of this fence are strong and simple and although it is of light construction it has considerable strength due to the use of the diagonal strips. The diagonals are fitted together with double groove joints (see the section on joints) and all cuts are simple 45° angles at the ends, simple and easy in either mitre box or power saw set to proper angle.

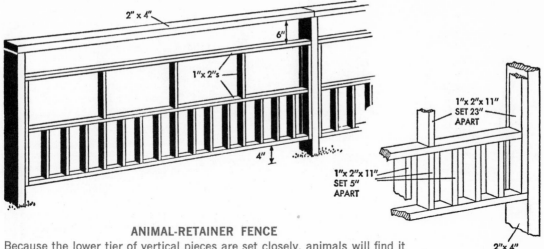

ANIMAL-RETAINER FENCE

Because the lower tier of vertical pieces are set closely, animals will find it hard to get out—or in—unless they are very small. Built on 2″ x 4″ posts and top rail, the other horizontals and verticals may be either 1″ x 2″s or heavier stock if you desire. Spacing of the verticals may vary according to what individual tastes or needs dictate.

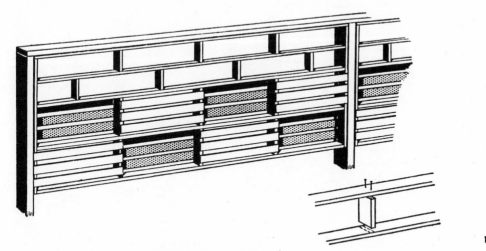

3-DIMENSIONAL OBLONGS

Long narrow oblongs form the two top tiers of this fence, wider oblongs the bottom two. The basic frame is: 2″ x 4″ posts, top rail 1″ x 4″ and oblong framing 1″ x 2″s, with the alternate in-and-out boxes of the lower two tiers faced with short trellis stock. The 3-dimensional quality of the in-and-out boxes creates interesting shadow play.

113

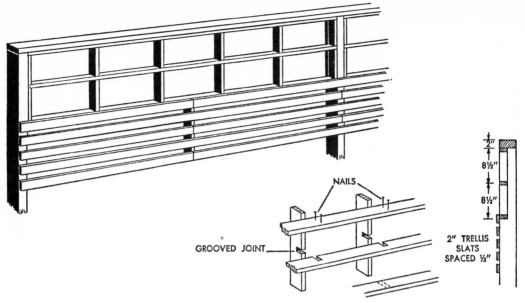

MAINLY HORIZONTAL

A good fence which accents horizontal lines but one which also has plenty of interest in the upper oblongs. One-by-twos fitted together with grooved joints (see section on joints) form the oblongs of the upper part as well as making the frame for the trellis slats used to make the lower part child- and dog-proof as well as producing long horizontal lines.

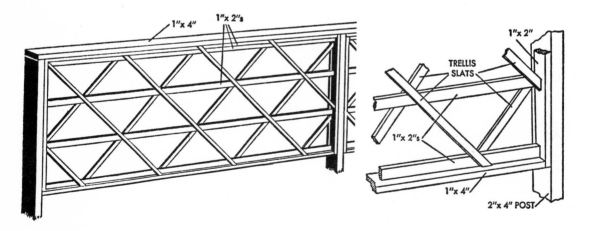

DIAGONAL SQUARES

The diagonals used in this fence give strength as well as producing a certain excitement of line. The 1″ x 2″ frame has the four 1″ x 2″s set in groove joints in the side pieces, and the diagonal trellis strips are nailed in place on either side of them. This fence would be a good one for Clematis or Roses or any of the good, lightweight vines to climb on.

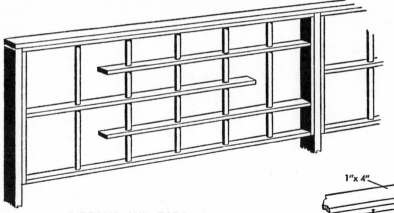

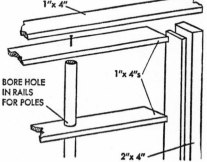

RIBBONS AND BARS

Somewhat offbeat and modern in effect, this fence is nevertheless simple enough to be used with any simple traditional home. The top rail is a 1" x 4", as is the framework between posts. Heavy dowel sticks or clothes poles are cut to the proper length to form uprights; the three horizontal 1" x 4"s in the center have holes bored to receive poles.

BORE HOLE
IN RAILS
FOR POLES

1"x 4"

1"x 4"s

2"x 4"

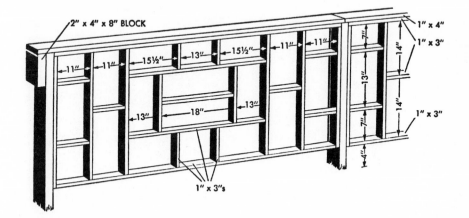

2" x 4" x 8" BLOCK

11" 11" 15½" 13" 15½" 11" 11"

13" 18" 13"

1" x 3"s

7" 14" 13" 14" 7" 4½"

1" x 4"
1" x 3"

1" x 3"

LONG AND SHORT OBLONGS

The space is divided into a series of harmonious oblongs in this fence, some upright and some horizontal in direction. Posts need be only 2" x 4"s, doubled below ground, the remainder of the stock (except for 1" x 4" top rail) being 1" x 3"s. Adapt posts to 4" x 4", make top and bottom rails 2" x 4"s, with 1" x 4" intermediate divisions for strength.

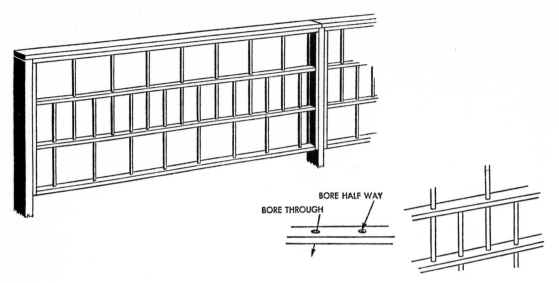

MODERN SPINDLE FENCE

Relatively simple to construct because it is possible to cut several pieces at once and to bore the necessary holes for the dowel sticks used for the spindles, this attractive fence is particularly interesting because of contrasting round and squared stock. It may be adapted to heavier construction by using 2" x 4"s and round clothes-pole stock.

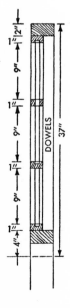

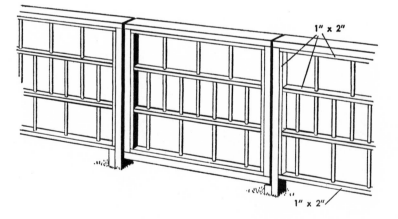

CONSTRUCTION OF GATE

The framework of the gate is all 2" x 4"s, with the framework holding the dowels made of 1" x 2"s with holes bored through where dowel penetrates, bored ⅜" to ½" deep to admit dowel on alternate spaces, a nail or screw being used through remainder of rail to hold dowel firmly in place. Use wood preservative over all, especially cuts.

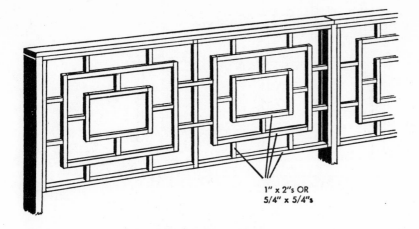

1" x 2"s OR
5/4" x 5/4"s

CONCENTRIC OBLONGS

Restful but with considerable interest inherent in its design, this fence uses oblongs to good effect. By centering the short connecting pieces on the center oblong and placing the short pieces on the outer oblong even with the sides of the inner oblong, variety is introduced with no sacrifice of design stability. Easily built because of straight joints.

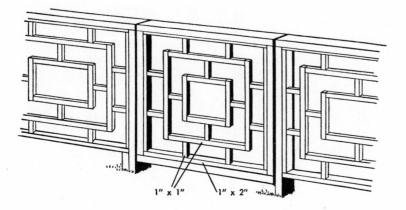

1" x 1"   1" x 2"

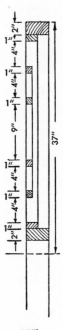

CONSTRUCTION OF GATE

This gate is constructed similarly to that above, with a 1" x 2" frame inside containing fretwork made of 1" x 1"s or 5/4" x 5/4"s. All fretwork is made with butt joints held by nails or screws and butted to frame, too. Set frame flush with front of fence.

117

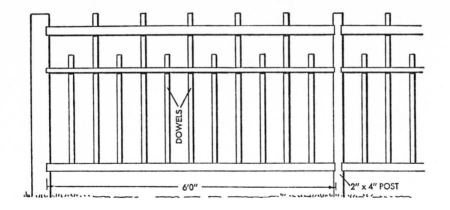

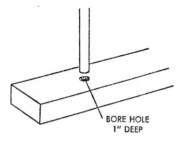

BORE HOLE
1" DEEP

LONG AND SHORT DOWELS

Pickets with a difference that will fit in with the most traditional house and go well with the intermediate modernized styles, too, are seen in this fence. Posts are 4" x 4"s for corners and every other post, intermediate ones 2" x 4"s with top and bottom rails 2" x 4", and 1" x 4" center rail dadoed in. Holes are bored in all rails to admit 1" dowels to penetrate.

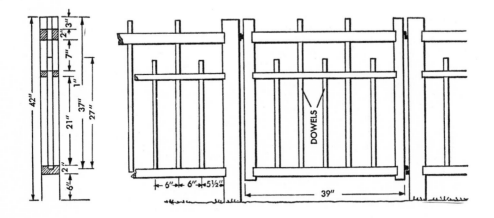

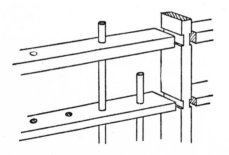

GATE CONSTRUCTION

Framework is all 2" x 4"s with center rail a 1" x 4". All rails are dadoed into sides of the gate. Dowels can be set closer together if desired, but they should not be placed at wider intervals or the fence and gate will look too stringy and proportions be in bad scale.

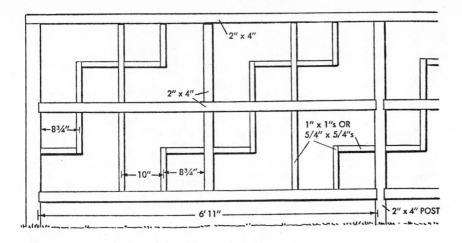

WITH DIAGONAL MOVEMENT

Modern but not bizarre, there is character and stability in the design of this fence. Posts at corners and ends are 4″ x 4″s; intermediate posts, top, bottom, center rails and 2″ x 4″s, the two latter dadoed into posts. All of the other verticals and horizontals are 1″ x 1″s, or $\frac{5}{4}$″ x $\frac{5}{4}$″s butted at corners, double dadoed at crossings, and dadoed into all rails.

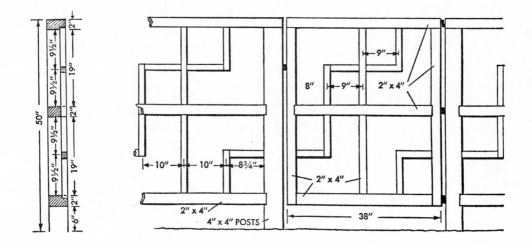

GATE CONSTRUCTION

Frame is made of 2″ x 4″s, butt-jointed, with center rail dadoed into side pieces. Other construction follows that of the fence. Note that 1″ x 1″s or $\frac{5}{4}$″ x $\frac{5}{4}$″s are dadoed $\frac{1}{2}$″ into the side, top and bottom rails, and aluminum or other nonrusting nails or screws used to secure pieces to the rails and posts. Make sure gate has swinging clearance.

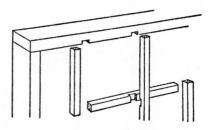

FENCES

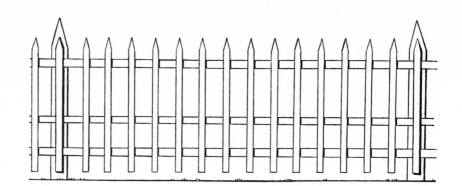

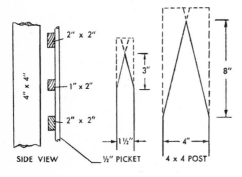

SIDE VIEW   ½" PICKET   4 x 4 POST

4" x 4"

2" x 2"

1" x 2"

2" x 2"

3"

8"

1½"

4"

PEEK-A-BOO PICKETS

Lightweight fences give an effect of airy charm because of the flowers seen peeping through them. They protect but do not interfere with the garden view, the wide spacing of the pickets giving plenty of ventilation area, too. The 2" x 2" frame is dadoed into the post, crossrail between being nailed to the front of the post. Joinings of the rails on the post are hidden by pickets on posts. Note that pickets echo "tent" cut of posts.

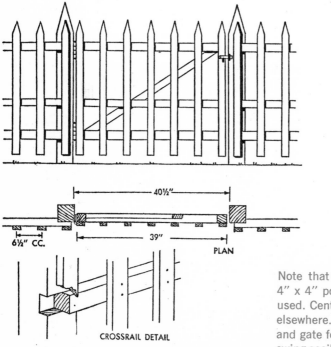

40½"

6½" CC.

39"

PLAN

CROSSRAIL DETAIL

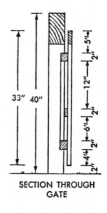

33" 40"

5¼"

2"

12"

2"

6¼"

2"

4¼"

2"

SECTION THROUGH GATE

GATE CONSTRUCTION

Note that 2" x 2" frame is not dadoed into 4" x 4" posts at gate. Instead, half-lap cut is used. Center rail is nailed to front of posts as elsewhere. Allow some space between posts and gate for hinges and clearance for frame to swing easily.

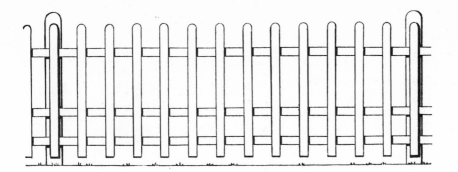

ROUND-TIP PICKETS

A pleasant variation in the top cutting of pickets is this semicircular topped style, the posts also being given a "barrel" cut. To make posts, draw semicircle with compass and then saw to rough shape, rasp file to finished size and sandpaper smooth. Cut all pickets to rough size, allowing extra wood for waste. Clamp several pickets together, draw semicircle, rough saw, rasp and sandpaper smooth.

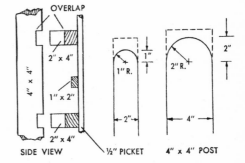

SIDE VIEW   ½" PICKET   4" x 4" POST

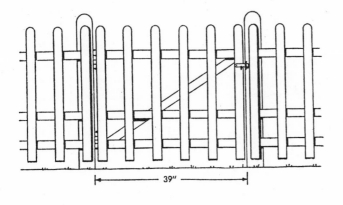

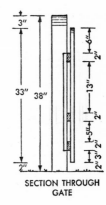

SECTION THROUGH GATE

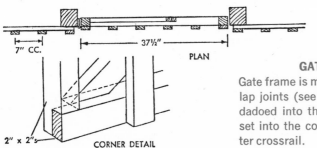

PLAN

7" CC.   37½"

2" x 2"s   CORNER DETAIL

GATE CONSTRUCTION

Gate frame is made of 2" x 2"s with corner half-lap joints (see detail, left) and with crossrails dadoed into the frame. The diagonal brace is set into the corners, crossing behind the center crossrail.

FENCES

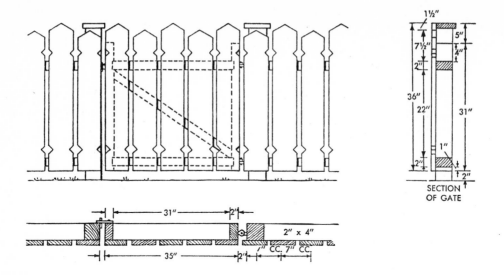

SECTION
OF GATE

2" x 4"

GATE AND FENCE OF WIDE PICKETS WITH CUTOUTS

A simple but effective pattern is achieved with little work, the "pickets" being 1" x 6" boards on a 2" x 4" framework, the posts being 4" x 4"s. All boards on both fence and gate may be made of even height or they may be of two heights, alternating as shown here. Note framework pattern in dotted lines on the gate, all frame being 2" x 4"s.

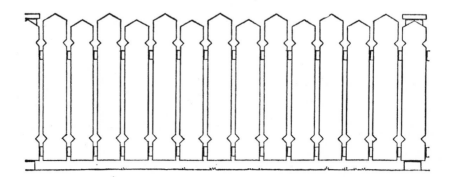

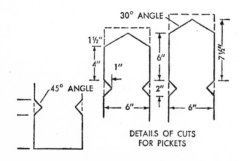

DETAILS OF CUTS
FOR PICKETS

HOW TO MAKE PICKETS

Cut all boards for the two sizes needed, allowing about 1/2" extra in length for each. By clamping several boards together and sawing them simultaneously the work can be cut down and precision maintained. Use one of the boards already cut for pattern to trace on cuts for all the rest of the pickets.

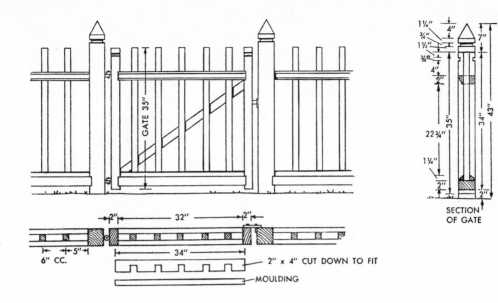

GATE 35"

2" ← 32" → 2"
5" →
6" CC. 34"
2" x 4" CUT DOWN TO FIT
MOULDING

SECTION OF GATE

1¼" 4"
¾"
1¼"
¾"
4"
½"
7"
35"
22¾"
34"
43"
1¼"
2"
2"
2"

POINTED POSTS, WIDELY SPACED PICKET FENCE

An open design which defines the extent of the property beautifully, utilizes $\frac{5}{4}$″ x $\frac{5}{4}$″ pickets set on a 2″ x 4″ bottom rail, with the tops held by moldings (see next page). The well-braced gate is made of 2″ x 4″'s and carries the lines of the fence through. Note details of the top rail above, and that the gate sides have been dadoed on front.

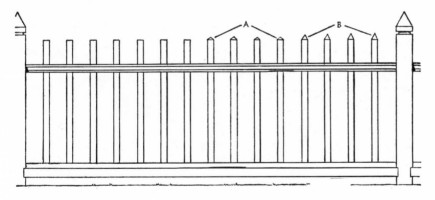

A   B

HOW TO ASSEMBLE PICKETS

Pickets may be square on top or pointed to conform with the posts, cut at 60° angle as in Picket B or at a 45° angle as in Picket A. By cutting and clamping several pickets together angle cuts may be done on bench saw to cut down labor. Note detail at right of the fence assembly, dadoing of bottom moldings firmly in place against post.

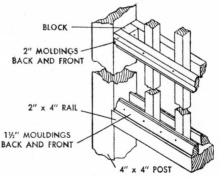

BLOCK

2" MOLDINGS BACK AND FRONT

2" x 4" RAIL

1½" MOLDINGS BACK AND FRONT

4" x 4" POST

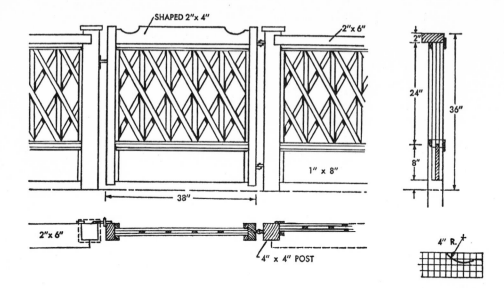

A DIGNIFIED DIAMOND PATTERN FENCE

Some bulky houses need a weighty fence to echo their massive quality and tone them down. Of particular interest to traditional homeowners will be this fence and matching gate. Gate's top rail is a shaped 2" x 4" between 2" x 4" uprights; bottom rail is a 1" x 8". The rails are not dadoed into the uprights. Upright pickets, diagonals are 1" x 1"s.

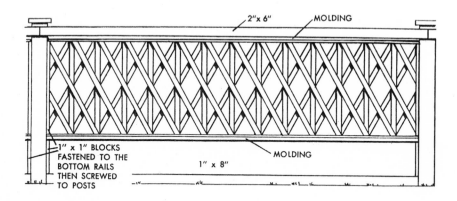

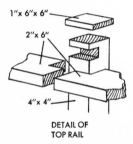

DETAIL OF TOP RAIL

CONSTRUCTION DETAILS

Top rails are 2" x 4"s cut to center on 4" x 4" posts. A 1" x 4" x 4" cap block with a 1" x 6" x 6" cap block topping it crowns post. Gate posts have top rail dadoed in. Bottom rail is a 1" x 8" with forward diagonal picket in front, rear diagonal behind it, and center upright picket on top of it. Bottom molding masks overlap of pickets and the top moldings hold the pickets in place.

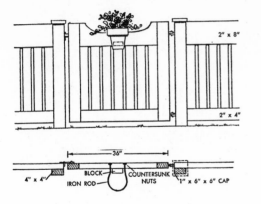

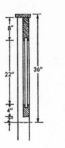

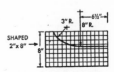

DETAIL OF RADII
FOR TOP RAIL SHAPE

AN INTERESTING DOWEL PICKET FENCE

Strongly constructed of 2″ stock with holes bored to admit dowels, this gate and fence have a directness of design that will go with many kinds of homes. The gate has a top rail of a shaped 2″ x 8″, with an iron rod bent into shape to hold a pot of flowers, the rod being threaded and held in place by nuts countersunk into the back side of rail.

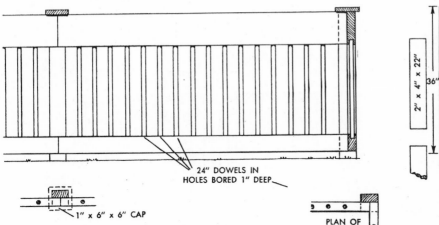

24″ DOWELS IN HOLES BORED 1″ DEEP

1″ x 6″ x 6″ CAP

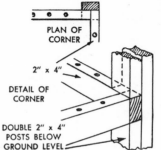

PLAN OF CORNER

DETAIL OF CORNER

DOUBLE 2″ x 4″ POSTS BELOW GROUND LEVEL

HOW TO BUILD THE FENCE

All posts are 2″ x 4″'s, doubled below bottom rail, the 2″ x 4″ bottom and 2″ x 8″ top rails being set in place, then a length of 2″ x 4″ fitted between to make a strong 4″ x 4″ post. Note detail of corner post of fence. Cap blocks are 1″ x 6″ x 6″ centered on the posts.

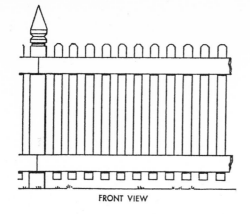

FRONT VIEW

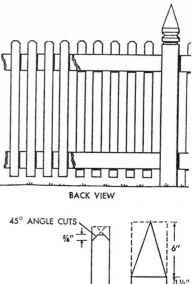

BACK VIEW

PUNCTUATED REGULARITY

A straight picket fence with no variation of height or spacing can be deadly dull. By emphasizing the posts, shaping the tops in an interesting way, and breaking the line of the fence, dullness is avoided. Top and bottom rails are overlapped on fronts of posts, fitted between on rear, pickets being held securely between the pairs of doubled rails.

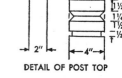

45° ANGLE CUTS

5/8"

6"

1½"
1¼"
1½"
½"

2"

4"

DETAIL OF POST TOP

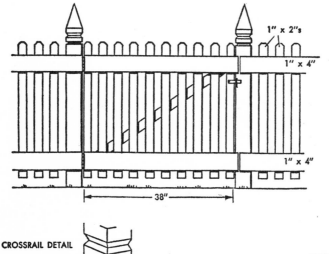

1" x 2"s

1" x 4"

1" x 4"

38"

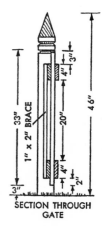

3"

1" x 2" BRACE

33"

46"

3"
4"
20"
4"
2"

SECTION THROUGH GATE

CROSSRAIL DETAIL

OVERLAPS ON FRONT OF POST

1" x 3" BLOCK

TOENAIL TO POST

BACK VIEW OF GATE POST

GATE CONSTRUCTION

Gate is constructed in the same manner as the fence, with the diagonal brace applied on the back side of top and bottom rails (see detail above). Note that rails overlap on one side of gate to prevent gate from swinging both ways. Use either concealed hinges shown here or barn-door type below.

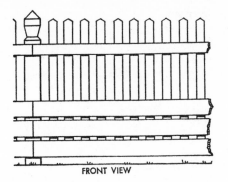

FRONT VIEW

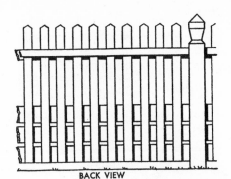

BACK VIEW

SLIGHTLY PUNCTUATED

A variation of punctuated regularity uses posts with interestingly cut tops that do not protrude very much above the fence line. Also, the rails have a variation in size as well as in number. The top rail is doubled here, and bottom rail may be doubled or not, since the three rails will give fence rigidity.

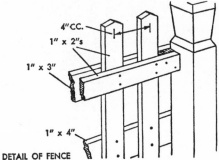

4" C.C.

1" x 2"s

1" x 3"

1" x 4"

DETAIL OF FENCE

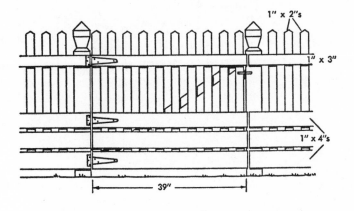

1" x 2"s

1" x 3"

1" x 4"s

39"

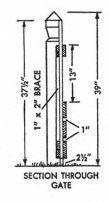

37½"

13"

39"

1" x 2" BRACE

1"

2½"

SECTION THROUGH GATE

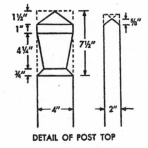

1½"

1"

4¼"

¾"

7½"

⁵⁄₈"

4"

2"

DETAIL OF POST TOP

GATE CONSTRUCTION

Again the gate follows in general the same construction as the fence. Note, however, that the top and bottom rails are doubled, using the same size wood as front rails, the diagonal brace being cut to fit between the rails. Either barndoor hinges, concealed, or more decorative wrought-iron hinges with a matching latch may be selected instead.

127

FENCES

FENCE ENDERS

Today, not all fences are total enclosures, Some merely run beside the house to be lines of demarcation between properties, giving visual division. Such fences usually just stop, but they can end in such a way as to enhance your property, providing real visual interest, furnishing a place for a vine or perhaps giving an architectural pattern.

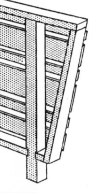

ANGLE ENDER

Give the fence a forward thrust by bringing out the top rail and angling back the end of the fence to the bottom rail. The evenly spaced thick and thin strips of the fence boards are cut back to follow the end angle.

PAINTED SPACE DIVISION

This fence-ender uses a square framework of 2" x 4"s like that of the fence, backed by 1" x 8" vertical battening. Short 2" x 4"s divide the ender into areas which are each painted in different colors for modern effect.

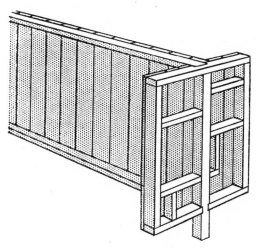

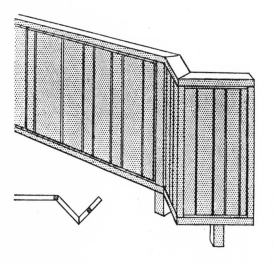

FENCE WITH A WAGGLE

A double-angled ender makes a feature of a fence. Simple and good-looking, it makes use of several random widths of vertical batten boards —1" x 4"s, 8"s, 12"s—for facing the fence, including the angles at the end.

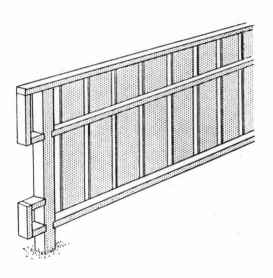

RAIL TERMINALS
The top, bottom and intermediate rail lines are extended, the top two forming a boxy oblong, the bottom line finishing with a boxy square. Scraps of 2" x 4"s are used. One-by-eights spaced 2" apart face the fence.

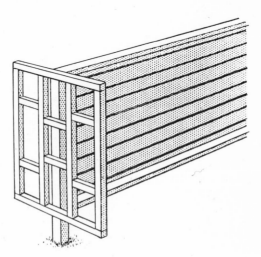

TRIPLE TRELLIS TERMINAL
A fence of grooved plywood used horizontally (1" x 4"s could be used instead if spaced ½" apart) can be ended with a simple trellis in three parts, built of 2" x 2"s, which is most ideally suited to pillar roses.

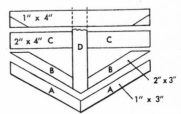

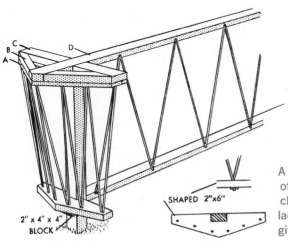

TAPERING TRELLIS
A good vine fence is made on a simple frame of 2" x 4"s laced with plastic-coated wire clothesline. The triangular trellis ender, also laced with clothesline, tapers to the base and gives good verticals for climbers.

129

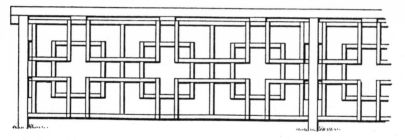

INTERLOCKING SQUARES

The fascination of pattern which is regular in its repetition, yet irregular in division of space is quite apparent in this design. Light in weight and airy in feeling, it is a fence which proclaims protection of an area, yet does it in a most gracious way. The design could be adapted to heavier weight materials than the 1″ x 1″ strips and frame of 1″ x 2″s used here, but would be less airy in effect.

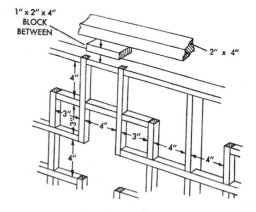

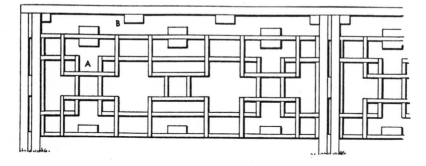

INTERLOCKING OBLONGS

Gracefully interlocking, the oblongs give the appearance of regularity, yet do not actually repeat, for the center oblongs vary from the end ones. Note, too, that the posts are doubled 2″ x 4″s with 1″ x 4″ blocks between, giving still further interest to the design from the resulting slots. Blocks of 1″ x 4″s are centered on the 1″ x 2″ frames in the middle of the oblongs and upper area.

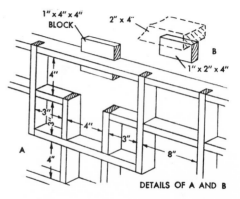

DETAILS OF A AND B

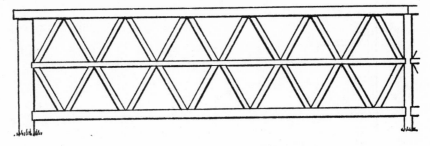

HARLEQUIN PATTERN

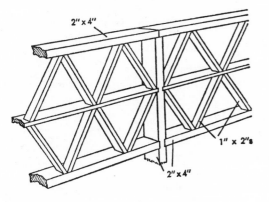

Bisecting the diamond pattern is a 1″ x 4″ horizontal piece which gives the fence more stability. Top and bottom rails are 2″ x 4″s as are the intermediate posts; corner and end posts are 4″ x 4″s. For the diagonals of the diamonds use either 1″ x 4″s or 1″ x 2″s, the 1″ edge facing the spectator. Note that the pattern can be varied to make it fit a higher or lower fence, the diamonds made thinner or fatter by varying the angle of the cutting.

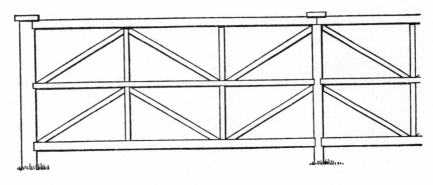

CHEVRON DESIGN

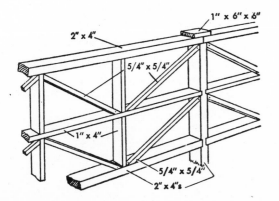

Using diagonals of a rather low angle in a chevron pattern, the fence can be made to seem longer and lower. In this case, rails at top and bottom are 2″ x 4″s, center rail is 1″ x 4″ with vertical strips also 1″ x 4″s. Diagonal strips may be 1″ x 2″s, 1″ x 1″ or $\frac{5}{4}$″ x $\frac{5}{4}$″ stock centered on the horizontals and verticals. Use a 1″ x 6″ x 6″ cap on all posts. For a taller fence, add a third row of chevrons and verticals with a second horizontal.

131

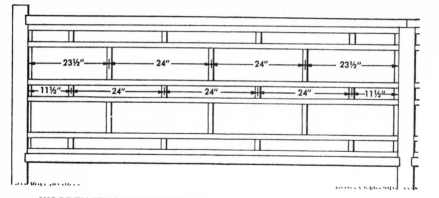

WOODEN FENCE, BRICK STYLE

Although this fence is built of wood it takes its design cue from brick bonding, the bands of wide and narrow oblong openings being set between top and bottom 2" x 4" rails dadoed into the 4" x 4" posts, the fretwork of the fence behind made from 1" x 2"s, 1" x 1"s or $\frac{5}{4}$" x $\frac{5}{4}$"s. Horizontal bars are dadoed into stock at ends, which is nailed to the posts.

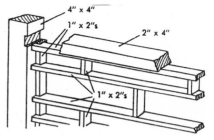

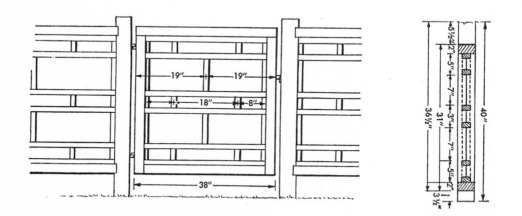

GATE CONSTRUCTION

Top and bottom rails of this gate are butt-joined to the sides, with the interior fretwork being constructed in all particulars in the same manner as that in the fence itself. Gate posts are 4" x 4"s raised like the corner posts $3\frac{1}{2}$" above the level of the top rail. Should it be felt necessary, a diagonal may be inserted behind fretwork to act as a brace.

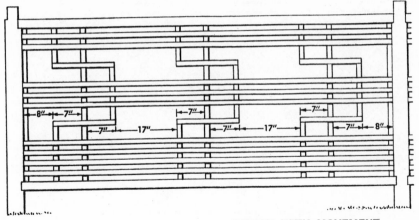

DESIGN WITH MOVEMENT

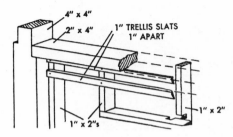

Although the various members of this fence are of necessity static, there seems to be considerable movement due to the design. Posts are 4″ x 4″s, rails 2″ x 4″s dadoed into the posts, and the interior members are 1″ x 2″s with butt joints except for horizontal crossbars which are double dadoed into upright members. The horizontals are 1″ trellis slats.

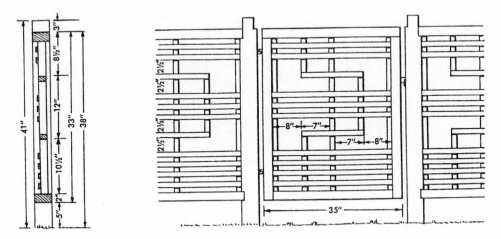

GATE CONSTRUCTION

One unit of the fence design is adapted to the gate, the construction being exactly the same as that used for the interior members of the fence. Use butt joints for the rails and sides of the gate frame. Dado the fence rails into the 4″ x 4″ gate posts, notching the tops of all posts as shown, the extenders being held 3″ above the level of the top rail of the fence.

133

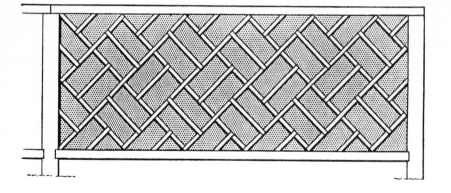

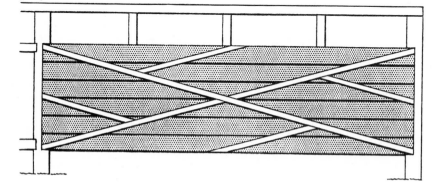

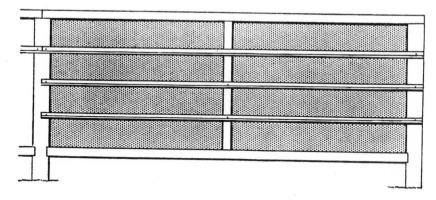

USE IMAGINATION AND MAKE AN INTERESTING FENCE

TOP: Narrow trellis slats form the basis of this design, set at 45-degree angles and overlapped to give added shadow effects. Paint the frame, slats white, background a soft, bright color.

CENTER: Sometimes fences may be faced on the near side. Here grooved plywood or matched boards face the fence, open top showing brightly-painted 2" x 4" framework.

BOTTOM: Heavy full-rounds or half-round moldings are nailed or screwed to the near side of the frame, the far side of the fence faced with plywood or transite for privacy.

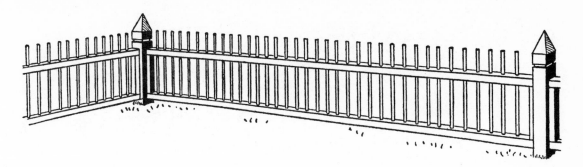

DOWEL PICKETS, POINTED POSTS

Posts are 4″ x 4″s or 3″ x 3″s pointed and dadoed. The framework of the fence is all 2″ x 2″, with holes bored through top rail for ¾″ dowels, bored part way into bottom rail to contain them. Screw frame to posts; use wood preservative over all, particularly cuts.

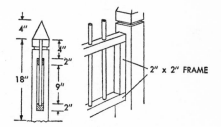

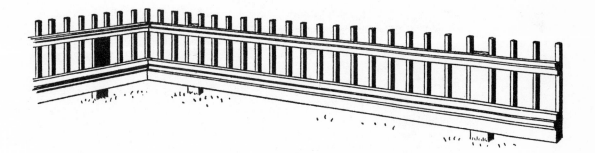

SQUARE PICKETS AND MOLDINGS

Moldings dress up top and bottom rails of this dooryard fence which features square pickets. Trellis slats, 1″ x 2″s or other shaped pickets may be adapted for use. Note that top rail and molding are narrower than bottom. All posts are set flat side to fence.

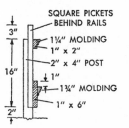

135

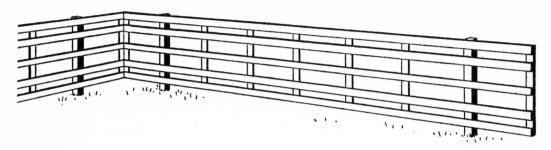

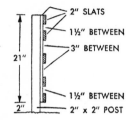

2" SLATS
1½" BETWEEN
3" BETWEEN
21"
1½" BETWEEN
2"
2" x 2" POST

LIGHTWEIGHT DOORYARD FENCE
On 2" x 2" posts, 2" trellis stock makes an attractive pattern, spaced as shown. It is not only lightweight, but also inexpensive, and is intended as a definition of the area rather than a strong, protective fence.

LOW FENCES FOR DOORYARDS
Grandmother's dooryard garden, a charming old-fashioned idea, is having a considerable success today in modern form. It is a tiny garden beside a front or back door, enclosed by a low fence. Concentrating bloom (and garden work) in one spot allows gardeners to keep the rest of the garden informal, and saves work. Shown here are a few fences; adapt others in the book to suit your own requirements and taste.

HOW TO MAKE PICKETS
While pickets may be simply square or flat boards, many traditional houses seem to demand shaped pickets. Most of these can be made by a bench saw or sawed by hand, but pickets B and D require a scroll- or jigsaw to shape their curves. Cut picket boards to approximate length and clamp several together so you can saw them simultaneously. Bore holes while clamped (B, C) and after sawing and boring, file with a rasp and sand down to remove unevennesses. For C, a board a little more than twice the width may be used, holes bored, then ripsaw to finished widths, reclamp, saw angle end.

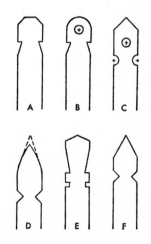

136

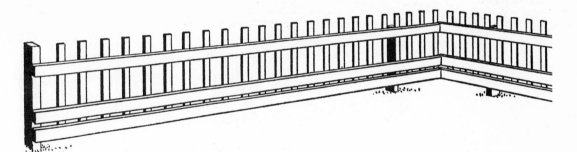

SQUARE PICKETS

Inexpensive, and little trouble to construct, this dooryard fence may be built of 1″ x 2″s or trellis slats, pickets the same or 1″ x 1″s, $\frac{5}{4}$″ x $\frac{5}{4}$″s. Posts are 2″ x 4″s or 2″ x 3″s driven into the ground or set in concrete. Note how corner joins without corner post.

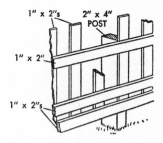

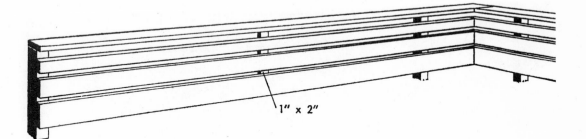

LONG LINES, GRADUATED BOARDS

Another easy-to-make dooryard fence uses 2″ x 2″ posts, 1″ x 4″ cap board and graduated boards—1″ x 2″, 1″ x 3″, 1″ x 4″—set apart 1″ for horizontal effect. Center supports are 1″ x 2″s set between the posts.

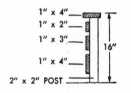

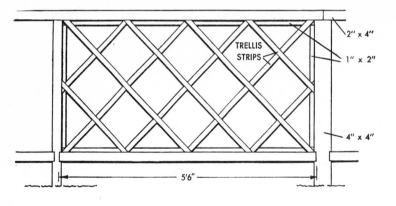

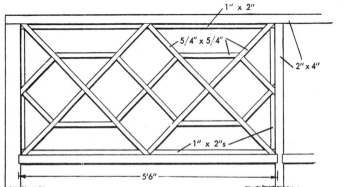

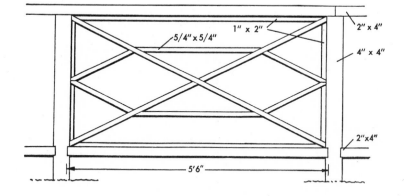

ADAPTED FROM DESIGNS OF THOMAS JEFFERSON
The stark formality of classic architecture in 18th Century America was relieved by the addition of gallery balustrades, roof railings and porch railings similar to these which are adapted from some which Thomas Jefferson designed for the University of Virginia. They fit well with the Colonial style, but

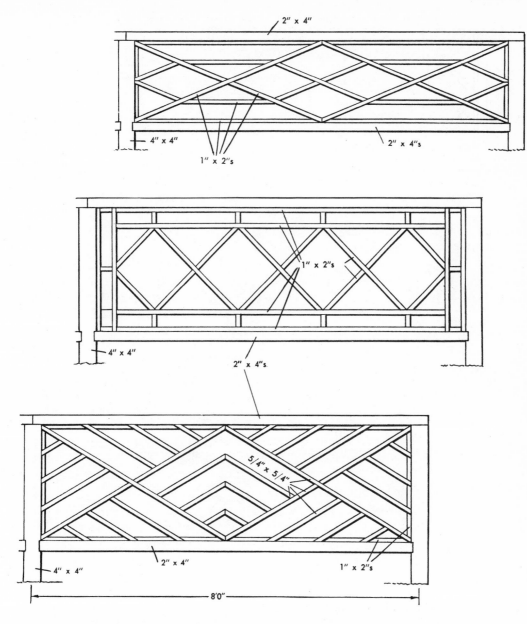

because they were adapted by him from Oriental sources, many, especially the simpler ones, will make good fences for modern homes. They may be varied to fit longer or shorter spans, turned up on end and used as trellises. Vary angles of diagonals as necessary, add design units to fill out and fit space you use.

FENCES

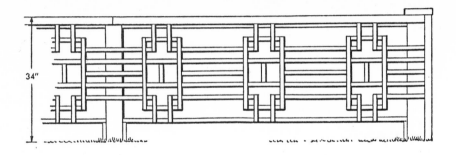

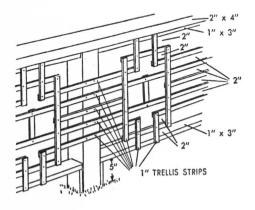

2" x 4"
2" x 1" x 3"
2"
2"
1" x 3"
2"
5"
1" TRELLIS STRIPS

INTERRUPTED HORIZONTALS

Although it may look intricate, this fence is actually rather easy to make, especially for the possessor of a power saw. It is neat and gives the effect of being rather strong, yet is actually light in weight. Posts are 4" x 4"s but might be 2" x 4"s except for corners and end or gate posts. Bottom rail might be a 2" x 4" if desired, reversing the order of top rail and setting it in a groove joint in post.

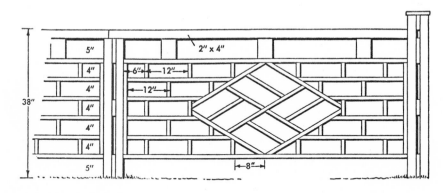

5"
4"
4"
38"
4"
4"
4"
5"
6" — 12"
12"
2" x 4"
8"

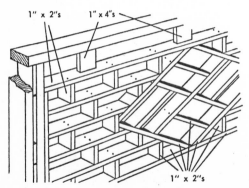

1" x 2"s
1" x 4"s
1" x 2"s

DIAMONDS AND OBLONGS

Strong horizontals lend serenity to this fence, but the diamond pattern lends it sufficient interest to keep it from being dull. The frame is entirely constructed of 2" x 4"s with doubled posts held apart by 1" x 4" blocks to make interesting slots. Although this is less easy to construct than that above, it is not beyond the amateur's scope, particularly if angular cuts are made by preset power saw.

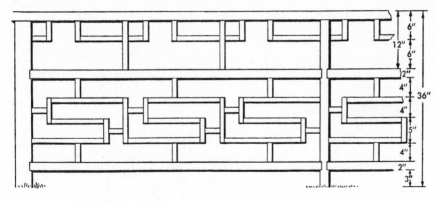

WHIRLING OBLONGS

Suitable for either a modern or traditional home, this fence has a lower part which will keep out animals and keep children in, but the upper part is kept open. The effect is light, airy, and off-beat enough to be most interesting. The end and corner posts are 4″ x 4″s with intermediate posts either the same or 2″ x 4″s; all posts are set in concrete. Top, bottom and intermediate rails are 2″ x 4″s set in groove joints in posts. The fretwork shown here is made of $\frac{5}{4}$″-square stock, but might be made of 1″ x 2″, 2″ x 2″ or even 1″ x 1″ stock.

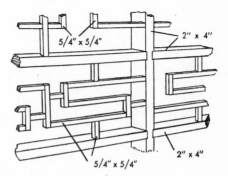

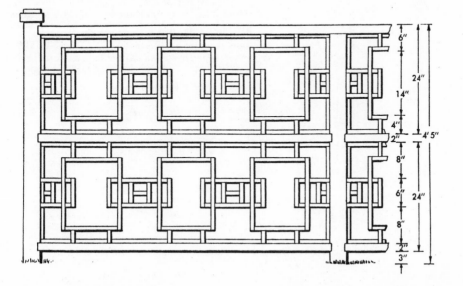

SQUARES AND OBLONGS

A somewhat more elaborate fence which would go well with a traditional or modern house where extra decorative effect is desired can be prefabricated in frame, set in.

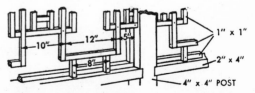

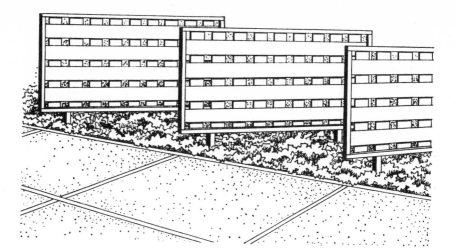

COMPLETELY BAFFLED

Where the view from the house or terrace includes some undesirable feature, but where ventilation or passage is needed, a baffled fence may be the answer. The sight lines, indicated on plan below by dotted lines, are blocked by the fence, which makes a very decorative background for plants or which may be under planted with ground covers as in our picture. Vines may be trained on it, too, but whatever the plantings, it will be an effective year-round baffled screen. If you wish, you may adapt other fence designs, shown elsewhere in the book, in place of this up-and-down-and-across pattern which is detailed.

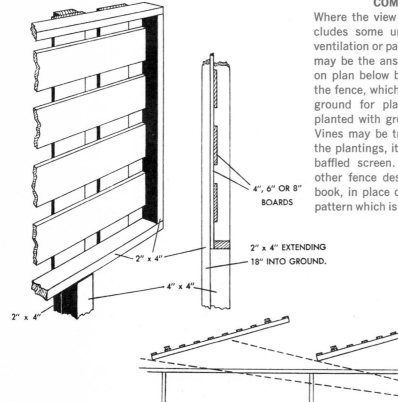

4", 6" OR 8"
BOARDS

2" x 4"

2" x 4" EXTENDING
18" INTO GROUND.

4" x 4"

2" x 4"

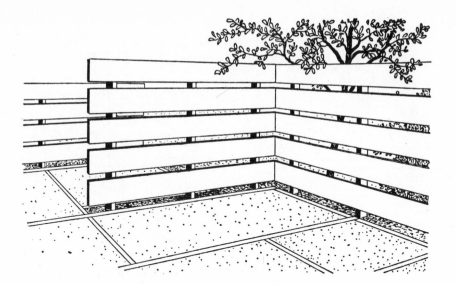

CORNERED AND BAFFLED

A terrace which is all too public can be kept from prying eyes of passersby with a fence which is balled at the entrance as shown here in a horizontal stripe pattern. If you want a completely horizontal pattern, paint the posts a dark green or black, the fence boards a light color and watch how the posts disappear. Boards may be spaced more closely or more widely according to how you want to use the leaf patterns of shrubbery planted on the back side of the fence. By adapting a design of some other fence in this book you can achieve a totally different visual effect while retaining the baffle fence privacy.

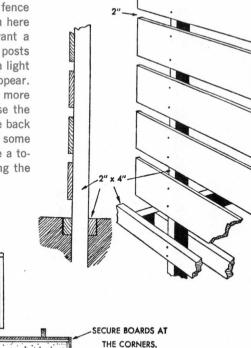

8"
2"

2" x 4"

SECURE BOARDS AT
THE CORNERS.

8'-0"
1'   3'   3'

3'0"

PLAN

FENCES

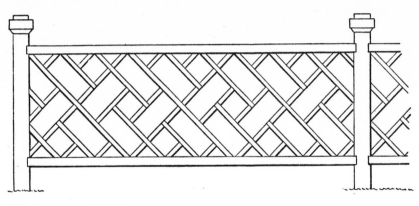

INTERLOCKING BLOCK PATTERN

A varied pattern, featuring broken diagonals, makes a handsome yet simple fence. Since all cuts are 45-degree angles, ease of making is guaranteed. 1" x 1"s, $\frac{5}{4}$" x $\frac{5}{4}$" or 1" x 4"s may be used for the center, top and bottom rails being 2" x 4"s, posts 4" x 4" with caps 6" x 6" x 2" supported by moldings. Trellis strips may be used if a light fence is desired, nailing them to back of rails or using a plywood backing with the trellis strips applied.

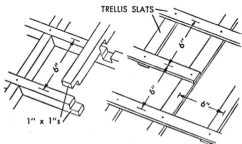

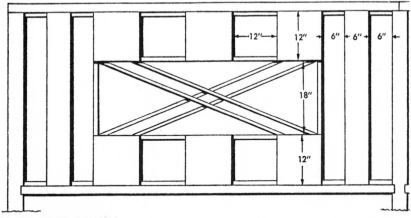

ORIENTAL PEEKABO FENCE

Adapted from the Japanese, this fence has posts of 4" x 4"s, rails 2" x 4"s with 1" x 2"s used as nailing strips for vertical members. The "window" is framed with 1" x 2"s, diagonals being narrow-cut trellis strips nailed to the back of the frame. If complete privacy is desired, baffle the fence by applying vertical members in all the openings (on the opposite side of the fence) and do not use "windows" or use them only for some exceptional view.

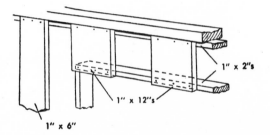

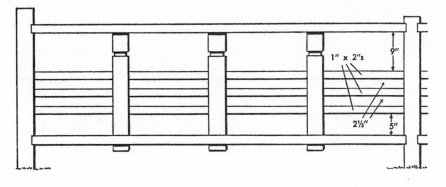

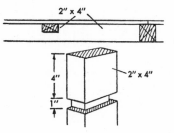

STRIP AND STRIPE

Sturdy, yet with delicacy of pattern, this fence fits well with modern, but is conventional enough to enhance traditional homes as well. Intermediate verticals are 2" x 4"s with a decorative dado cut; horizontals are 1" x 2"s screwed or nailed to the back of all uprights, including 4" x 4" end and unit posts. Both the top and bottom rails are 2" x 4"s.

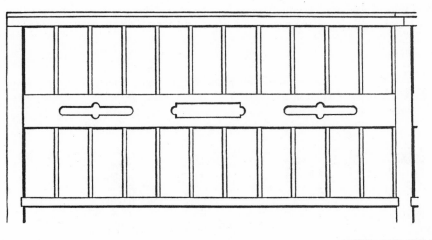

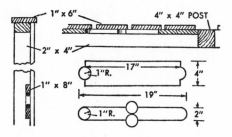

BAND AND BAFFLE

Although the effect of this fence is rather oriental due to the center band with its cut-out pattern, a plain wood band or two or three strips of 1" x 2"s or trellis stock would give the same effect and perhaps allow it to fit in with more kinds of houses. The 1" x 6" uprights are fastened to backs of 2" x 4" rails and are protected from the weather by 1" x 6" caps.

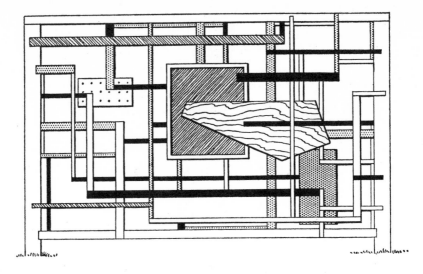

FEATURE FENCES—FOR MODERN, ABSTRACT EFFECT

Relieve the starkness of a modern house by making a Feature Fence beside the terrace. Oddments and scraps of wood, ranging from bits of trellis slat to 2" x 4" ends, plywood, moldings, and even metal may be utilized. Paint the various parts in different colors as indicated here by varieties of shading or leave the woods to mellow into natural greys and browns. For the best effect, cover the back of the fence with a piece of plywood or Transite painted white so that design will show up well and shadows will produce still greater effect. Allow some pieces to overlap the frame and get as much 3-dimension as possible. Shown above is a Feature Fence designed to stand alone, while that below ends a somewhat conventional fence where it overlaps the terrace. Fence frame is painted in one of the colors of the feature, with a second color used on the uprights. Corner and end posts are 4" x 4", other posts, top and bottom rails, 2" x 4"s.

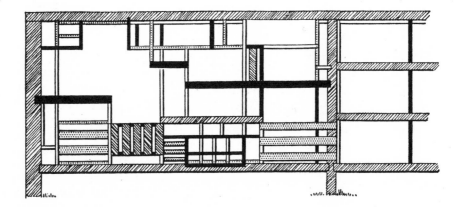

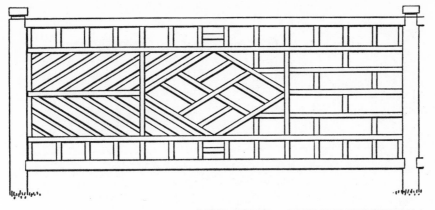

SAME FENCE, DIFFERENT TREATMENT

Don't be alarmed. We do not advocate using two unrelated patterns in the same section of any fence. What we are attempting to show is that even though the basic framework may be the same, the whole effect of the fence can be changed by varying some portion of the interior design. On the left is an exciting treatment with diagonals parallelling the center diamond, while the one on the right is rather quiet and static and could be used in more places, because it would draw less attention to itself than vivid pattern might.

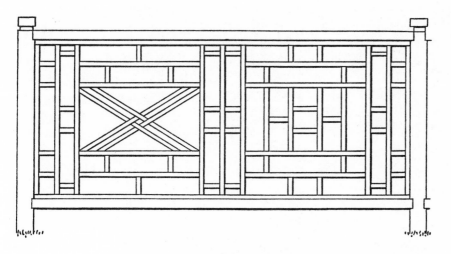

RIGHT-HAND OR LEFT-HAND DESIGN?

Although everything except the small central oblong is the same in each half of this fence, the effect could hardly be more different. In the left-hand side the "window" has interwoven 1" trellis strips, crossing it diagonally, breaking the space and giving excitement to the area, although it is a quiet excitation. On the right we have the same segment filled in with fretwork which follows the verticals and horizontals of the framework surrounding it. Similarly, you can adapt other fence designs, varying them as necessary.

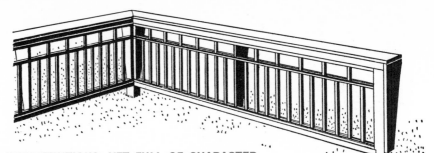

UNPRETENTIOUS, YET FULL OF CHARACTER

Although this fence is built of the simplest of materials, it has a certain character which will enable it to be used with the most traditional of houses, or with modern ones. The shaped end boards and center braces, the wide overhang of the top rail, the square pickets and center rails combine to make it an easy fence to build too. The openings of the top can be varied (see below) with every other or every third picket cut full height.

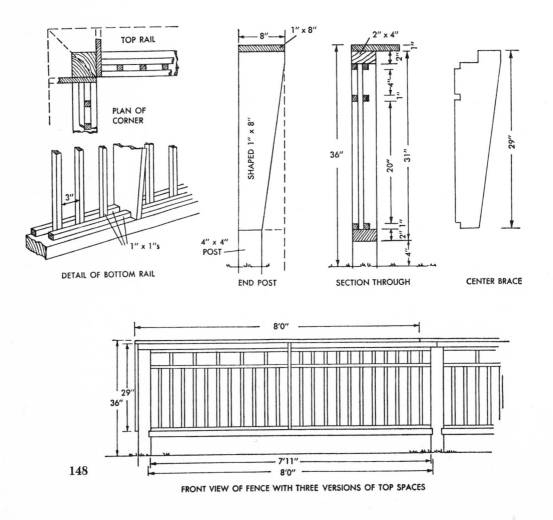

TOP RAIL

PLAN OF CORNER

3"

1" x 1"s

DETAIL OF BOTTOM RAIL

8"   1" x 8"

SHAPED 1" x 8"

4" x 4" POST

END POST

2" x 4"

36"   20"   31"

SECTION THROUGH

29"

CENTER BRACE

8'0"

29"
36"

7'11"
8'0"

148

FRONT VIEW OF FENCE WITH THREE VERSIONS OF TOP SPACES

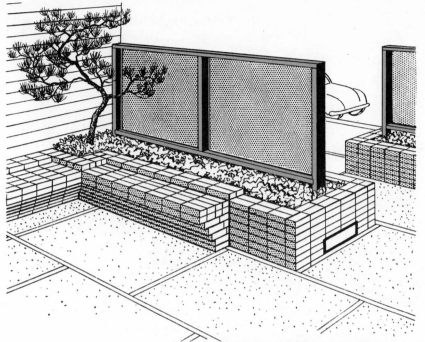

BAFFLE FENCE PLANT BOX

A terrace may be very close to the street or the neighbors and still be private if a low plant box is built incorporating a seat and baffle fence. Note the light fixture set in box end to guide night traffic safely across the terrace. Vines on the side street give a warm welcome, while a picturesque pine and a carpet of flowers dress up terrace side.

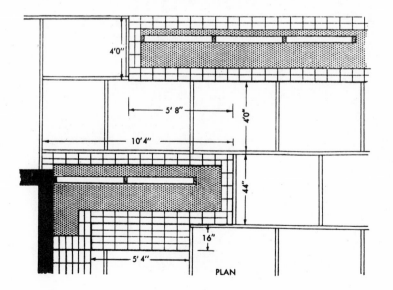

PLAN

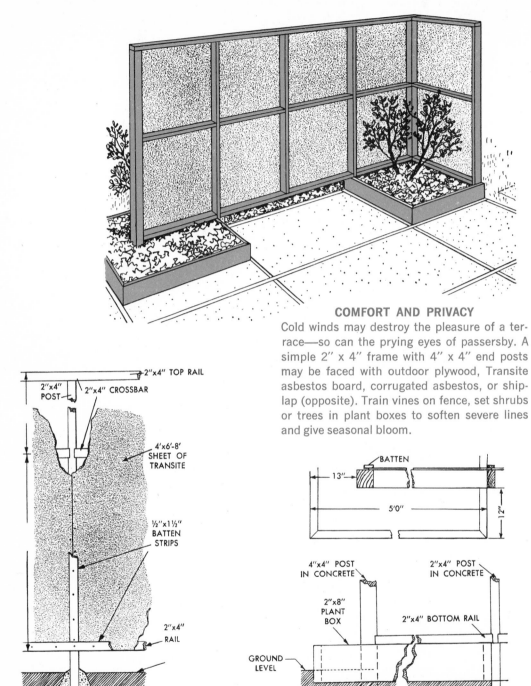

COMFORT AND PRIVACY

Cold winds may destroy the pleasure of a terrace—so can the prying eyes of passersby. A simple 2″ x 4″ frame with 4″ x 4″ end posts may be faced with outdoor plywood, Transite asbestos board, corrugated asbestos, or shiplap (opposite). Train vines on fence, set shrubs or trees in plant boxes to soften severe lines and give seasonal bloom.

2″x4″ TOP RAIL

2″x4″ POST

2″x4″ CROSSBAR

4′x6′-8′ SHEET OF TRANSITE

½″x1½″ BATTEN STRIPS

2″x4″ RAIL

BATTEN

13″

5′0″

12″

4″x4″ POST IN CONCRETE

2″x4″ POST IN CONCRETE

2″x8″ PLANT BOX

2″x4″ BOTTOM RAIL

GROUND LEVEL

FENCES

A VARIETY OF MATERIALS CAN BE USED TO CLOTHE THE FENCE SHELTERS

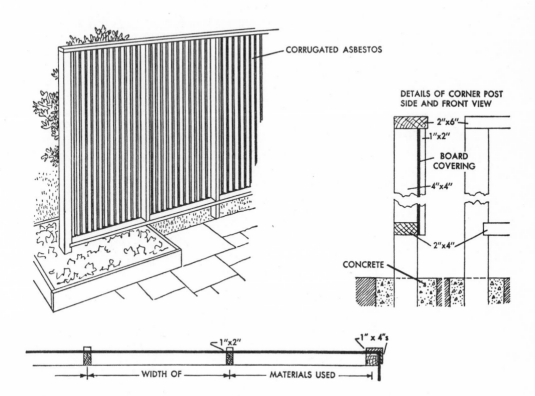

CORRUGATED ASBESTOS

DETAILS OF CORNER POST
SIDE AND FRONT VIEW

2"x6"

1"x2"

BOARD
COVERING

4"x4"

2"x4"

CONCRETE

1"x2"

WIDTH OF

MATERIALS USED

1" x 4"s

151

8

Every Garden Needs a Shelter

Today we find that the life of many a house has overflowed into the garden, making it a roofless outdoor room, so that the case for the need of a shelter is a good one. Most home owners will welcome some sort of shelter to add comfort to outdoor living. In moist climates they not only add a place outdoors in which to escape from sudden showers, but they also provide a place in which to sit of an evening and be protected from falling dew.

There are many shelters to choose from today. No longer need they be formally placed on the axis of the principal room of the house and have surroundings planted with overwhelming dignity, as in the past. Then the effect was impressive and beautiful but hardly invitingly cosy. Today's shelters are placed where they are *needed* and are usually casual and informal. Sometimes shelters will be found at the end of the garden, sometimes attached to the house, and sometimes freestanding along one side of the garden border—in fact, they will be placed wherever they are found to be most useful. They are practical, then, built for use as well as beauty. That is, perhaps, the outstanding thing about today's shelters that is universal.

Some of them bring a certain architectural significance to an ordinary development house, extending its lines and making it look larger, while giving it a distinctive appearance which divorces it from the factory-made appearance of its fellows on the street.

Sometimes the shelters are solidly roofed; sometimes they have crossbars for a roof (the so-called "egg crate" style); and sometimes they are roofed with laths or trellis slats, with sunbreakers of snow-fencing, or bamboo shades which can be rolled up in the autumn and stored indoors or in the garage during those months when light and warmth in the house are needed. Increasingly, however, you will find shelters becoming half-and-half structures—partly roofed and partly open or covered with sunbreakers. This will provide shelter from sudden storms under the roofed part, while the sunbreaker or open part will be pleasant for those who like the sun, either tempered a bit or full strength. Or it may be that the terrace paving will extend out beyond the structure of the shelter so that it is possible to move furniture into the sun when that is desirable and to pull it quickly under the sunbreaker or solid shelter when protection is wanted.

A revival of the old time gazebo seems to be taking place. These garden shelters were sometimes roofed, sometimes covered with slats or *treillage,* a crisscrossing of trellis slats that leave openings that are square, diamond-shaped, or oblong. There is much to be said for the gazebo—it is a pleasant and private place to retire to for a small gathering, to sit in and read or converse with a friend, to have a small luncheon or supper party of four or six, according to the size of the structure. In areas that are plagued with insects, the gazebo can be screened and a screen door fitted to give protection from these pests. An electric light can also be installed so that one may enjoy the outdoors and summer cool in the evenings, as well as in the daytime. See photograph of a modern type.

Also some of our designs will readily adapt to use for basic structure for gazebos; note the Rose Arbor on page 169 which will adapt by being doubled in size or made more compatible with the space you have to use by adjusting the dimensions and by paving or gravelling the floor. The Lath House Pavilion on page 168 might be used in its original dimension or another module added on each side and the roof rafters increased accordingly; the laths could be more closely spaced or crisscrossed in

A garden shelter need not be elaborate—it may be as simple as you like. Consider this A-frame, made of redwood planks set in concrete embedded in the soil. Latticed roof uses 1″ x 1″ strips to shade the brick floor set on sand. Railroad ties form the rustic steps leading to it.

treillage fashion if greater privacy were desired. However, the classic gazebo was open at least on the sides where a good view was to be obtained, with windows or other openings giving out on that view. To shelter the interior from wind or from view, certain sides were enclosed. The Shelter in a Corner on page 186 is perhaps closest to the old-time gazebo and could easily be adapted. Note the seats, a desirable built-in feature which could be incorporated in any of the other designs mentioned above. All in all, the gazebo idea can be most successfully adapted to modern outdoor living by using one's imagination and ingenuity.

In the designs herewith we have tried to provide a wide range so that you will find one which will harmonize with your house or which can be adapted to it. Even old houses are being brought up to date these days, with the addition of a garden house or shelter attached to the house or garage, or placed elsewhere in the garden, rejuvenating the whole site. A little remodeling of the garden to bring its plan up to date, to open it up, and to simplify it so that it is easier to take care of, and perhaps also the installing of a picture window to look out upon this garden-with-a-new-look, will help to make an old house more pleasant to live in as well as more saleable, should that necessity arise.

By taking into account your hobbies, your family life, your climate with its prevailing winds and other pertinent features, and the situation of neighboring houses, you can place your shelter where it will do the most good and add to your use of your garden, even improve your plants' chances for survival. For instance, if you live in areas where strong sun makes it difficult to grow camellias or other broad-leaved evergreens, you can give them the needed protection by placing your lath-roofed shelter so that its shadow is cast over adjoining beds. This will enable you to plant your favorites alongside your outdoor entertaining room, making a real conversation piece of this exhibit of your hobby plants.

By using a slat-roofed shelter, too, you will achieve privacy from above, where neighbors have second-floor windows overlooking your terrace or where a street or path overlooks it from a hill. (See the sketch on p. 182 on how to achieve privacy.) A solid roof offers complete protection, of course, but it is surprising how much protection a slatted sun-breaker will give, how it can hide the view, especially when reinforced with vines in summer. It will not cut off all the light from the rooms of the house, should the terrace be placed alongside it; and, if you want the warmth of the sun and more light in the house in the winter, the slats can be placed on a frame which is removed each winter for storage and replaced before the sun gets too hot, or as soon as you begin to use the terrace in the spring. They may be alternated in direction if you wish.

If you want protection from the eyes of observers but like to sit in the sun, the entire shelter need not be covered: part of it may be solid, another part slatted, and part of it left open so that you have three choices; and the divisions need not always be rectangular, either, but may be at an angle if you wish. Snow-fencing, bamboo, and basswood porch shades or roll-up blinds can also be used for slatting, being rolled up for storage in winter and held in place by 1" x 2"s or other light wood strips screwed in place on the shelter during the summer. This will prevent their being blown off or damaged by summer winds.

WHICH SHELTER FOR YOU?

To assist you in choosing the kind of shelter most suitable for you, how big it should be, where to put it, and so on, we suggest that you study our chapter on Terraces. This will aid you in making your deci-

sions, for a shelter is usually used in connection with a paved terrace, and many of the same considerations must enter into the planning. It is logical to place the shelter on a paved terrace, because it will be used so much that the maintenance and repair of a lawn beneath it would be a factor to reckon with. Also paving will make it more usable in damp and rainy weather.

There is a preliminary consideration, however, which you must take into account. What do you need shelter from? Is it the sun, the wind, the prying eyes of neighbors and passersby? Once you have decided this question you will have a very good idea of where it would be best to place the shelter and what you will need in the way of screening.

Next you must decide on how permanent you want the shelter to be. Is it to be a freestanding one which will be a stand-in for a young tree still too small to give shade and visual privacy from upstairs windows or adjacent houses? A lightweight but well-built, sturdy structure would be indicated so that it would last until the tree took over.

Is it to be a really permanent shelter? Then there must be even more care given to its planning and construction so that it will withstand the stress of all-year weather and storms, and it should be well-designed so that it will be a permanent asset, a credit to both the house and the garden. Don't rush things at this point, for the shelter must stand the test of future years, which makes the preliminary thinking of primary importance. The permanent shelter should have permanent paving; its posts should be firmly embedded in concrete; and if they are of wood they should be treated with wood preservative before being set. All parts of wood structures outdoors will last indefinitely if they are thoroughly treated with wood preservative. (See the section on Wood Preservatives.) Proper advance consideration and lucid thought will give you the kind of sturdy structure you will use with pleasure and no regret in future years.

A fanciful and pleasing octagonal garden shelter that is a modern descendant of the Victorian gazebo is open on two sides, windows are seen on two others and the slatted sides give it textural interest, provide climbing support for vines to give privacy.

SIZE AND HEIGHT

In general, with today's homes being built long and low, it will be a good plan to echo these proportions in other structures. The average height of indoor rooms today is 8'0". Outdoor rooms may not *need* to be higher than that, particularly those with open or slatted roofs, but, architecturally speaking, most shelters attached to the house or another building look better when they have their roofs or trellises placed at, or just below, the eaves of the adjacent roofs, usually 9 to 10 feet above ground level. Attach the stringer enough below the line of the gutter so that crossbars or other superstructure may be placed on top and it will still come below the gutter. If this is too high, any normally visible architectural division of the house—the eaves of a low ell, the strong line of a tall picture window or door—may be your cue for placement. If not, then place the shelter so that the lower side of the rafters will clear the doors and windows by at least 6 inches or even a foot.

PLACEMENT OF POSTS

Similarly, in deciding on the placement of posts or other uprights, it is obvious that they should never be placed in front of a door or window to bisect it, obscuring the view of the garden from indoors or impeding direct entry into the house. It is also apparent that in attaching a shelter to a building it should not end inside the vertical lines of a window or a door opening, but should come at least to the edge or, better still, should extend beyond it by 6 inches or a foot, if possible.

Frequently there is an ell or a jog in the house wall which will make it convenient to place your shelter in the corner formed by the two walls, provided that this location will work out from the practical standpoint of use, and also that it will look well in the garden. If it gives you privacy and protection from wind, and if it is where its use at night will not disturb sleeping children or elder members of the family, you will find that the use of two existing walls will cut down on the labor required for building and also the expense of materials for construction, saving the cost of several posts and their setting. Jogs or breaks in roof lines also make good places to start or end shelters.

OTHER FEATURES

Frequently two or more useful functions may be combined in one structure. For instance, it may be possible that in building your garden shelter you will make a screen for a service yard where compost heaps, clotheslines, trash and garbage cans, and other un-beautiful but necessary adjuncts of modern life are congregated. This can be done by merely adding a solid wall to the back of the structure, if it is free-standing in the garden, or else the wall may be backed up by a tool shed, an outdoor potting bench, or a garden storage house. The shelter wall may also be used as a place to display and shelter summering house-plants if shelves are built for the pots. They should be conveniently placed for necessary watering but also with an eye to showing off the potted plants in a decorative manner. The plants will thrive beneath a lath roof where they get enough light but are protected from the burning rays of the sun.

A shelter is a good play-place for children, too. With their vivid imaginations it can become a castle, a pirate ship, a prairie schooner, or the most modern of space ships. They can play in its shade all day long during the dog days when heat stroke stalks the open lawn. They will remember it with pleasure in adult years as a place where lunching and dining outdoors during the green seasons helped to knit the family together and make it a unit. When the children have gone to bed, it becomes a refuge for the adults on pleasant summer nights.

MATERIALS

In general, any wood which is used structurally in building houses may be employed for outdoor shelters. (See also section On Choosing Lumber.) Certain woods are favored because they are less susceptible to decay than others when used outdoors. Cedar, cypress, and redwood are the most prominent on the list, but almost any good, sound wood, well treated with wood preservative before it is painted, would last for many years if given yearly inspection and repaired and repainted as often as needed.

In some sections bamboo poles are cheap and available. They may be used as crossbars or set closely together as slats for view-breakers, or

A garden-house-gazebo of unique design is constructed throughout of long-lasting Western red cedar. Two-by-six posts are bolted to U-straps of steel set in concrete footings. Two-by-twos are centered on each side of the posts. Seats and backs are 2″ x 4″s; flooring, as well as framing, 2″ x 6″s.

they may be used in conjunction with bamboo porch shades. It should be realized, however, that bamboo is not noted for its long-lasting qualities and that it will need periodic renewal. Reed mats, available through nurserymen who use them as cold-frame coverings, are also available in many places. They may be used as view-breakers on fences or as shelter roofs, and may be rolled up and stored over winter. They do not cost very much, and it may be that their cheapness will make them attractive enough to compensate for their not lasting more than a few seasons.

Trellis slats in conventional 1½- to 2-inch widths, ¼- to ½-inch in thickness, will last for many years if painted and properly prepared. Cut the slats to size. A laborsaving device is to clamp a half dozen or more together and saw them all at once. Then unclamp them and stand them in a can containing about 4 inches of wood preservative. Soak the other ends too, letting them stand in the preservative for two to four hours or more; then remove them and let them dry well. Paint them with undercoat on all sides and ends and with a final coat of outdoor gloss paint. When that has dried install them on the trellis using aluminized or other non-rusting nails, and give the entire shelter a second coat of outdoor gloss paint, paying particular attention to filling well any joints or holes with paint. If there are any larger gaps, knotholes, etc., fill them with putty to prevent moisture from entering and then paint them. Use nails or screws which won't rust. This will preserve the shelter from rust stains and will cut down on the number of times it must be painted to keep it fresh-looking and to prevent further rusting.

SPARE THAT TREE

If one of your prized large trees should die, don't feel that you must immediately have it cut down. Instead, as demonstrated on a page of the designs for shelters, remove as many limbs and branches as may be necessary to prevent their breaking off or being blown off in storms and causing trouble. Plant a small tree nearby to give shade in future years, and then bolt long rafters of 2-inch lumber to the trunk of the dead tree (at least 2″ x 6″ lumber should be used), using 6- to 10-inch bolts to secure them. Keep rafters level and parallel to each other, using blocks if necessary to keep them equidistant. Eyebolts in the trunk at a distance of several feet above the rafters will hold cables to support outside corners of the structure. If you want to plant your new tree almost on the spot of the old one, support the middle of the rafters on the trunk and at the far end on posts.

A vine, either perennial or annual, or a combination of the two until the perennial vines grow large enough to mean something, will quickly cover the trellis and add shade-giving qualities, and will even climb the trunk and remaining limbs of the tree, preventing it from looking quite dead. Peeling the bark from the dead tree will prevent termites from doing too much damage to it. If you want to paint it a soft silver grey or a pale pastel color it can become a feature of your garden. Or, if you want to minimize it, paint it black or a soft dark green. Either way it will add distinction to your garden until the replacement tree is large enough. Then the shelter can be removed for its shade will no longer be needed.

To digress from the use of a tree as a shelter, perhaps you will want to use it in the following way if your dead tree has a picturesque shape which will compose well with a fence, a shelter, or some other feature: use it as a trellis for a vine. We have seen in this country large dead trees ablaze with wistaria bloom and, later in the season, feathery and lovely with the pale green leaves of that vine. In England and in the forests of France we have seen dead trees spreading their limbs in interesting patterns, their trunks and limbs clothed with the glossy leaves of ivy. It takes some years for ivy to grow that big, of course, and it won't grow in many places in this country; but other vines can be substituted, even annual ones, to make the dead tree a center of interest in the garden. At the base of the trunk a trellis may be built to support the vines and help them to reach the lower limbs.

USING YOUR SHELTER

Once your shelter is finished and planted with vines (if you have decided that you want such a leafy green roofing) you will begin to use it. We urge that you should not think of it as *finished,* however. As you use it, always be on the lookout for any way in which you can improve it, perhaps using some of the demountable coverings which we have advocated, perhaps developing some of your own methods of making sun- or view-breakers. Perhaps by hanging a roll-up blind on the sunny side of a necessarily small shelter you can outwit the broiling sun during the hot part of the day and then roll it up for evening or late afternoon. Probably some of the other hints shown in trellis or fence sections of this book will give you some ideas to adapt to your own problems or ideas of construction.

But when your shelter has finally come to its completion and you are satisfied with it, enjoy it to the fullest and make the most of it. You will find that your new outdoor room has broadened and deepened the interest you take in your garden and in the outdoor living which it will make possible.

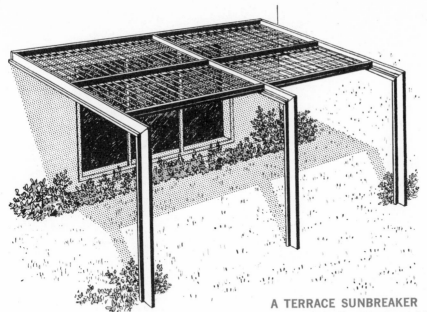

A TERRACE SUNBREAKER

Many picture windows are so placed that the glare of the sun cuts down their usefulness. A simple framework with posts securely placed on concrete and with the top rails attached to the house can hold snow-fencing, as shown here, which breaks the glare and heat in the summer and can be rolled up, stored in winter, when the sun will be welcome. Bamboo roll-up blinds, cut to fit, may also be used, or a more permanent trellis of slats, fiberglass, plastic aluminum screen, dowel rods, placed on basic frame.

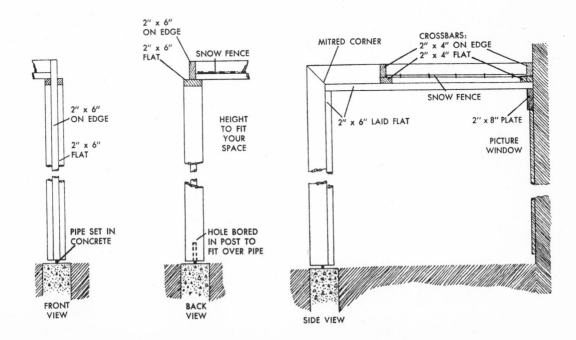

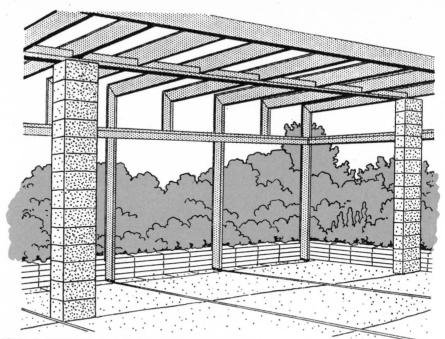

MAKE A TRELLIS ROOM BESIDE THE HOUSE

A good-looking and sturdy trellis covering a terrace alongside the house will add an outdoor room to your living space. Piers of blocks support the center stringer; uprights alternate full-length pieces with short uprights resting on the horizontal crossbar. A low plant bed constructed of flat blocks surrounds the terrace, giving further sense of enclosure. In our sketch, note that the back stringer is attached to the house, but it is possible to build a free-standing room by using upright construction on both sides.

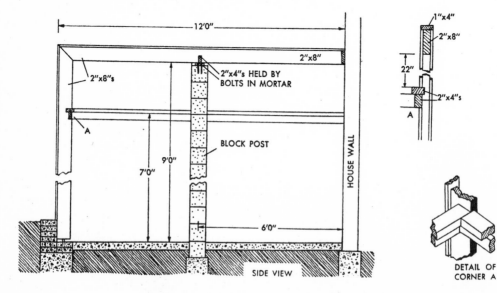

SIDE VIEW

DETAIL OF CORNER A

CONSTRUCTION DETAILS

The block piers should be given good strong footings so that they will not settle and fall out of line. Dig down at least 18″ in ordinary soil, deeper if the soil is light and sandy, or if you do not pave the terrace (which would protect foundations from frost). Footings should extend 5″ on each side of the block; but if the soil is very light, increase them to as much as 10″– 12″ on all sides to give a proper base. The holes in the top blocks should be plugged to prevent moisture damage in future. Put a wadded-up newspaper well down in the block and then fill to the top with mortar. Bolts, with which to secure the stringer, can be inserted in the mortar while it is still wet, and secured.

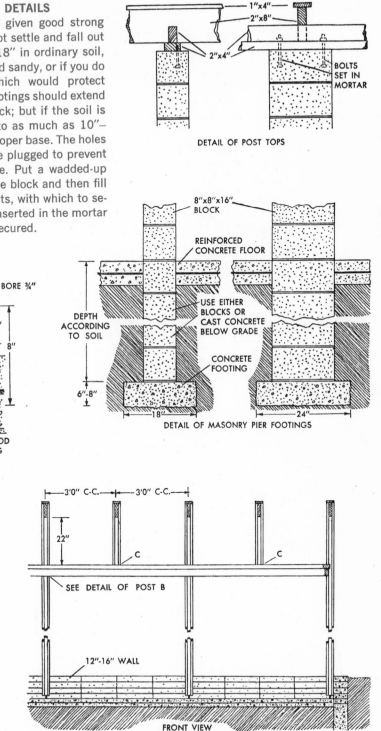

DETAIL OF POST TOPS

DETAIL OF MASONRY PIER FOOTINGS

DETAIL OF WOOD POST SETTING

DETAIL OF SHORT MEMBERS C

DETAIL OF POST B

FRONT VIEW

SHELTERS

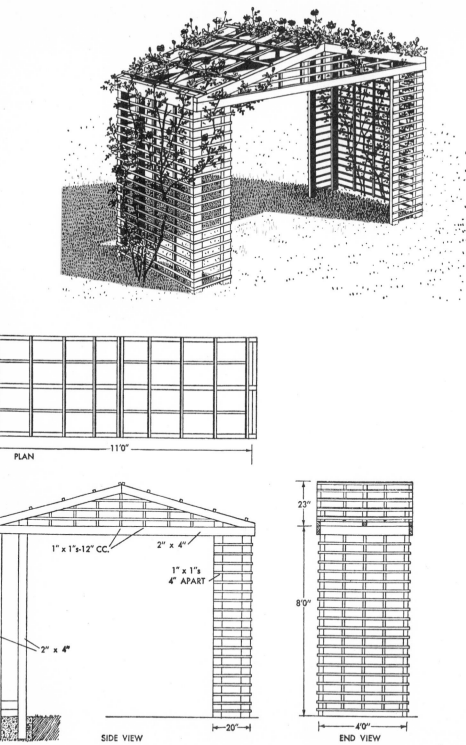

PLAN

⟵————————— 11'0" —————————⟶

1" x 1"s-12" CC.

2" x 4"

1" x 1"s
4" APART

2" x 4"

SIDE VIEW

⟵— 20" —⟶

END VIEW

23"

8'0"

⟵——— 4'0" ———⟶

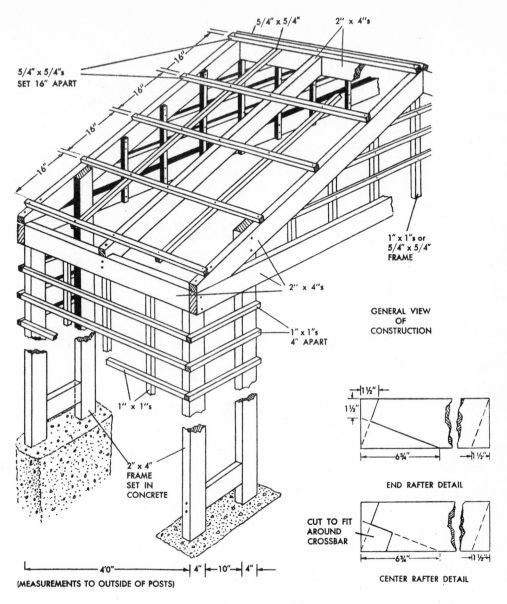

5/4" x 5/4"

2" x 4"s

5/4" x 5/4"s
SET 16" APART

16"

16"

16"

16"

16"

1" x 1"s or
5/4" x 5/4"
FRAME

2" x 4"s

GENERAL VIEW
OF
CONSTRUCTION

1" x 1"s
4" APART

1" x 1"s

2" x 4"
FRAME
SET IN
CONCRETE

1½"

1½"

6¾"

1½"

END RAFTER DETAIL

CUT TO FIT
AROUND
CROSSBAR

6¾"

1½"

CENTER RAFTER DETAIL

4'0"

4" 10" 4"

(MEASUREMENTS TO OUTSIDE OF POSTS)

A MODERN CLASSIC ROSE ARBOR

Adaptable to many uses, this arbor will go well with many traditional houses,
yet it has a modern flavor, too. It can be set in the open as shown here, used
in connection with a fence of similar design, be attached to a house or a
garage to roof a terrace, with suitable regard for harmonizing the roof lines
with those of the building. Its charm lies in the airiness of its proportions,
its form being geometric yet open enough to complement the wayward,
natural curvature of the climbing roses or vines, and at the same time sturdy
enough to bear their weight when, in future years, they will need firm support.
Use preservative on edges before painting.

167

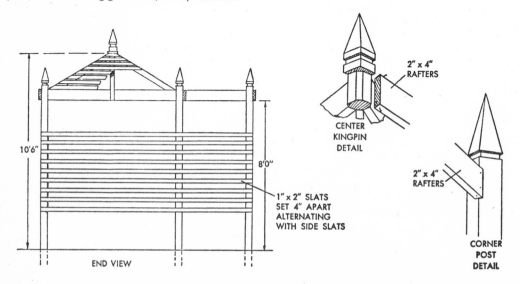

THE LATH HOUSE PAVILION

A garden house which combines the better features of both the modern and traditional styles will give any garden a real focus as well as providing a useful outdoor living area. The corner "house" part is made more interesting by by the use of 4" x 4" posts cut to taper to a point and rabbeted for decorative effect. The flat portions of the structure may be left open or roofed with laths, like the hip-roofed corner house. Note how laths are used on sides as view-breakers and give privacy to the corner where they overlap. The entire area of the terrace under the structure may be paved, or only that under the house part, the rest being gravelled, or kept in lawn.

10'6"

8'0"

1" x 2" SLATS
SET 4" APART
ALTERNATING
WITH SIDE SLATS

END VIEW

2" x 4"
RAFTERS

CENTER
KINGPIN
DETAIL

2" x 4"
RAFTERS

CORNER
POST
DETAIL

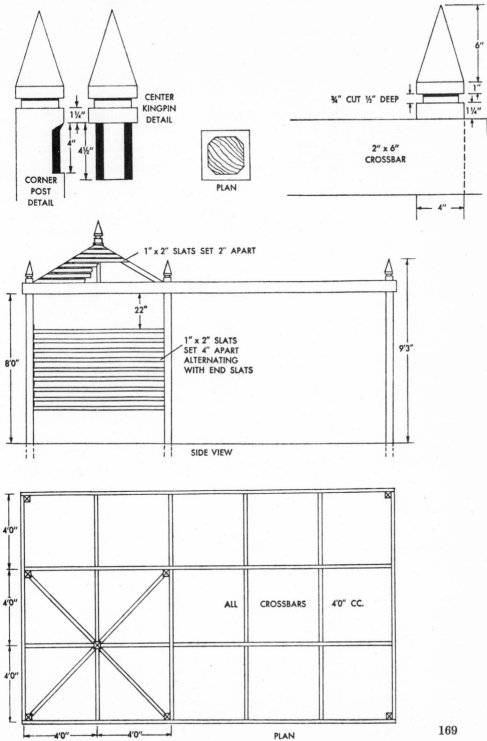

CENTER
KINGPIN
DETAIL

1¼"

4"

4½"

CORNER
POST
DETAIL

PLAN

¾" CUT ½" DEEP

6"

1"

1¼"

2" x 6"
CROSSBAR

4"

1" x 2" SLATS SET 2" APART

22"

1" x 2" SLATS
SET 4" APART
ALTERNATING
WITH END SLATS

9'3"

8'0"

SIDE VIEW

4'0"

4'0"

4'0"

ALL   CROSSBARS   4'0" CC.

4'0"   4'0"

PLAN

169

A TRELLIS ROOM BESIDE A WINDOW

A little "room" with leafy walls and ceiling outside a window adds to the attractiveness of the house, both outside and inside. Privacy is also achieved, and shade for windows placed where the sun produces unwanted glare during the summer months; yet in winter, when sun is needed for heat and light, the leafless trellis will admit it.

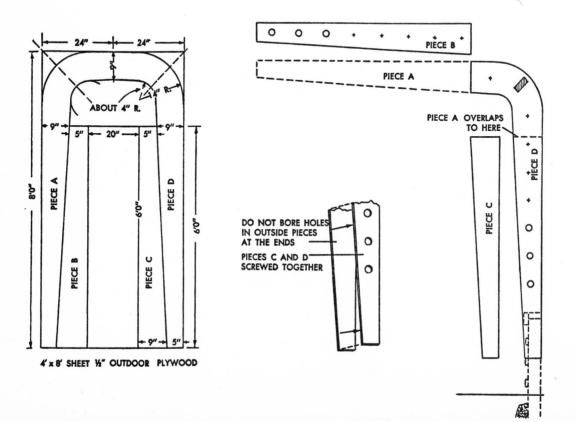

24" 24"

9"

ABOUT 4" R.   4" R.

9" 5" 20" 5" 9"

8'0"

60"

6'0"

PIECE A

PIECE D

PIECE B

PIECE C

9" 5"

4' x 8' SHEET ½" OUTDOOR PLYWOOD

DO NOT BORE HOLES IN OUTSIDE PIECES AT THE ENDS

PIECES C AND D SCREWED TOGETHER

PIECE B

PIECE A

PIECE A OVERLAPS TO HERE

PIECE D

PIECE C

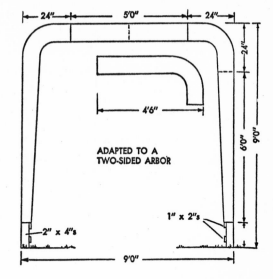

CONSTRUCTION DETAILS

Pieces A and B, C and D (below left) are cut from a single piece of 2″ outdoor plywood, left-over pieces being utilized for other projects requiring outdoor plywood. One-by-twos and 1″ x 4″s make the fence at trellis base, cross-bars being clothespoles inserted through holes bored through the plywood (after it has been put together with aluminum screws) on each side of the angle. Two-by-fours make frames for the fence, being doubled to make the posts, set in concrete; and are also used for spacer bars at the angle of plywood posts.

ADAPTED TO A
TWO-SIDED ARBOR

2″ x 4″s

1″ x 2″s

24″
5'0″
24″
24″
4'6″
6'0″
9'0″
9'0″

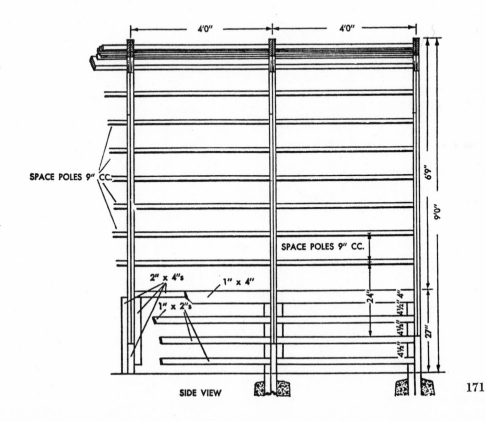

SPACE POLES 9″ CC.

SPACE POLES 9″ CC.

2″ x 4″s
1″ x 4″
1″ x 2″s

4'0″
4'0″
6'9″
9'0″
24″
4″
4½″
4½″
4½″
27″

SIDE VIEW

171

SLIDING SCREEN WITH BAMBOO

Bamboo porch shades are cheap to buy, can be fastened to basic framework as shown below (bosswood detail), finished off with moldings. When sun or more view is desired, screens slide aside easily on marbles in a rabbeted channel cut into the bottom piece.

DETAIL OF MARBLE TRACK

1" x 3"

END VIEW

1" x 1"

2" x 4"s

END VIEW (SEE DETAIL)

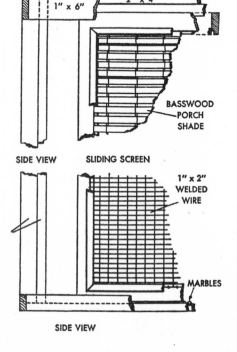

1" x 6"   2" x 4"

BASSWOOD PORCH SHADE

SIDE VIEW   **SLIDING SCREEN**

1" x 2" WELDED WIRE

MARBLES

SIDE VIEW

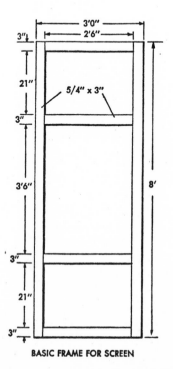

3'0"

2'6"

3"

21"

3"

5/4" x 3"

3'6"

8'

3"

21"

3"

BASIC FRAME FOR SCREEN

SLIDING SCREEN FOR PRIVACY

Picture windows bring problems of privacy and frequently too much sun. One way to solve both problems is to use sliding screens mounted on a sturdy framework. They can be painted to contrast with or to match the house, stained or allowed to weather. Many materials can be used—bamboo or basswood porch shades, wire hardware cloth, trellis slats, snow fence, corrugated fiberglass or plastic—and many other coverings can be applied on the basic framework shown below. With bamboo and basswood shades, a groove should be cut deep enough to admit the shade, which is tacked to the frame and a molding applied to give a neat finish to the edge. With corrugated plastic, hardware cloth or wire, material can be stretched firmly over frame, molding applied on top if a finish is desired. An interesting new variation on the permanent slat idea is the horizontal slat set straight, which shades and does not impede the outward view, but breaks the inward view. Permanent slats can also be set at an angle, as seen in lower version. The molding gives a neat-looking finish to the dadoed slat inserts.

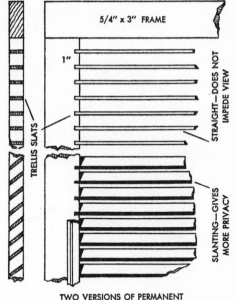

TWO VERSIONS OF PERMANENT
SLATS IN SLIDING SCREEN

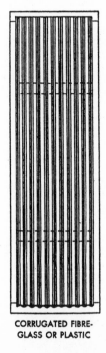

SNOW FENCE

CORRUGATED FIBRE-
GLASS OR PLASTIC

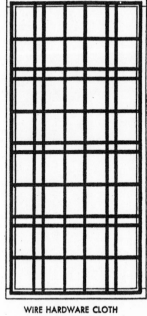

WIRE HARDWARE CLOTH
AND WOOD SLATS

HALF-AND-HALF, SUN AND SHADE

Roofed with a 1″ x 2″ trellis supported by a 2″ x 4″ frame and sturdy posts set in concrete, this terrace cover provides shade on hot days in summer, or warmth of the sun on cool days in spring and autumn. On two sides the neighbors are kept at bay by a raised fence faced with either Transite or plywood, louvers or spaced trellis slats.

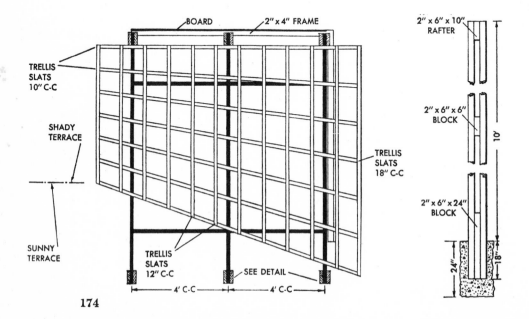

SHADE FOR WHEN A TREE DIES

Regrettable as the loss of a large shade tree may be, it needn't rob you of shade completely. It is possible to utilize the tree as a source of shade for some years more until its replacement grows big enough to contribute sufficient protection. Cut off all limbs and branches likely to be dislodged by the wind, then bolt the center 2″ x 6″s to the trunk, then attach the 2″ x 4″ frame members securely and apply the trellis strips. For complete support, use steel cables through eyebolts to four corners. A wisteria or some other permanent vine planted at the trunk and trained up to cover the shelter will eventually clothe it and even the trunk and limbs of the tree. Meantime plant quick-growing annual vines (see Chapter 6) for thick cover, yearly shade. You may wish to peel the bark off and paint the tree a pleasant, soft color or stain soft green or brown.

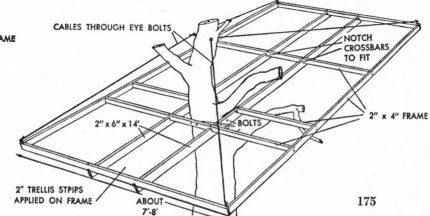

DETAIL OF TRELLIS FRAME

CABLES THROUGH EYE BOLTS

NOTCH CROSSBARS TO FIT

2″ x 4″ FRAME

2″ x 6″ x 14′

BOLTS

2″ TRELLIS STPIPS APPLIED ON FRAME

ABOUT 7′-8′

175

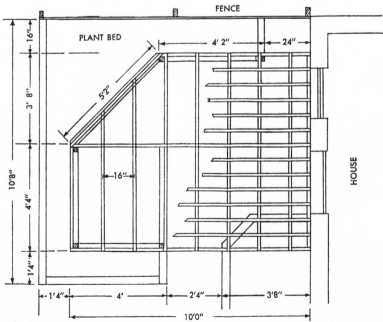

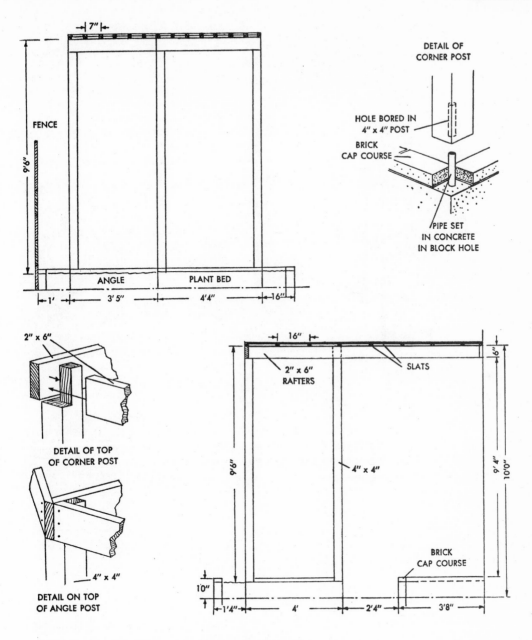

DETAIL OF
CORNER POST

HOLE BORED IN
4" x 4" POST

BRICK
CAP COURSE

PIPE SET
IN CONCRETE
IN BLOCK HOLE

FENCE

9'6"

7"

ANGLE

PLANT BED

1'   3'5"   4'4"   16"

2" x 6"

DETAIL OF TOP
OF CORNER POST

4" x 4"

DETAIL ON TOP
OF ANGLE POST

16"

6"

2" x 6"
RAFTERS

SLATS

9'6"

4" x 4"

9'4"

10'0"

BRICK
CAP COURSE

10"

1'4"   4'   2'4"   3'8"

A NEW ANGLE ON TRELLISES

Even though a terrace is quite tiny, it can still have charm—in fact, the
smaller it is the more it *needs* charm. An unconventional trellis which fits
well with a traditional house, such as this one, lends piquance to the home
scene. Blocks form a low plant bed, give year-round definition to the terrace
which in summer is graced with potted plants; annuals are used with shrubs
in the beds. The trellis breaks the sun's rays, providing a place for annual
vines to climb on. The clapboard fence continues lines of the house.

177

AN EYRIE FOR EATING—OR DREAMING

Although shown as a platform extended over a declivity, this corner arrangement might be built equally well on the level. The seats are hung from the supporting members of the trellis, as is the table on one side. Outdoor plywood may be used for seats as well as the table top if a sturdy frame of 2″ x 2″s or heavier were used for support.

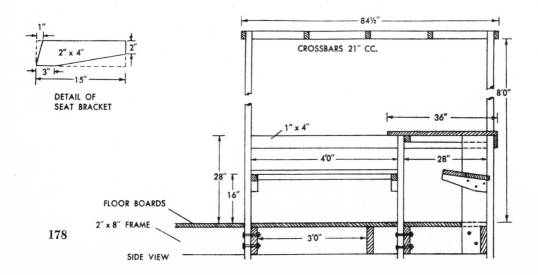

1″

2″ x 4″

2″

3″

15″

**DETAIL OF
SEAT BRACKET**

84½″

CROSSBARS 21″ CC.

8′0″

1″ x 4″

36″

4′0″

28″

28″

16″

FLOOR BOARDS

2″ x 8″ FRAME

3′0″

SIDE VIEW

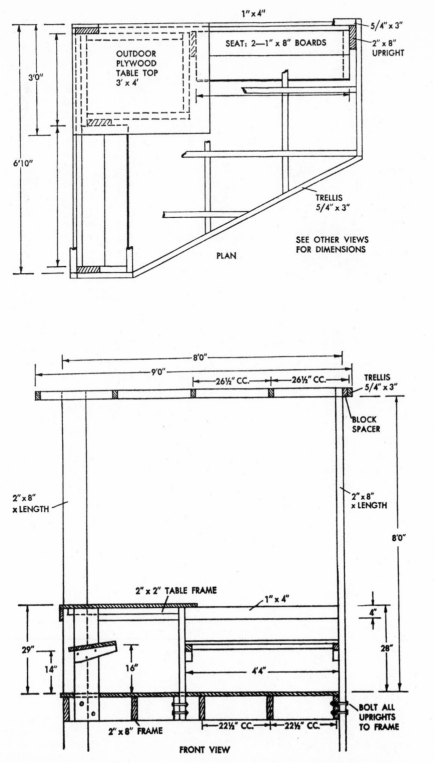

1″ x 4″

5/4″ x 3″

SEAT: 2—1″ x 8″ BOARDS

2″ x 8″
UPRIGHT

OUTDOOR
PLYWOOD
TABLE TOP
3′ x 4′

3′0″

6′10″

TRELLIS
5/4″ x 3″

SEE OTHER VIEWS
FOR DIMENSIONS

PLAN

8′0″

9′0″

26½″ CC.

26½″ CC.

TRELLIS
5/4″ x 3″

BLOCK
SPACER

2″ x 8″
x LENGTH

2″ x 8″
x LENGTH

8′0″

2″ x 2″ TABLE FRAME

1″ x 4″

4″

29″

14″

16″

28″

4′4″

2″ x 8″ FRAME

22½″ CC.

22½″ CC.

BOLT ALL
UPRIGHTS
TO FRAME

FRONT VIEW

179

A SHADOWPLAY SHELTER

This quadrangular shelter is a little off the beaten track, with its seat roofed over and its trelliswork casting shadows of ever-changing patterns on the gravelled terrace. It adapts itself to many kinds of gardens.

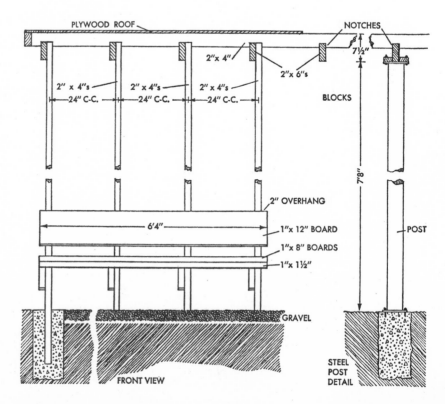

PLYWOOD ROOF

NOTCHES

2" x 4"

7½"

2" x 4"s

2" x 4"s

2" x 4"s

2" x 6"s

BLOCKS

24" C-C.

24" C-C.

24" C-C.

7'8"

2" OVERHANG

6'4"

1"x 12" BOARD

POST

1"x 8" BOARDS

1"x 1½"

GRAVEL

180

FRONT VIEW

STEEL POST DETAIL

DETAILS OF POST AND SEAT

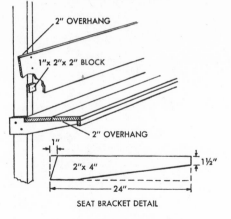

2" OVERHANG

1" x 2" x 2" BLOCK

2" OVERHANG

1"

2" x 4"

1½"

24"

SEAT BRACKET DETAIL

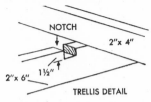

NOTCH

2" x 4"

2" x 6"

1½"

TRELLIS DETAIL

FURTHER DETAILS

Basically the shelter is simplicity itself, consisting of four upright wooden posts set in concrete on one side, supporting trelliswork and seat, while on the other side three steel basement columns, bolted to a concrete base, hold up the front, giving unobstructed views of the garden. Wooden posts (4" x 4") could be substituted if desired. Only the area over the seat is roofed over, either outdoor plywood or Transite being used, but both need sloping from front to back to provide quick drainage and thus preserve the roofing material. The 2" x 4" frame for the roof is notched to fit over the trelliswork frame, and nailed in place. Vines which climb on the trellis give adequate shade in the summer; and when they are leafless in spring and autumn, sufficient sun enters the trellis to make it possible to sit outdoors.

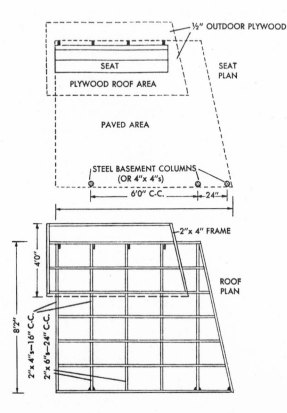

½" OUTDOOR PLYWOOD

SEAT

SEAT PLAN

PLYWOOD ROOF AREA

PAVED AREA

STEEL BASEMENT COLUMNS (OR 4" x 4"s)

6'0" C-C.   24"

2"x 4" FRAME

4'0"

8'2"

2" x 4"'s—16" C.-C.

2" x 6"'s—24" C.-C.

ROOF PLAN

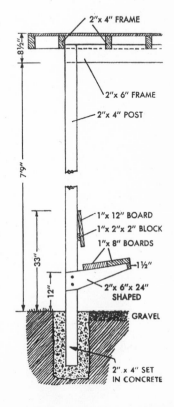

2"x 4" FRAME

8½"

2"x 6" FRAME

2"x 4" POST

7'9"

1"x 12" BOARD

1"x 2"x 2" BLOCK

1"x 8" BOARDS

1½"

2"x 6"x 24" SHAPED

33"

12"

GRAVEL

2" x 4" SET IN CONCRETE

GARDEN HOUSE WITH A MODERN FLAVOR

Informality is the keynote of this structure with its seat along the back wall and the poles on the sides giving it strength and a feeling of enclosure without restricting outward view. Annual vines can clamber up the poles toward the shingled roof to make dining under cover a delight during the summer. The paving is set into the soil here, but a concrete or a brick- or stone-on-concrete platform would be equally acceptable. Similarly, in place of the dwarf trees in pots, plant beds could be let into the paving and made gay with annuals.

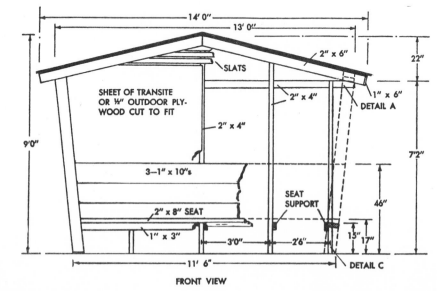

FRONT VIEW

CONSTRUCTION DETAILS

Shingles for the roof should be heavy, thick-butt type, applied on outdoor plywood roof. Rafters are notched to admit the supporting rails and the bracing members "B" are either nailed or bolted to post. Place 65" members first, then cut angle supports accurately to fit spaces, secure them to post and ridgepole.

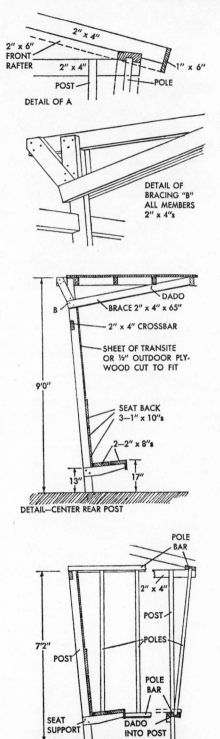

2" x 4"

2" x 6" FRONT RAFTER

2" x 4"

POST

1" x 6"

POLE

DETAIL OF A

DETAIL OF BRACING "B" ALL MEMBERS 2" x 4"s

B

DADO

BRACE 2" x 4" x 65"

2" x 4" CROSSBAR

SHEET OF TRANSITE OR ½" OUTDOOR PLYWOOD CUT TO FIT

SEAT BACK 3–1" x 10"s

2–2" x 8"s

9'0"

13"   17"

DETAIL—CENTER REAR POST

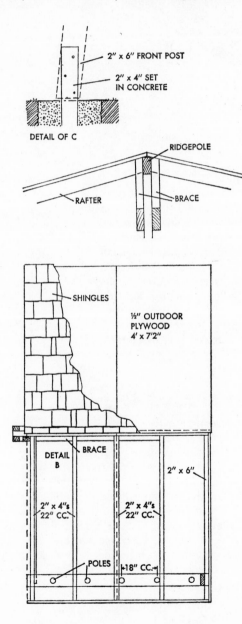

2" x 6" FRONT POST

2" x 4" SET IN CONCRETE

DETAIL OF C

RIDGEPOLE

RAFTER   BRACE

SHINGLES

½" OUTDOOR PLYWOOD 4' x 7'2"

DETAIL B   BRACE

2" x 6"

2" x 4"s 22" C.C.   2" x 4"s 22" C.C.

POLES   ⊢18" C.C.⊣

PLAN

POLE BAR

2" x 4"

POST

POLES

POLE BAR

7'2"

POST

SEAT SUPPORT

DADO INTO POST

DETAIL—RIGHT REAR POST—FRONT VIEW

A TREE HOUSE FOR THE CHILDREN

The first requisite of any tree house is, of course, the tree. Choose a good strong one with a sufficient number of limbs to give good solid support to the framework. Keep the frame as light as is consistent with safety (always the prime consideration where children are concerned) and brace the platform adequately. In the sketches are shown two types of flooring, either of which may be chosen. Note that the upright posts are securely bolted to the frame, rails being fastened by nails. Hinged "gate" rail is lowered when tree house is in use to prevent accidents. Adapt the frame to the limb structure, securing it to the trunk only—limbs are likely to wave about in the wind. Holes in floor permit them to move without endangering the platform. Crossbars are bolted to the tree trunk with 6" lag screws, braces being notched to fit over cross-bars and nailed.

DETAIL OF GATE AND RAIL

FLOOR DETAILS

Conventional floor boards, 1" x 6" or 1" x 8", may be utilized, or $\frac{1}{2}$" outdoor plywood laid on diagonal subflooring of rough or scrap lumber. Alternate flooring (below) may require more work but it will last indefinitely, making a rigid framework and providing perfect drainage in wet climates. Use 2" x 4"s for its outer framework, 1" x 3"s separated by 1" x 3" x 3" blocks, all bolted together with steel tie-rods through holes bored through boards and frame, giving rigidity.

2"x4" FRAME

STEEL TIE-RODS

1"x3" BOARDS

1"x3"x3" BLOCKS

DETAIL OF ALTERNATE FLOORING OF 1"x3"s

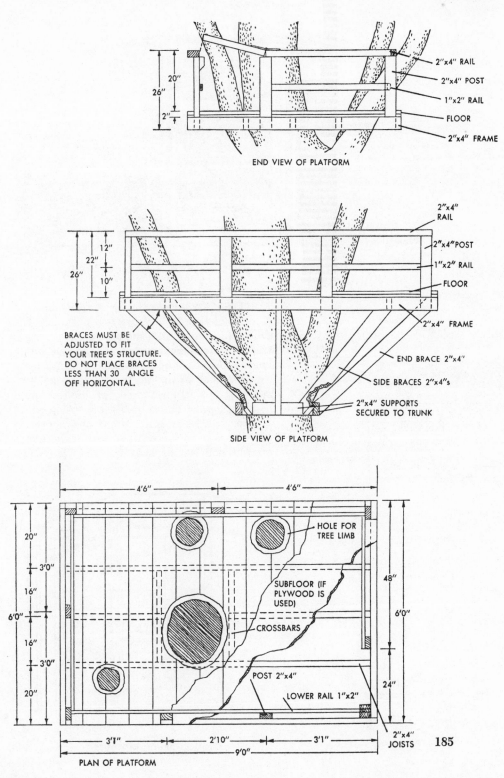

END VIEW OF PLATFORM

2"x4" RAIL

2"x4" POST

1"x2" RAIL

FLOOR

2"x4" FRAME

20"

26"

2"

2"x4" RAIL

2"x4" POST

1"x2" RAIL

FLOOR

2"x4" FRAME

12"

22"

26"

10"

END BRACE 2"x4"

SIDE BRACES 2"x4"s

2"x4" SUPPORTS SECURED TO TRUNK

BRACES MUST BE ADJUSTED TO FIT YOUR TREE'S STRUCTURE. DO NOT PLACE BRACES LESS THAN 30 ANGLE OFF HORIZONTAL.

SIDE VIEW OF PLATFORM

4'6"

4'6"

HOLE FOR TREE LIMB

SUBFLOOR (IF PLYWOOD IS USED)

CROSSBARS

POST 2"x4"

LOWER RAIL 1"x2"

2"x4" JOISTS

20"

3'0"

16"

6'0"

16"

3'0"

20"

48"

6'0"

24"

3'1"

2'10"

3'1"

9'0"

PLAN OF PLATFORM

185

ATTRACTIVE SHELTER IN A CORNER

A corner of a garden may create a most difficult problem in the small-home grounds. By using shiplap or scored outdoor plywood to give long lines to the fence and by echoing those lines with the slats of the shelter, by using boldly opposing vertical lines to support the roof, we create a garden house both good-looking and unique. The seats add comfort and a place for dining, while the plywood part of the roof provides both sun and shelter.

LAYOUT FOR CUTTING SEATS FROM ONE 4′ x 8′ SHEET OF PLYWOOD. LETTERS KEY SEATS TO PLAN BELOW

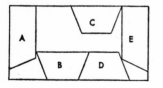

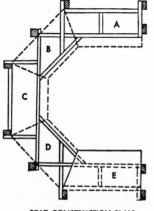

SEAT CONSTRUCTION PLAN

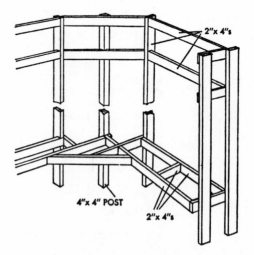

2″x 4″s

4″x 4″ POST

2″x 4″s

CONSTRUCTION OF FRAMEWORK

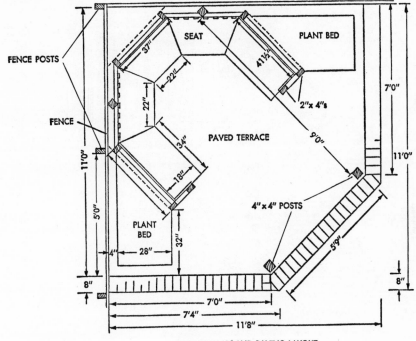

FENCE POSTS

SEAT

PLANT BED

37"

41½"

22"

2" x 4"s

7'0"

FENCE

22"

PAVED TERRACE

9'0"

11'0"

11'0"

3'4"

5'0"

18"

4" x 4" POSTS

PLANT
BED

32"

5'9"

4"

28"

8"

7'0"

8"

7'4"

11'8"

GROUND PLAN AND PAVING LAYOUT

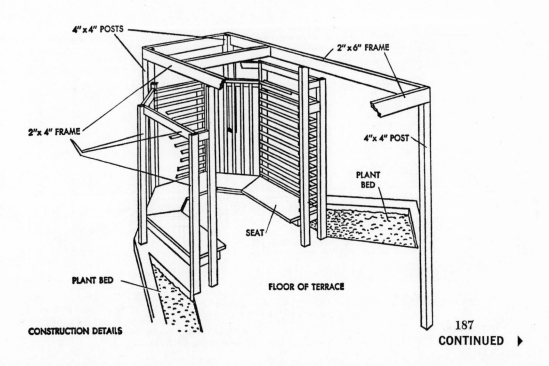

4" x 4" POSTS

2" x 6" FRAME

2" x 4" FRAME

4" x 4" POST

PLANT
BED

PLANT BED

SEAT

FLOOR OF TERRACE

CONSTRUCTION DETAILS

187

CONTINUED ▶

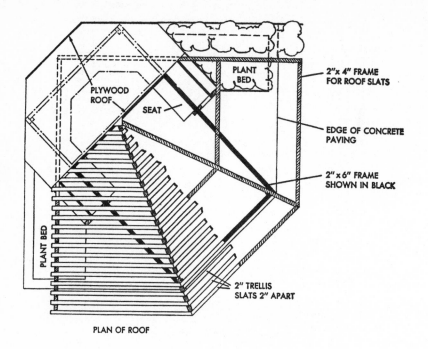

PLYWOOD ROOF

PLANT BED

SEAT

2" x 4" FRAME
FOR ROOF SLATS

EDGE OF CONCRETE
PAVING

2" x 6" FRAME
SHOWN IN BLACK

PLANT BED

2" TRELLIS
SLATS 2" APART

PLAN OF ROOF

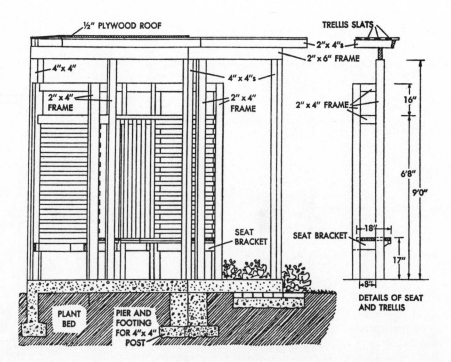

½" PLYWOOD ROOF

TRELLIS SLATS

2" x 4"s

2" x 6" FRAME

4" x 4"

4" x 4"s

2" x 4"
FRAME

2" x 4"
FRAME

2" x 4" FRAME

16"

6'8"

9'0"

SEAT
BRACKET

SEAT BRACKET

18"

17"

PLANT
BED

PIER AND
FOOTING
FOR 4"x 4"
POST

8"

DETAILS OF SEAT
AND TRELLIS

Well integrated into a shelter and fence, a tool storage house built of grooved exterior ⅝″ plywood with interior grade inside in ¾″ and ¼″ thicknesses provides a place for everything needed in the garden and even a storage place for the barbecue between uses.

A shelter with off-the-ground storage is built from exterior-type plywood on a sturdy wood frame. The 2″ x 4″ columns are firmly held by metal U-supports set in deep square concrete footings. Door shelves and hangers store small tools, supplies. Note slatted 2″ x 4″ seat.

Everything under one roof in a triangular garden house: tool storage, potting shelf, even a tiny "greenhouse" taking advantage of waste space. A veritable gardener's dream. Framing is 2" x 4"s throughout; siding and roofing is Masonite exterior hardboard.

Drainholes are located in the hardboard floor. Glazed frames lift off.

A flat-roofed exterior-plywood-covered storage unit houses not only the garden tools but also furniture and barbecue equipment, bicycles and other possessions. It even has a place for logs and refuse cans. Shaped built-up roof beams are made of doubled 2″ x 8″s.

A garden storage house or a playhouse for the kids is designed like a barn and painted red. Built on a stout wood frame faced with Masonite Panelgroove hardboard (outdoor type) with pegboard inside for hanging tools and equipment, it is topped with wood shingles.

A neat and unobtrusive garden house has space not only for everything needed for entertaining outdoors but wide-opening front is composed of doors with door shelves similar to those on refrigerators that, with inside shelves, make furniture, dishes, implements available.

Who would think that what appears merely a pleasant jog in a fence could be a complete tightly-roofed tool storage house? The same boards and spacing of fence are used on 4" x 4" and 2" x 4" framing; padlocked doors keep everything safe.

9

Build Your Own
Outdoor Furniture

With life becoming more informal every day and with outdoor life an accepted part of the American scene, there are many good reasons for building your own furniture. Use it for sitting outdoors, for entertaining on your terrace—take full advantage of all the possible delights offered—and have furniture which is of a different character from that run-of-the-mill stuff one sees so much of today. While there are many kinds of inexpensive outdoor furniture which can be purchased for use on the terrace, some of it rather good in design, there is an increasing need being felt for good *permanent* tables and for seats which are strong and may be left outdoors the year round. Also, with our small quarters and changeable climates, a need for a good *demountable* table has been expressed with frequency.

Therefore we present in this section a number of simple designs which almost any amateur craftsman will find easy to construct, for furniture which will add to the joy you take in your garden. Some is modern in character, so it will fit more comfortably with the modern home in good taste than will commercial furniture with its shrieking colors, its machine-made curves and angles. Other pieces will fit well with traditional homes, and several of our designs will integrate with either modern or traditional styles.

One of the main drawbacks to commercial furnishings is the imper-

194

manence of the pieces offered. Cheaply constructed, they are so light in weight that they won't last long, are never satisfactory to use, and when the final analysis is made of costs, they prove not to have been so inexpensive after all. The furniture shown here is planned to be able to withstand ordinary weather, so that it can be left outdoors; it need not be whisked under cover at the approach of every storm. It will take its beating from children's (ab)use and still look all right, for its simple, good-looking lines, its rugged character are *designed for use*. And it will age well. If you should ever need to replace any parts for any reason, you will find them either in your wood scrap box or at your nearest building supply house. Can commercial furniture offer this advantage?

SPECIAL FEATURES

Note that the seats and tables we call "demountable" have tops which can be taken off for storage in the garage or cellar during the inclement months, leaving only the masonry parts to bear the brunt of ice and snow during the winter. Those made entirely of cinder blocks and stone slabs can, of course, be left outdoors and used on those few bright days of winter when the sun lures you out for a quick picnic in the sparkling air. These, too, can be called "demountable" because they can be moved from one spot in the garden to another if you wish—from the open terrace into a sun-trap in a corner, from the terrace alongside the house to another place where you may have an outdoor fireplace for cooking. But they are rather heavy and you won't want to move them too often.

Consider the tables which fold down from a porch or terrace railing and those which lower from a kind of outdoor "breakfront." With one of them you can have a good-looking piece of furniture which does not take up a great deal of space when it is not in use.

In addition to these features you will find a number of designs for seating pieces. Note in the shelter section how seats have been built in as integral parts of the shelters, and how, in the wall section of this book, low walls are shown used as seats, too. These will "come in very handy," as my mother used to say, when you are entertaining the family or large numbers of people on the terrace for luncheon or for cocktail parties, or other gatherings.

The primary consideration in choosing outdoor furniture, whether you build it yourself or buy it, is that it should be practical and useful. Secondarily, it should be good-looking and simple, so that it fits well into its surroundings and is a part of the general picture—an unobtrusive part. You will find that the natural qualities of unfinished wood, stone, and building blocks used for furniture will help to keep it in the background and in harmony with the house and natural parts of the garden. If the materials must be painted, we urge that soft colors be employed— never bright or dominant ones, *never* hard reds or bright oranges, or even that unnatural green one so often encounters.

BENCH FOR GOOD COMPANIONS

Backless garden benches are coming more and more into use. This bench can be used alone as shown in the bottom version, or can be used to angle off to follow a terrace boundary or one of the part-shade, part-sun trellises now so popular. It will adapt itself to any site and can be portable or permanent. Built of 1" x 2" x 2" blocks it has apron boards and two center boards of 1' x 6" shaped at ends to meet end aprons of 1" x 4" stock. Supports consist of two 2" x 8"s shaped as shown resting on 2" x 4" feet which protect them from rot. (Permanent version is set in concrete after treatment with a wood preservative.) Note that boards are butted at corners, nailed or screwed together securely, as center blocks are, also.

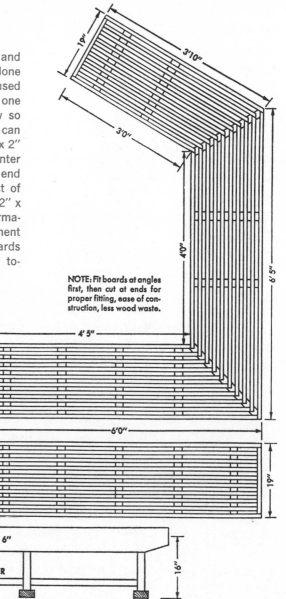

NOTE: Fit boards at angles first, then cut at ends for proper fitting, ease of construction, less wood waste.

END VIEW   SIDE VIEW

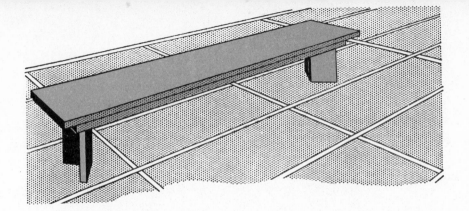

BENCH—PORTABLE OR PERMANENT

A bench of simple lines, built of sturdy 2″ stock and painted to harmonize with the house or stained a dark, woodsy color, will be a useful addition to any terrace. If you wish to place it permanently, treat the posts with wood preservative, set them in concrete beside the terrace or in whatever location you have chosen for the bench to be placed.

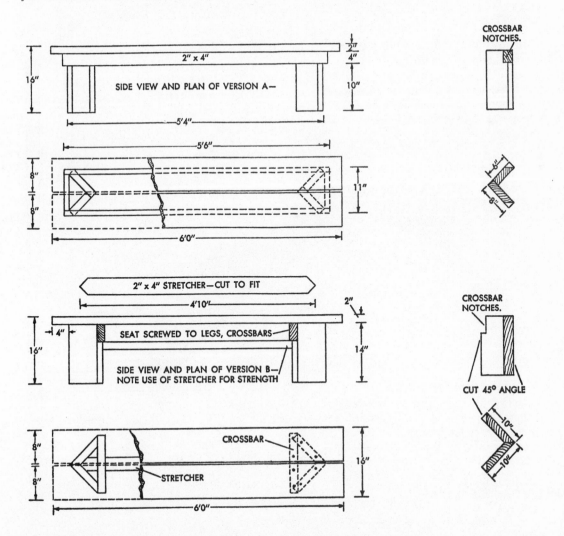

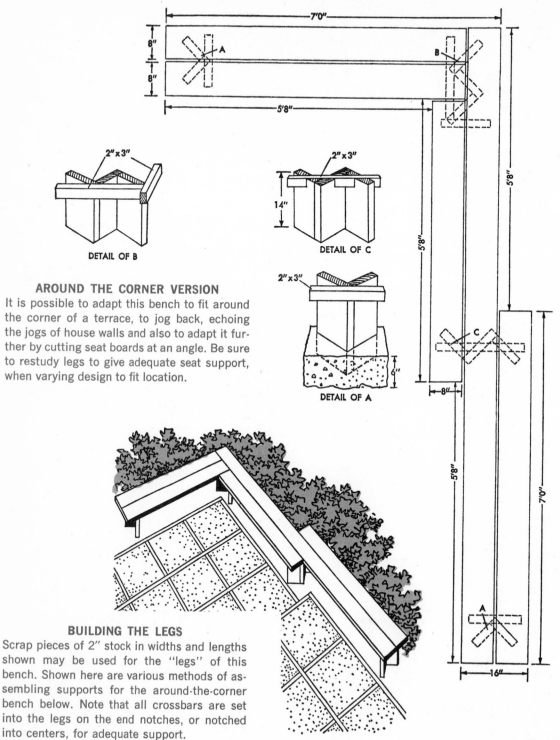

7'0"

8"

8"

A

B

5'8"

5'8"

5'8"

2"x3"

DETAIL OF B

2"x3"

14"

DETAIL OF C

2"x3"

C

6"

DETAIL OF A

8"

5'8"

7'0"

A

16"

AROUND THE CORNER VERSION

It is possible to adapt this bench to fit around the corner of a terrace, to jog back, echoing the jogs of house walls and also to adapt it further by cutting seat boards at an angle. Be sure to restudy legs to give adequate seat support, when varying design to fit location.

BUILDING THE LEGS

Scrap pieces of 2" stock in widths and lengths shown may be used for the "legs" of this bench. Shown here are various methods of assembling supports for the around-the-corner bench below. Note that all crossbars are set into the legs on the end notches, or notched into centers, for adequate support.

**DETAIL OF LEG
HINGE AND HOOK**

FOLDAWAY TABLE ON A BALUSTRADE OR WALL

A table which folds away when not in use, leaving precious terrace space free, is an asset to be prized. Hinges attach the 2″ x 2″ frame to balustrade, fence or wall; hinged legs fold up inside frame to be held by turn-buttons on a stationary block. Legs are secured by a $\frac{1}{4}$″ steel rod bent into a double hook, fitted into holes bored in leg and frame. Use $\frac{1}{2}$″ outdoor plywood for table top and for legs, too, if desired. Vary the dimensions to suit your needs. This table will seat five adults with good elbowroom.

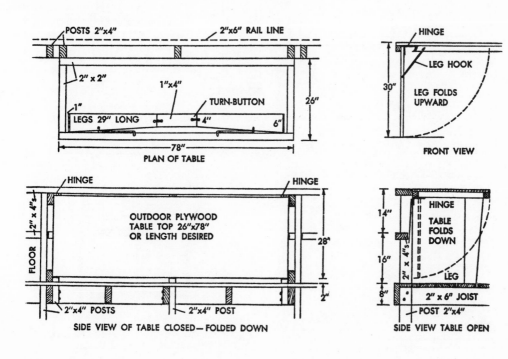

POSTS 2″x4″   2″x6″ RAIL LINE

2″ x 2″

1″x4″

1″   TURN-BUTTON

LEGS 29″ LONG   4″   6″

26″

78″

PLAN OF TABLE

HINGE

LEG HOOK

30″   LEG FOLDS UPWARD

FRONT VIEW

HINGE   HINGE

2″ x 4″s

FLOOR

OUTDOOR PLYWOOD
TABLE TOP 26″x78″
OR LENGTH DESIRED

28″

2″

2″x4″ POSTS   2″x4″ POST

SIDE VIEW OF TABLE CLOSED—FOLDED DOWN

14″

HINGE

TABLE
FOLDS
DOWN

2″ x 4″s

16″

LEG

6″

2″ x 6″ JOIST

POST 2″x4″

SIDE VIEW TABLE OPEN

FOLDAWAY DEMOUNTABLE TABLE

For winter storage, take off the table top made of 1″ x 6″ boards mounted on 2″ x 4″ frame, which fits over leg and stretcher construction. A long pivot bolt permits folding up for carting and storage. Rot-block feet may be replaced as necessary. Table top stays on by own weight if frame is carefully fitted to legs. Vary top size to suit needs.

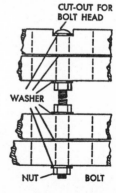

CUT-OUT FOR BOLT HEAD

WASHER

NUT   BOLT

BLOCKS

CENTER BOLT DETAILS

4′ 1½″

6″

LEGS CUT FROM 1″ BOARD OR ¾″ PLYWOOD

17″

28″

11″

SIDE VIEWS

3″

2″

END VIEWS

13″

4′ 4½″

4′ 6″

2″ x 4″ APRON

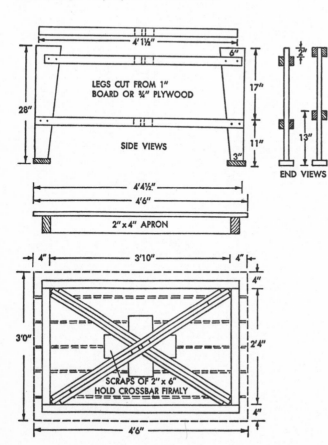

4″   3′10″   4″

4″

3′0″

2′4″

4″

SCRAPS OF 2″ x 6″ HOLD CROSSBAR FIRMLY

4′6″

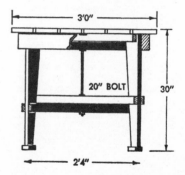

3′0″

20″ BOLT

30″

2′4″

FOLDING LOVESEAT OR BENCH FOR THE TERRACE

A chair and a loveseat of similar design are easy to build, easy to fold up and take in for the winter or for quick shelter from a summer shower. The boards used for seats may be $\frac{5}{4}$" or 2" (the lighter weight may make them more transportable) with 1" x 4" stock used for backrest boards. Legs may be cut from 2" x 10"s or heavy outdoor plywood.

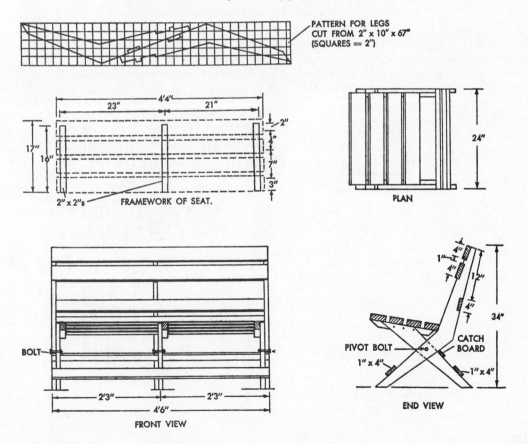

PATTERN FOR LEGS
CUT FROM 2" x 10" x 67"
(SQUARES = 2")

4'4"
23" — 21"
17"
16"
2"
4"
7"
3"
2" x 2"s — FRAMEWORK OF SEAT.

PLAN
24"

FRONT VIEW
BOLT
2'3" — 2'3"
4'6"

END VIEW
PIVOT BOLT
CATCH BOARD
1" x 4"
1" x 4"
1" x 4"
4"
1"
4"
12"
4"
34"

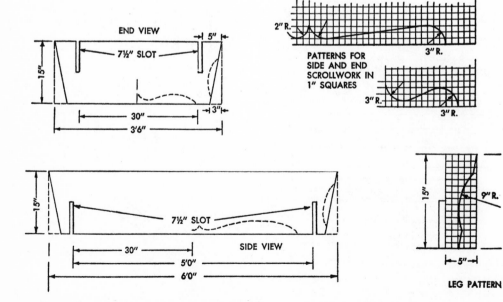

LARGE COFFEE TABLE FOR OUTDOORS

For informal lunches, suppers and for snacks at any time a large coffee table built of outdoor plywood, protected by wood preservative, and painted in a gay color is a good solution. This one may be made as shown in the sketch, with shaped apron and the corner "legs" cut out, too, or it may be made perfectly wedge shaped as in the sketches of the parts below. The feet may be changed, should they decay from contact with moisture, waste blocks being used for them. Or if the table is used on a porch where it is dry the feet may be dispensed with.

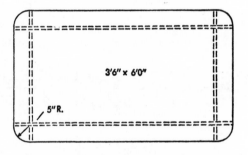

3'6" x 6'0"

5" R.

END VIEW

5"

7½" SLOT

15"

30"

3"

3'6"

2" R.

3" R.

PATTERNS FOR
SIDE AND END
SCROLLWORK IN
1" SQUARES

3" R.

3" R.

15"

7½" SLOT

30"

SIDE VIEW

5'0"

6'0"

15"

9" R.

5"

LEG PATTERN

CUPBOARD WITH LET-DOWN TABLE

When a garden house or terrace has limited space, the need is acute for a dining table which will fold away. This table, combining storage space for plates and other necessary equipment, is hinged to let down, the rear resting on the jutting lower cupboard space, the front supported by a hinged leg which is a part of the decorative frame when not in use. Potted plants grace the shelves on either side. Lower cupboards may have two doors to divide into center and two side compartments.

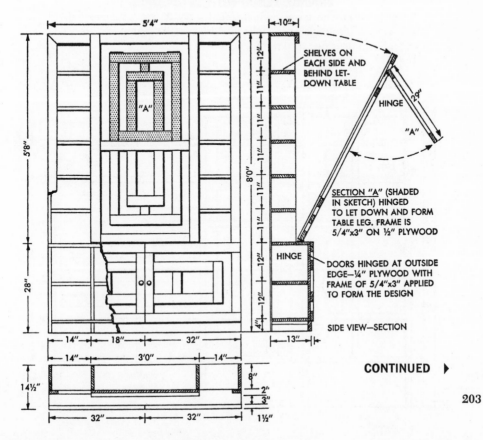

SHELVES ON EACH SIDE AND BEHIND LET-DOWN TABLE

HINGE

29"

"A"

SECTION "A" (SHADED IN SKETCH) HINGED TO LET DOWN AND FORM TABLE LEG. FRAME IS 5/4"x3" ON ½" PLYWOOD

HINGE

DOORS HINGED AT OUTSIDE EDGE—¼" PLYWOOD WITH FRAME OF 5/4"x3" APPLIED TO FORM THE DESIGN

SIDE VIEW—SECTION

CONTINUED ▶

203

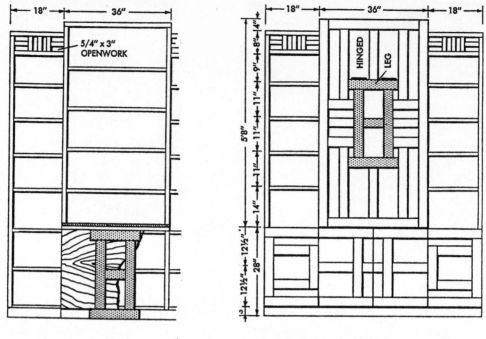

FRONT VIEW OF LET-DOWN TABLE

BREAKFRONT LET-DOWN TABLE SHOWN CLOSED

PRIVACY FENCE WITH LET-DOWN TABLE

Fence table has no lower compartment, only 2″ x 6″ shelves behind table which continue alongside to make open shelves for plants, etc. Vary cupboard and table dimensions to suit.

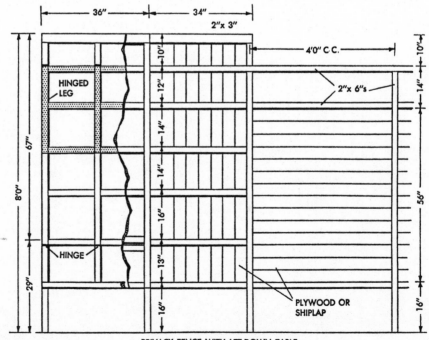

PRIVACY FENCE WITH LET-DOWN TABLE

204

BREAKFRONT FOR PLANTS

A breakfront with adjustable shelves that may be used indoors or on a covered terrace to summer houseplants outdoors also provides storage space below for pots, plant foods, vases, and other equipment. The shelves are made of 1″ x 2″s fitted together as shown in the details. The triangular cuts in the projecting ends fit over the dowel rod supports and prevent shelves from sliding.

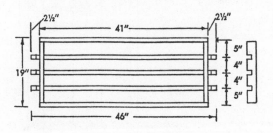

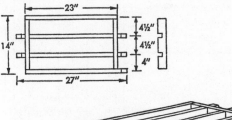

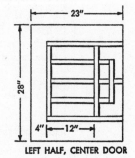

LEFT HALF, CENTER DOOR

CENTER SHELF DETAIL

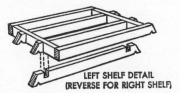

LEFT SHELF DETAIL
(REVERSE FOR RIGHT SHELF)

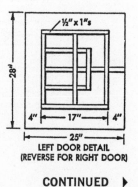

LEFT DOOR DETAIL
(REVERSE FOR RIGHT DOOR)

CONTINUED ▶

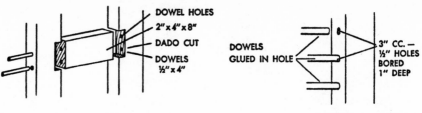

DOWEL HOLES

2" x 4" x 8"

DADO CUT

DOWELS
½" x 4"

DOWELS
GLUED IN HOLE

3" CC. —
½" HOLES
BORED
1" DEEP

DETAIL OF END SECTION "A"

DETAIL OF SHELF SUPPORT

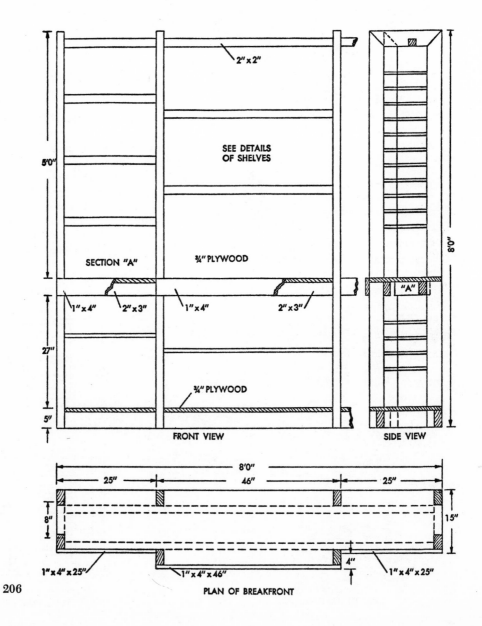

2" x 2"

SEE DETAILS
OF SHELVES

5'0"

SECTION "A"

¾" PLYWOOD

1" x 4"   2" x 3"   1" x 4"   2" x 3"

27"

¾" PLYWOOD

5"

FRONT VIEW

8'0"

"A"

SIDE VIEW

8'0"

25"   46"   25"

8"

15"

4"

1" x 4" x 25"   1" x 4" x 46"   1" x 4" x 25"

PLAN OF BREAKFRONT

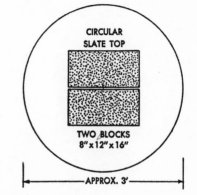

YOU CAN MOVE THE TABLE FOR LUNCH

A few chimney flue blocks, a slab of cut stone (outdoor plywood could also be utilized) and you can move your table around to enjoy various parts of your garden or terrace as the sun and blossoms may dictate. Chimney flue blocks form the support here for the table and make the seats when paired. Cushions will make them more comfortable, but are not necessary.

COFFEE TABLE OF BLOCKS AND SLATE

A slate top cut round or in irregular slab style set on a pair of wall blocks makes a most acceptable low cocktail or coffee table. It can be easily picked up and moved, and because it is impervious to weather can be left out all winter as a bird feeding table.

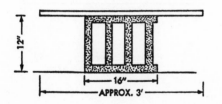

12"

16"

APPROX. 3'

CIRCULAR SLATE TOP

TWO BLOCKS
8" x 12" x 16"

APPROX. 3'

SLATE COFFEE TABLE

In most sections it is possible to buy a slate top cut in a circle. The table shown here is either demountable or permanent and the supports are simple to assemble. A pair of half flue blocks are stacked as shown inside a full square chimney flue block. If you want to make them permanent, mortar them on a secure foundation, mortar the slate top securely in the center, and your table is done.

TWO INFORMAL TABLES

A table which is demountable yet is impervious to weather can be built by stacking up three chimney blocks and placing a slab of slate or other stone on top. It may be cut stone or irregular in shape as shown above. Or, by using several half-blocks placed to support the slab, a larger piece may be used for a low table. Half blocks with cushions made to fit make low seats for terrace use.

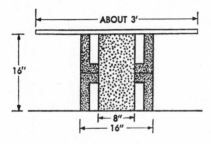

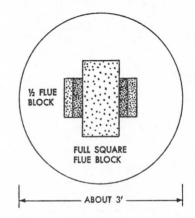

DEMOUNTABLE PLYWOOD AND BLOCK FURNITURE

An outdoor plywood top screwed to a simple framework which fits around a pedestal of cement blocks mortared to a concrete foundation makes a most useful outdoor dining table. Benches on block supports are also sturdy, simple and good looking, built of planks on a framework which also fits around blocks. Note that both the table and bench tops may be removed and taken indoors for storage or for winter protection.

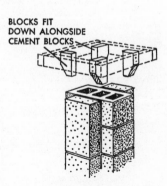

BLOCKS FIT DOWN ALONGSIDE CEMENT BLOCKS

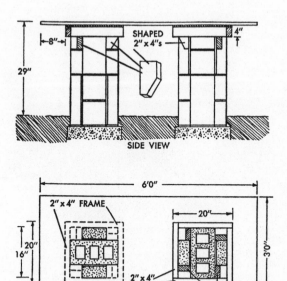

SIDE VIEW

SHAPED 2" x 4"s

29"

8"

4"

PLAN

6'0"

2" x 4" FRAME

20"

16"

20"

3'0"

2" x 4" FRAME

SHAPED 2" x 4"s

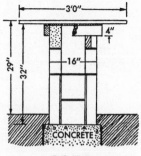

END VIEW

3'0"

29"

32"

16"

4"

CONCRETE

CONTINUED ▶

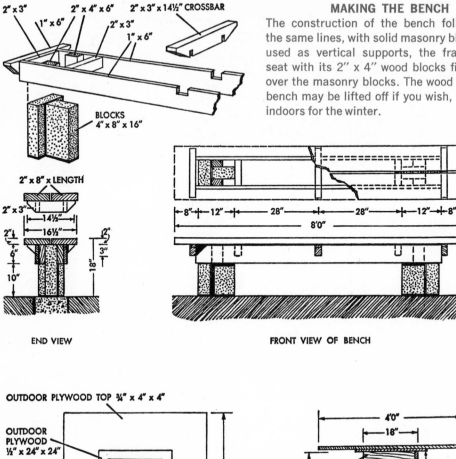

2" x 3"
1" x 6"
2" x 4" x 6"
2" x 3"
1" x 6"
2" x 3" x 14½" CROSSBAR

BLOCKS
4" x 8" x 16"

2" x 8" x LENGTH
2" x 3"
14½"
2"
16½"
2"
6"
3"
18"
10"

END VIEW

8" 12" 28" 28" 12" 8"
8'0"
16½"
18"

FRONT VIEW OF BENCH

MAKING THE BENCH

The construction of the bench follows much the same lines, with solid masonry blocks being used as vertical supports, the frame of the seat with its 2" x 4" wood blocks fitting down over the masonry blocks. The wood part of the bench may be lifted off if you wish, for storage indoors for the winter.

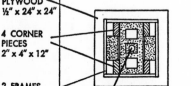

OUTDOOR PLYWOOD TOP ¾" x 4" x 4"

OUTDOOR
PLYWOOD
½" x 24" x 24"

4 CORNER
PIECES
2" x 4" x 12"

2 FRAMES
1" x 4"s

CEMENT PIPE IN
CENTER HOLE TO HOLD
UMBRELLA

4'0"

PLAN OF TABLE AND PEDESTAL

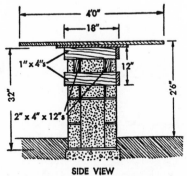

4'0"
18"
1" x 4"s
12"
32"
2" x 4" x 12"s
2'6"

SIDE VIEW

A DEMOUNTABLE SQUARE TABLE

A square table built of outdoor plywood fits on a single pedestal of cinder or concrete blocks. To keep it solid and stable, longer wooden blocks and a double banding of wooden strips are used to fit down around the masonry pedestal. Screw a 24" square of plywood to the top of the framework, then mount plywood and framework to table top with aluminum screws, thus avoiding holes in the top surface of the table. One-by-three stock can be used to frame under table edges for strengthening it, if you should wish to do so.

FURNITURE

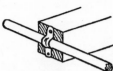

AXLE DETAIL

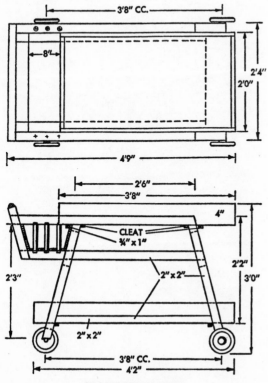

3'8" CC.

8"

2'4"
2'0"

4'9"

2'6"
3'8"

4"

CLEAT
¾" x 1"

2" x 2"

2'3"

2'2"
3'0"

2" x 2"

3'8" CC.
4'2"

SIDE VIEW (WITH TRAYS)

ROLL-AWAY SERVICE CART

Push this attractive cart up to the kitchen door, load it with food, dishes, bottles - all the appurtenances of outdoor eating and roll it out to your picnic table. You'll find outdoor eating work halved, cutting trips back and forth to one each way. Return the cart with dirty plates and the debris of a meal to the kitchen door, then lift off the tray on top (you may also want one for the bottom, shown in side view (below), and you can take it all into the house. Rubber-tired wheels bought together with a suitable axle at the hardware shop (or possibly adapted from a child's discarded wagon) roll easily. Note that trays have cleats on bottom to fit over crossbars and that bottoms of trays may be either plywood or metal mesh. Bottles stay secure in the space just below the pushing bar at the end of cart.

CONTINUED ▶

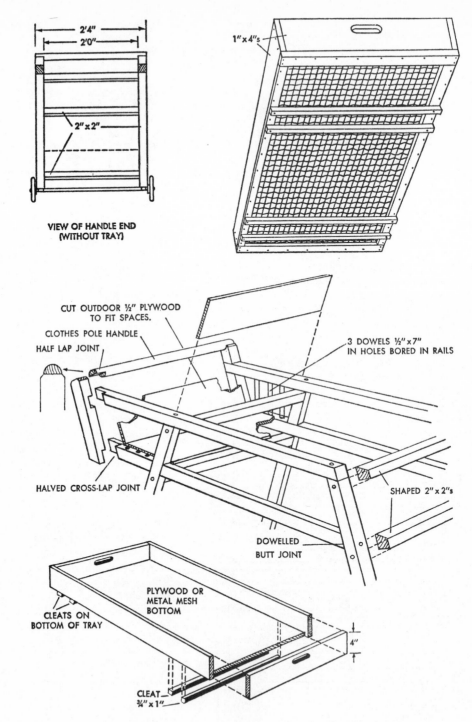

2'4"

2'0"

1" x 4"s

2" x 2"

VIEW OF HANDLE END
(WITHOUT TRAY)

CUT OUTDOOR ½" PLYWOOD
TO FIT SPACES.

CLOTHES POLE HANDLE

HALF LAP JOINT

3 DOWELS ½" x 7"
IN HOLES BORED IN RAILS

HALVED CROSS-LAP JOINT

SHAPED 2" x 2"s

DOWELLED
BUTT JOINT

PLYWOOD OR
METAL MESH
BOTTOM

CLEATS ON
BOTTOM OF TRAY

4"

CLEAT
¾" x 1"

Table of Measurements and Heights:
Furniture for Adults

SEATS:	HEIGHT
Floor to seat	16″–18″
Floor to shoulder rest	36″–37″
Seat to head clearance	2′ 11″ min.
Width, side to side, minimum	15″
Depth, back to front, minimum	15″
Clearance, top of seat to table	11″
Clearance, top of knee to table	5″
Clearance, seated people, elbow to elbow	27″ min.
Clearance, kick space from stationary seat to table upright	19″–20″
Clearance, space in front of knee	7″ plus 12″ foot-space
Clearance behind heel, legs vertical	7″
Stationary seats should extend no more than 2″-3″ under table line.	
TABLES:	
Width, table used one side only	22″–24″ min.
Width, table used on both sides	28″–20″ min.
Table height, top to floor	29″
Clearance for knees, seated, min. height	24″
Clearance for seated people, elbow to elbow	2′ 3″ min.

Children's Furniture Heights

Height of Child	Seat Height	Table Height
3′ 3″	1′	1′ 10″
4′ 0″	1′ 2″	2′ 1″
4′ 4″	1′ 3″	2′ 2½″
4′ 8″	1′ 4″	2′ 4″
5′ 0″	1′ 5″	2′ 5″
5′ 4″	1′ 5″	2′ 5″
5′ 6″ and over	1′ 6″	2′ 5″

NOTE: All the above measurements can be varied to suit the uses of the child and the adult. If you want to experiment to find out what is the best height for you, measure various tables and sitting pieces in your home, using the measurements of those pieces you find most comfortable. The above chart was compiled on the basis of average measurements: that is, tall- to average-sized people.

POTTING BENCHES

Although perhaps potting benches are not strictly furniture, they are a part of the garden picture, and the ones we show here can also be built into basements, garages, or tool sheds. The rolling potting bench will be trundled out on the terrace frequently, so it may be included as a part of the furniture section.

Every gardener has difficulty with storage—where to put stakes, where to put hand tools and all the various useful things which make such clutter when left about the house or garage or out in the garden. These benches will help to organize things and make them orderly and useful. For the indoor gardener, a potting bench is the answer to a prayer when it has storage bins for soil, peatmoss, sand, etc., lots of shelves, and plenty of storage space for pots. Outdoors, too, where people are using more potted plants to get quick color, such a potting bench with a shelter built over it will be most welcome in many a garden.

With this section, as we have urged for other sections, we suggest that you look at our designs and, if they do not exactly meet your needs, adapt them as you see fit. Change merely for the sake of change is not what we mean, of course, but change for improving the use of the piece so that it adapts to fit your specific needs will give you a taste of the creative experience which is one of the most satisfying joys of craftsmanship. Your furniture can express *you* as much as any other part of the garden does.

An odd corner behind a curving fence is put to good use as a potting bench. The trellis from the opposite side is extended over the waist-high 2″ x 4″s that form the counter. Large boxes below this contain soil, peatmoss and sand for use in potting up.

A workmanlike potting bench provides a potting surface, adequate bins for soil, sand and peatmoss, shelves for pots and other needs. Exterior-type plywood is used in ¾″ thickness, mounted on a sturdy framework of bolted 2″ x 4″s. Note guards on top and sides.

A permanent bench seating several has back members attached to pergola columns for stability. Framework of 2″ x 6″s takes cue for angle cuts from slant of back, echoes it in front. Seat and back are made of spaced 2″ x 4″s. Paint bench to match pergolas.

A curved seat is supported at ends by bolted multiple 2″ x 4″s with others flat against the curving back at the corners. Saber-saw a hefty wide plank in the desired curve, place the two straight sides together and attach them to end cleats fastened to end supports.

Portable painted bench has a seat formed of three 2″ x 8″s screwed to a 2″ x 6″ crossbar and supported by three 2″ x 8″ legs that are overlapped 2 inches each side of center space. Note how legs are angle-cut for tight fit, how crossbar is beveled to meet leg line.

A bench may be used for sitting or to display a bonsai tree. Made of redwood—2″ x 8″ planks for seat, invisibly attached with epoxy-glued dowels set in holes bored in seat plants and in crossbars that, like legs, are 4″ x 4″s. Dado crossbars 1½″ into legs.

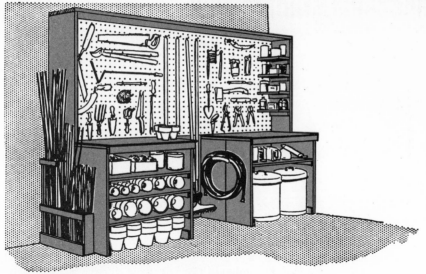

A PLACE FOR EVERYTHING

It's always easier to put a tool back when you've finished using it if there is really a place to put it. This simple pegboard arrangement adapts to the tools you now have, can be adjusted to those acquired later. Work tables for potting, etc., are built above shelves for pots and garden equipment. Table sides are put to work as storage for stakes and hose. Note upper shelves for dusts and sprays. Garbage cans provide covered storage for soil, sand and peatmoss gardeners need for use in pots, flats.

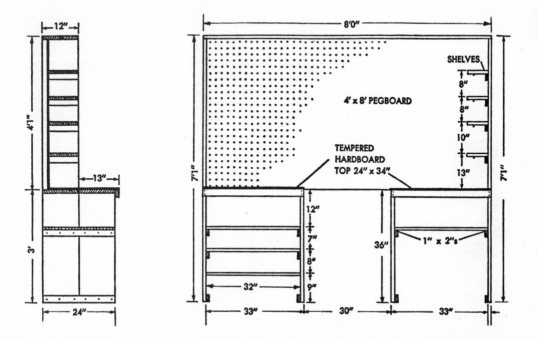

PUSHAROUND POTTING BENCH

Sometimes it is most convenient to have a portable potting bench which can be pushed from one area to another and also be easily stored when not in use. The removable tray on top makes it still more portable—for a small garden perhaps the tray would be sufficient. Peatmoss, sand and soil are stored in the bins below, pots kept on the shelf. Note that the outer parts are made of plywood, the frame and caster supports are of 2″ x 4″ stock. If desired, use tempered hardboard for topping the bench and tray.

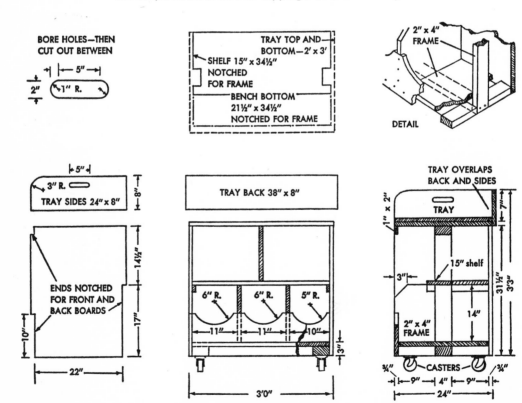

BORE HOLES—THEN CUT OUT BETWEEN

5″
2″ 1″ R.

TRAY TOP AND BOTTOM—2′ x 3′
SHELF 15″ x 34½″
NOTCHED FOR FRAME
BENCH BOTTOM 21½″ x 34½″ NOTCHED FOR FRAME

2″ x 4″ FRAME

DETAIL

5″
3″ R.
TRAY SIDES 24″ x 8″
8″

TRAY BACK 38″ x 8″

TRAY OVERLAPS BACK AND SIDES

1″ x 2″ TRAY 7″

15″ shelf
3″
31½″ 3′3″
14″

14½″

ENDS NOTCHED FOR FRONT AND BACK BOARDS

17″
10″

6″ R. 6″ R. 5″ R.
11″ 11″ 10″

2″ x 4″ FRAME

3″

22″

3′0″

¾″ CASTERS ¾″
9″ 4″ 9″
24″

FRONT VIEW

SIDE VIEW

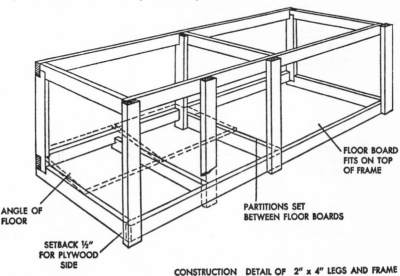

A POTTING BENCH WITH MANY OTHER USES

Gardeners fortunate enough to possess greenhouses or those who garden on a fairly large scale outdoors will have need for a potting bench that is sizeable. This one also provides storage bins for peatmoss, potting soil, sand, and has shelves for pot storage and for insecticides, fertilizers, etc. The top is made of hardboard, outdoor quality, which wears well and is relatively impervious to water. Bench may be used for other purposes in between its uses as potting bench, for cutting and arranging flowers, as an auxiliary carpentry bench and in many other ways. The collar around the back and sides prevents soil and debris from falling off, permits putting ingredients of potting soil in a corner for mixing. Note, too, that as the contents of the bins lower from use, front boards may be removed to facilitate remainders being reached. Adapt this bench to a smaller size if you wish, to fit your space, and alter it to suit your own specific needs.

27"   35"

**DETAIL OF
REAR LEG**

**ANGLE OF
FLOOR**

**SETBACK ½"
FOR PLYWOOD
SIDE**

**PARTITIONS SET
BETWEEN FLOOR BOARDS**

**FLOOR BOARD
FITS ON TOP
OF FRAME**

CONSTRUCTION DETAIL OF 2" x 4" LEGS AND FRAME

CONTINUED ▶

FURNITURE

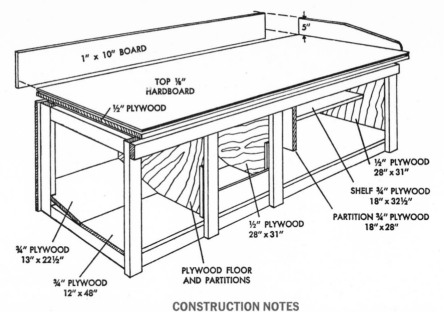

1" x 10" BOARD
5"
TOP ⅛" HARDBOARD
½" PLYWOOD
½" PLYWOOD 28" x 31"
SHELF ¾" PLYWOOD 18" x 32½"
PARTITION ¾" PLYWOOD 18" x 28"
½" PLYWOOD 28" x 31"
¾" PLYWOOD 13" x 22½"
¾" PLYWOOD 12" x 48"
PLYWOOD FLOOR AND PARTITIONS

CONSTRUCTION NOTES

Note how bottoms of bins are inclined, to throw soil, etc. toward front of bench. Bins may be lined with aluminum or other sheet metal to permit storage of damp materials, if desired, and top may also be finished with metal instead of hardboard. Outdoor plywood in ¾'' or ½'' thickness is shown, but boards of various widths may be used instead, if top is covered and bins are lined.

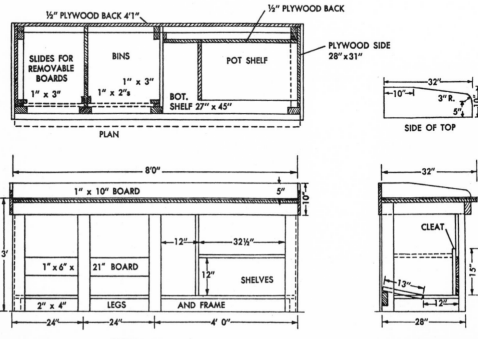

½" PLYWOOD BACK 4'1"
½" PLYWOOD BACK
PLYWOOD SIDE 28" x 31"
SLIDES FOR REMOVABLE BOARDS
1" x 3"
BINS
1" x 3"
1" x 2"s
POT SHELF
BOT. SHELF 27" x 45"
PLAN

32"
10"
3" R.
5"
SIDE OF TOP

8'0"
5"
1" x 10" BOARD
10"
3'
1" x 6" x 21" BOARD
2" x 4" LEGS
12"
32½"
12"
SHELVES
AND FRAME
24"
24"
4' 0"
FRONT VIEW

32"
CLEAT
13"
12"
15"
28"
SECTION THROUGH BINS

II

MASONRY

10

Tools for Masonry

Any sort of building work is made more difficult by not having the proper kinds of tools to work with. No matter how small or how simple the job may seem it will go more quickly and be done better if you have the right tools and know how to use them properly.

In masonry work, as in any other kind of project, tools made specifically for the job will be required for certain operations. In other parts of the work, carpentry tools may be used, or perhaps even improvised tools, if you want to take the time and trouble to make them. It is possible to build most of the masonry shown in this book, and to get good results, by using only three or four basic tools, as many thousands of do-it-yourselfers have proved on similar undertakings. Other projects and some of the more ambitious designs in this book may require certain other tools.

The implements most needed are few: a good trowel, a mason's brick hammer, a good level, and a pointing trowel. With these tools you can make do in producing most kinds of masonry work. You will also find the following tools useful when you are able to add them to your

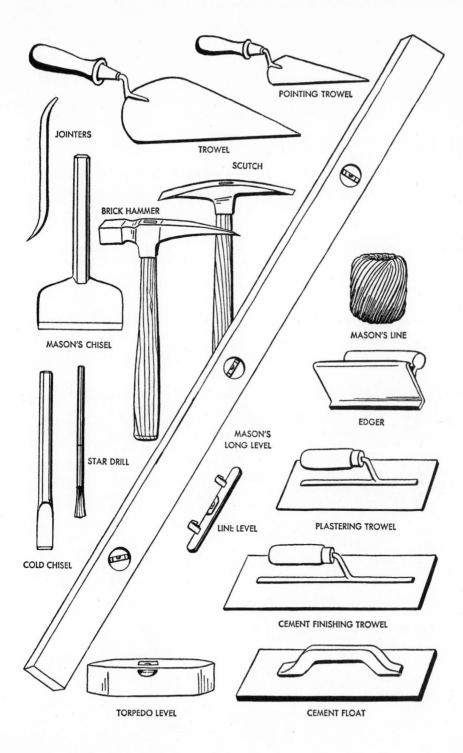

POINTING TROWEL

JOINTERS

TROWEL

SCUTCH

BRICK HAMMER

MASON'S CHISEL

MASON'S LINE

EDGER

STAR DRILL

MASON'S
LONG LEVEL

PLASTERING TROWEL

LINE LEVEL

COLD CHISEL

CEMENT FINISHING TROWEL

TORPEDO LEVEL

CEMENT FLOAT

223

collection: a mason's chisel for cutting bricks, a ruler (either a folding one or a rollup tapeline), a ball of mason's line or cord, and a steel framing square such as carpenters use. You will probably want to add a set of jointing tools and perhaps a scutch (a pointed hammer), a pointed mason's chisel, and probably a standard iron cold chisel 1 to 1¼ inches wide. If you are laying up a wall, you will need a plumb bob to make sure that the wall is kept vertical.

The trowels you'll need are of the two above mentioned types: a large one with a pointed, somewhat flexible blade about 10 to 12 inches long for use with mortar when you are laying bricks or blocks; and a smaller one called a pointing trowel with a 4- to 6-inch blade, which is most useful when pointing up joints in constricted space and for various other small jobs. Both should have the handle shank brazed to the trowel blade or made in one piece with the blade, the wooden handle being set firmly on the shank. Good blades are scientifically heat-treated to harden the steel and good handles are made of durable hardwood. (See the sketches of tools in this chapter for types.)

The mason's brick hammer should be well balanced so that it is easy to use, be made of good hardened steel, and weigh about 1½ pounds. The head of the usual type is square at one end and either tapers to a broad, chisel-like point or tapers on all four sides to a sharp point.

A mason's scutch tapers to both ends, one end being a chisel-like point and the other tapering on all four sides to a sharp point. Both ends are useful for various jobs of chipping away bricks, blocks, and stones to make them fit into the required spaces.

Made of wood, the mason's level differs from a carpenter's level in that it is usually longer, averaging 42 to 48 inches in length, or even more. It probably will have three spirit bubbles, one of which either is set permanently at a 45-degree angle or can be adjusted to various angles up to that. In a pinch, or for small jobs, it is possible to use a short carpenter's level, adapting it to use by binding with masking tape or friction tape, a good straight board, 42 to 48 inches in length, to the side of the level; keep the board even with its lower edge. Be *sure* that the board and level *exactly* coincide on the bottom sides in order to be used with assurance of accuracy. If, however, you plan to do much masonry work now or later, it will be to your benefit to buy a mason's level. In some narrow spaces you will need a shorter level and for this purpose a 12-inch or shorter one will do. A torpedo level (see sketch) about 9 inches long is a convenient size. It usually contains one or two spirit bubbles in the body of the tool in addition to the central horizontal one on the top side. A line level hooks onto the mason's line when it is stretched tightly between corners of bricks or blocks; it is used as a guide for the line so that courses of masonry can be kept level and straight. By using such a level you can be assured that the line is stretched properly to a true level. Actually, for most work detailed in these designs, only the large mason's level will be needed. Only when building a very long wall or some large structure will a line be used and a torpedo or line level be useful or necessary.

Plumb bobs are hung from a point on scaffolding or from a cleat on a high stake as a guide for keeping walls from bulging in or out, and for keeping corners plumb and truly vertical. Plumb bobs come in various weights, but we suggest that small ones are as accurate as large ones and are cheaper. Again we must qualify the recommendation by saying that a plumb bob will not be needed for most projects in this book. A mason's level can be adequate and frequent testing of the vertical wall with the level will suffice. However, some craftsmen find a plumb bob most useful for building tall masonry piers, as corners or supports for trellises and the like, so we felt it necessary to describe them.

Brick chisels are most useful for cutting bricks squarely. The wide one in our sketch has a blade about 3 inches wide, while the blade of the narrow chisel is 1 to 1¼ inches wide.

A mason's point and also a star drill will prove useful. The former is used for chipping where a scutch could not be used, and for various other things: for instance, starting a hole (where the star drill will be used later on) in cement or stone or in a masonry wall. Such a hole would be used for a bolt, an iron spike to hold a trellis, or for some other reason.

An electric or hand drill can also be used for boring holes. Use a carbide or tungsten-carbide-tipped drill made especially for boring holes in cement, concrete, brick, stone, or marble. When the carbide-tipped drill strikes a particularly hard piece of aggregate or a hard vein of stone, it is a good plan to stop drilling and to use a star drill and a hand hammer to crack the hard spot until it is penetrated or sufficiently broken up to permit use of the drill again.

The folding rule or roll-up tape line—in both cases the minimum length recommended is 6 feet—can also be used for carpentry projects, as can the steel square. The professional mason's square is somewhat smaller, but the large carpenter's square is usually adaptable for the amateur's use.

Mason's line or cord may be made of cotton, but nylon or linen cord is less likely to stretch excessively. The cord is used in laying out a project and making sure that the corners are squared up (explained in Chapter 12), as well as to keep courses straight, as we spoke of earlier in this text. A pair of mason's line blocks, either purchased or home-made, are used to hook around the built-up blocks or bricks at each end of a wall to keep courses straight, are held in place by the tension of a well-stretched mason's line between them. They are most useful.

Jointers, used for striking the mortar joints in various ways, are shaped for easy use in the hand. If you plan to make any of the fancier kinds of joints (see sketch of masonry joints), you will find that a jointer of the right type will speed your job. They come in various shapes, each especially formed to make a certain kind of joint, sometimes combining two types in one S-shaped tool. You'll use them also in certain kinds of concrete work and for striking joints in bricks or blocks used for paving.

TOOLS FOR CONCRETE WORK

You will need some other tools if you are going to do concrete work. (See sketches.) You will want to have an oblong cement-finishing trowel, a plastering trowel, and a cement float, all made of good hardened steel and all with handles made of good hardwood. A possible exception may be the float, which is sometimes satisfactorily made of wood. An edger for finishing off the sides of cement on walks or drive-ways, terraces, etc., will complete the tool list you'll probably need for most cement and concrete work.

The cement-finishing trowel is oblong, about 4" x 14", good ones having the handle shank securely riveted or brazed to the blade, as does the plastering trowel. The latter is smaller about 4½" x 10" to 11". Handles of both should be of good quality hardwood. The cement float should be about 5" x 13" to 14", with a wooden handle secured at both ends to the blade. The edger and the groover are usually slightly curved at both ends so as not to pick up or drag the wet cement. Both are made of metal and have wooden handles. You may find some of the newer versions of these tools made from stainless steel which is non-rusting and consequently easy to care for. This type of tool will last indefinitely.

CARE OF TOOLS

It pays to buy quality tools even though you may not be able to buy many to begin with. Cheap tools can cost money if they buckle or break in the middle of a job. They may cause you to ruin a batch of concrete or cement, or prevent you from working on a weekend, when shops are closed and it is impossible to replace tools quickly.

Once you have purchased your tools, let us suggest that you be sure to arrange a place to keep them in orderly fashion, somewhere that is dry and out of harm's way. Don't just lay them on a cluttered shelf or put them in a corner of a damp basement where they will rust or get lost. Chisels and other bladed tools should be kept sharpened, and all iron or steel tools should be cleaned and wiped with an oily rag after each use to prevent rust. Well-kept tools are not only ready for use at a moment's notice, but they are the signature of a good craftsman, one who respects his implements. They are good to look upon and they will last for many years, thus protecting your investment.

11

Materials for Masonry

All masonry is basically the same: it is composed of units of certain sizes held apart, as well as joined, by a "bond" of mortar. The strength and durability of the wall will depend upon how good the mortar is and how carefully the units are laid. All horizontal points must *be* horizontal. The plumbness of corners and all vertical lines will depend upon the horizontals being level. Similarly, the looks of the wall depend upon the straightness and the levelness of the horizontals, the uprightness of the verticals, and the plumbness of the wall.

To avoid walls cracking or breaking from normal stress and strain, vertical masonry joints are "staggered," placed so that they do not fall immediately above each other. This added strength and durability has been achieved in various ways, often quite decorative in the patterns produced, as shown in the pages of illustrations of bonds. In Common Bond the units are placed with vertical points falling exactly in the center of the preceding units below. Staggering one-third or one-quarter is also possible and when units are laid as headers—facing end-on across the wall—the result is most engaging. See the Diaper Pattern, the English and Flemish Bonds, and Rolok.

In modern buildings Stacked Bond is often called for; that is, each unit is laid exactly above the preceding unit in the course below, with all vertical joints aligning with those below, resulting in emphasized vertical as well as horizontal lines in the pattern created by the joints. Raked joints and/or colored mortars will give added pattern and textural effects. In this case, in order to tie units together, metal tie-rods, tie strips, or metal mesh strips should be inserted to bridge the vertical joints and be held in place by the mortar. Stacked Bond is frequently employed in veneering, facing a block, concrete, or cheap brick wall with finished bricks, broken-texture blocks, etc.

Ancient bricks were hand-made and varied widely in size and even in shape. They were rough and probably rather difficult to lay, yet we find them in walls which are still standing after many centuries, so we can see how durable masonry can be. In Pompeii, in Rome, and in other cities of even greater antiquity, we see many walls of great beauty made with sun-baked brick. With today's machine-made bricks and other precisely formed masonry units which are almost identical in shape and size, varying only a tiny fraction due to shrinkage in baking or drying, we find it quite easy to obtain walls that are pleasantly uniform and which conform to exact measurements. In any batch of bricks which are delivered to your property you may find variations of up to $\frac{1}{8}$ inch; and in blocks, hollow tile, and other units you may find some variations of as much as $\frac{1}{4}$ inch or so. These are allowable and should cause you no alarm, because shrinkage is inevitable in drying out or in the baking and varies from batch to batch in the same brick-yard under identical conditions.

However, you may find it hard to match bricks at a yard other than that where they were made. Different factories use different machinery in making bricks and masonry units, for not all of them have as yet conformed to the modular size advocated by the industry. It will be well to check the sizes of the bricks in your present construction, if you can, before ordering to see if you can obtain new ones from the same source. In all our charts and plans, therefore, we give approximate measurements, the ones which the industry has approved as standard. If yours do not conform, make allowances as necessary.

KINDS OF BRICK

Common brick is made from ordinary clay with no coloring added, and no texturing or other processing. Bricks will vary in color and texture even in the same baking batch. This is something which some builders deplore and others prize, according to individual tastes and the purpose for which the brick is to be utilized. Many degrees of hardness and softness are to be found in the category embraced by the name "common brick." Those nearest to the baking fire are hardest, those farthest from it are softer, more likely to chip, and less resistant to weather and moisture conditions. The hard-burned bricks are sometimes called "clinker" in the trade nomenclature, or "hard-burned" in some sections of the trade. Where non-load-bearing walls are built with bricks sheltered from weather or from moisture and freezing, the cheapest common brick may be used. These are usually called "salmon" brick (although the color may be anything *but* salmon). For all other outdoor work and for load-bearing walls, clinker or hard-burned bricks are the best to use. Most common brick is rough in texture, relatively porous, and probably will have a slight bow in the lengthwise measure. Always place the bow on the *lower* side when laying up a wall.

Pressed brick is more uniform in size, with well-squared corners, and is altogether smoother and finer in texture. These bricks are usually used for facing walls, since they are more weather-resistant and also are likely to run more uniform in color. They may be used for facing any wall except one exposed to fire (such as in backing-up a fireplace) in which case use . . .

Firebrick, which is a specially-made unit somewhat larger than the standard-sized brick, running to a size of about 2½″ x 4″ x 8″. It is made of a specially selected fire-resistant clay, and it is coarser in quality and in finished texture than other bricks. Hearths and flues or fireplaces may be lined and floored with this brick. It will resist the hottest of fires.

Unbaked brick is a recent development. Because they are not baked but formed in molds of cement and sand and allowed to dry out until they are completely set, these are usually most precise in size and admirably square-edged. They are quite uniform in color, too, because the color can be controlled in the mixing, but to our mind they are less interesting for this. Some types are made as substitutes for baked fire-bricks and seem to be quite satisfactory. The common un-baked bricks are well suited for use in walls either as backing-up or as facing bricks. To some, their texture as well as the absolute uniformity of color is less pleasing than that of baked bricks, but, if the joints are well raked and if the wall is to be painted, this objection might be overcome.

Used bricks salvaged from razed buildings are perfectly good to use provided that they are well cleansed of mortar and the edges have not been too chipped or damaged. Such damaged bricks are all right for backing-up if they are still sound and cleaned. It is sometimes poor economy to buy salvaged bricks when the cost of new brick is balanced against the price of salvaged brick and against the effort required to clean, examine, and sort each brick for use. Of course if the brick comes from a building or wall on your own property which is being demolished, or from some structure being razed nearby where the only cost involved is in the hauling, it may save you considerable money to clean and re-use them.

The method of cleaning is this: grasp a piece of broken brick about half the size of a regular brick (this is called a "brickbat") and rap it smartly against the mortar on a brick held in the other hand. This chips and dislodges most of the mortar, and the rest can be rubbed off with the brickbat. We would be the last to say it is an easy job or an interesting one, but it can save money if the conditions are as detailed above.

Paving brick is much harder and usually larger than ordinary types of brick. All the edges are rounded and each face of the brick is smooth. It is sometimes possible to buy used paving brick from salvage yards for paving walks or traction strips in a concrete driveway. They will be too large and heavy to use for wall-building but are perfect for paving walks, terraces, and driveways.

Roman brick is longer in proportion to its height and width than the ordinary types of bricks and is being much used in modern construction because of the long, low lines it gives to walls. It looks particularly good as veneer in Stacked Bond with the joints raked and with a fairly rough-textured surface on the facing side.

SCR bricks are wider and longer than most other bricks. They can be used for garden walls, for house walls one-story high, and for various other projects. Because of their size—6″ x 2⅔″ x 12″—they can be laid quickly, reducing the time needed for the work and also saving a little on the cost of mortar. Vertical holes through the bricks permit mortar to penetrate, making a strong bond between courses. Metal clips may be inserted in the joints to hold furring strips; 2″ x 2″ stock gives space for insulation and furrs out to allow placement of wallboard, lath, plaster-board and the like to be applied for interior finishes. For outdoor use, the width allows use of a single-brick wall, particularly where frost is not a problem and especially for low walls in all regions.

Other types of bricks.   Many kinds of bricks are available by special order for particular use. Various shapes, such as round corner, coping in different shapes, molded face, hexagonal paving bricks or tiles, and many others, are made in most areas but must, as we indicate, be specially ordered. They are more expensive than common bricks but they give effects which are not obtainable in any other way. Check locally for types, prices, and availability.

Features found in some bricks.   Various surface textures are made, the wirecut surface being widely used. We do not, personally, favor this type of surface because it is so full of texture as to look shaggy. Furthermore, it is likely to collect dust, grime, and soot in city areas and will be very hard to clean. A good brick wall with slightly raked joints gives a most pleasing texture in itself and no further textural emphasis will be necessary.

Today many bricks, even common ones, are found with various depressions moulded into the top and bottom surfaces to hold mortar and thus give a better bonding. These depressions are usually of an oblong shape and run to within an inch or so of each edge. Other types have round holes bored through the brick at intervals so that when the bed joint is laid the mortar sinks into these holes and is pushed up into the holes in the next course, thus assisting in making a strong joint. Iron rods may be forced through these holes, too, making an exceptionally strong wall even though it may be only one brick in width. The mortar in this case is forced tightly around the iron rods, of course.

BLOCKS OF ALL KINDS AND OTHER MASONRY UNITS

Many types of building blocks are available in all sections of the country, some more prevalent according to local tastes or availability of materials than others. We have only to look about us to see what strong, durable and beautiful walls they make, how interesting in texture and architectural effects they can be.

Concrete blocks. Whether made of cement and any of a number of materials—cinders, scoria, pumice, Haydite, or gravel—all blocks are known by the familiar name of concrete blocks. Some are more finished looking than others, so it is well to check on the looks of the blocks before ordering them to be sure that you obtain what you desire. They come in an astonishing array of sizes and shapes, some molded for use on corners, for chimney construction, for surrounding window sashes and door frames and for other specific uses. Blocks which use cinders may contain iron particles or scraps which rust when blocks are used for facing walls, exposed to weather. Rust stains will appear, bleeding through paint and disfiguring the surface.

Blocks are made by mixing portland cement with the aggregates and pouring the mixture into molds which are then shaken well by machinery to settle the mixture and release air pockets. When set and fairly dry, the blocks are removed by releasing the molds and are stacked for curing and drying. Because most blocks are hollow-cored, they are relatively light and strong for their size. They may be used for load-bearing-walls, the size being selected according to the width needed for the load to be borne. Walls go up quickly and blocks are satisfying to work with. A complaint of some amateurs, that blocks are heavy and therefore tiring, isn't necessarily true. Many are made now with light-weight aggregates and the core area lightens the weight a good deal. Working with blocks is a very pleasant task and the fascination of seeing the project progress so quickly is part of the pleasure.

Split blocks.   Concrete blocks are cast, then broken. The uneven side is used, laid as the outer face of the wall. Most interesting textures result, a rather stone-like quality, rough-textured and pleasant in walls, breaking up large expanses in a most unusual way. Colored-aggregate and integrally colored blocks are often available in this type.

Slump blocks.   These are cast, and before they are really set, forms are removed allowing the block to "slump" or subside. The squashed effect is similar to that found in adobe brick walls, so typical in old buildings in the Southwest. Especially suited to homes with a Spanish or Latin-American look, they are also useful for garden walls, retaining walls, and other construction work where a different texture is desired. In modern design homes or where a conventional block might be cold and too smooth, this may be used with good effect.

Glazed blocks.   Mostly for indoor use, where a finished and formal look is desirable, these may also be used for facing patio or terrace walls. Their extra cost may also have to be considered as well as their suitability. However, they will need little or no maintenance.

Grille blocks.   This is a field that has expanded rapidly. Many ordinary blocks lend themselves to grilles. For instance, the *single-core blocks* laid horizontally or set vertically, overlapping at corners make a good checker-board patterned wall. *Square chimney blocks,* or those with *round cores* for use around cylindrical flue tiles are also effective, and easy to obtain, easy to lay. *Keyhole-shaped lintel blocks* as well as *regular lintel blocks* make unusual patterns used alone or combined with *single-core blocks.* A most effective diagonal pattern results when they are used alone, placed foot to foot, *double-corner blocks* placed on top and bottom courses between the *lintel blocks.* Combining 8″ x 8″ x 8″ *lintel blocks* with *partition blocks* 4″ x 8″ x 8″ and 4″ x 8″ x 16″ will give a stately effect if the longer partition blocks are placed vertically, allowed to project in alternate rows.

Corner blocks in *stacked bond,* laid with corner-up and corner-down, fitted into each other, give a most unusual and different grille. *Column blocks* (these are like square chimney blocks) as well as *half-*

column blocks can make handsome grilles, either alone—laid in stacked bond—or combined with *partition blocks* or with *thinner patio-paving blocks* laid between them in double rows, projecting on either side of the wall. Similarly, *single-core blocks* may be set vertically, with two *partition blocks* laid horizontally between them, to give a kind of open-and-closed checkerboard design. Many other combinations of blocks may be used to form grilles and a truly creative workman will take much pleasure in figuring out new patterns, designing his own grilles from the blocks available.

Shaped-face or sculptured blocks.   Coming in many styles and varied facings, these blocks give variety and textural interest to walls. Diagonal or diamond-shaped patterns fit together as blocks are laid to give directional patterns to the finished wall.

Colored blocks, aggregates selected.   More frequently than ever before blocks are now offered in integrally colored concrete or with selected aggregates that will make the finished wall anything but a cliché of blocks that are dull, that must be painted, stuccoed or otherwise decorated to give architectural significance to them. Often you will be able to select from blocks with interestingly varied aggregates set in colored concrete, for complete surface finish.

Ground-face blocks.   The most recent development, one that will probably be available about the time this book is offered, is the smooth-face block that has a look similar to that of a terrazzo floor. The finished block is placed in a machine that grinds down its surface (in a manner similar to being fed into a wood-sanding machine with an abrasive belt) to expose and also grind flat the aggregates. The finished blocks can be treated with a sealant, and this will make them attractive for indoor use, as well. Colors that contrast or blend with the aggregates used make them doubly attractive. The sealant makes them waterproof, too, I believe, so that they would be more useful than ordinary blocks for areas where moisture is a problem or for walls which would need to be washed down occasionally.

Clay tile is coming into wide favor for use as flooring and for facing pools, making wall designs, and so on. It had been out of fashion for some time, except for use in bathrooms and for floors in public places, but now it is definitely back in favor, with many exciting and beautiful new interpretations and many new uses being advocated. It is a most durable and permanent material, and we note in museums tile and mosaics from Roman and pre-Roman remains which are as beautiful today as they were when they were laid. Many beautiful designs can be worked out with small tiles and bits of tiles; mosaic work as fine as embroidery can be achieved by artists in this field. For tabletops, garden seats, and as inserts in walls, tiles which are glazed or unglazed, painted or molded or in plain colors are all excellent. For anyone interested in pursuing this line we suggest research in books on Spanish and Portuguese architecture and gardens to see how interestingly tiles were used. Some of the modern books on how to make mosaic may also give inspiration and assistance in the creation of new design approaches with ceramic tiles.

12

How to Lay Masonry

Because the first thing to be done after the wall units have arrived and you are ready to build your wall is to make ready the mortar, we shall consider that job first, assuming that suitable footings for the wall have been prepared. (See Chapter 14 on Concrete.)

Good mortar is strong mortar and therefore of the utmost importance. If it is carefully made, with the ingredients precisely measured so that the formula is correct, and the materials thoroughly blended, then the mortar should be as nearly perfect as it is humanly possible to make it. The perfect mortar should be wet enough so that the masonry units cannot rob it of too much moisture (this will be discussed later on); should contain enough lime so that it will spread well without losing its adhesiveness; and should be composed of *fresh* ingredients which are free from impurities. The water should not be heavily alkaline or contain salts, for either of these two elements will tend to reduce its strength.

MORTAR FORMULA

If you have only a small amount of masonry work to do, it may be to your advantage to buy Sakcrete or one of the other commercially prepared mixtures in which the sand and cement have been blended in the proper proportions. There are two or three sizes of bags available, allowing you to buy only what you need for small jobs, or if a larger quantity is needed, the more economical larger bags. For all except very large jobs, it is probably best to use these convenient mixes, because of the time saved in measuring out quantities and the fact that pre-blending has been done by the manufacturer. All that is needed is water in quantities specified on the package and mixing according to directions there, too. A strong, durable mortar results.

In cases where larger quantities of mortar are needed, it will be much more economical to buy sand and cement in the quantities needed and mix it according to the formula below. This is, of course, where quantities of more than a single bag of cement are required and the aggregates or sand exceed a half load. Use portland cement for the formula ingredient. It should be fresh and clean, with the bags unbroken, and should be stored in a dry place. Be sure to keep the bags *off the floor or ground* and in a *dry building*. Bags, although fairly moisture-proof, will absorb moisture from concrete floors or from the soil and this will cause the cement to harden or to deteriorate. Should you ever find any lumps in the cement which cannot be crushed with your fingers, it will be well to discard the cement and buy some fresh bags. The sand used should be free of soil, weeds, twigs, and other plant or animal matter. Un-washed beach sand should not be used, or the strength of the mortar will be affected. Beach sand itself is perfectly good to use, provided it has been *thoroughly washed* with fresh water to cleanse it of salt.

Lime putty or commercial hydraulic hydrated lime should be used to increase the workability of the mortar and also to augment its waterproof qualities.

Although many formulae for mortars used for specific purposes have been developed, the following formula will be found to serve very well for most masonry work and for laying bricks or tiles for paving. Be sure that your sand is dry, that cement and lime are free of lumps.

> 1 part portland cement
> 1 part lime putty or hydrated lime (Use up to 2 parts, if needed,
>   to increase workability and waterproof qualities of mortar.)
> 5 to 6 parts sand
> Sufficient water to give mortar consistency of putty

Mix the dry ingredients well until they have sufficiently integrated so that the mixture is a uniform brownish-grey color. Then, into a depression scooped out of the center of the dry heap, pour a *little* water; mix it in well; then add more and again mix well. *Never* add a large quantity. It may be too much and make the mortar soggy and too wet. By adding water bit by bit you can get it to just the proper point and then stop. The ideal mixture is wet enough to spread easily but not so wet that it is loose and will spread like cream. If it is the consistency of putty, stiff enough to stand by itself without collapsing yet wet enough to be easily spread and worked by a trowel, it will be just right. Observe professional masons at work, if you can, and take note of the look of the mortar consistency they are using.

Let us emphasize that not too much water be used. This will destroy the strength of the cement. And let us add that not too much mortar should be made up at any one time because it begins to set within two hours, and in hot summer weather or warm areas it will set even sooner. Therefore you should mix only enough to use for an hour or so of work (probably two bucketfuls will be enough until you see how your speed and proficiency work out). Once mortar has begun to set, never add water to soften it or you will find that the strength is decreased and the mortar probably ruined.

Mix the mortar in an old tub, in a bucket, on a wooden platform made of scrap lumber or a piece of outdoor plywood. Best of all, perhaps, is a metal wheelbarrow, which will not be harmed by the mixing and can be cleaned afterward quite easily. Use a hoe in mixing if the quantity will permit it, pulling it through the ingredients to distribute the moisture and to stir up dry particles scraped from the bottom of the heap. In measuring out the ingredients it is imperative to use the same container for *all* of them, wet or dry. Let us point out, too, that the mixed quantity will turn out to be a little less than the total of the dry quantities, the reason being that the small particles of cement and lime fit in between the sand particles when well mixed and liquefied. Keep the decrease in mind when mixing any specific amount, and try always to mix a little more than is needed so as to be sure to have enough.

WETTING BRICK OR TILE

All brick and clay tile should be thoroughly moistened before being laid so that a minimum of moisture will be extracted from the mortar, thus keeping it strong and durable. On the other hand, there should be no free water visible on the units when laid, either, or they may not bond properly with the mortar. Some pull between the two is desirable.

Stack the masonry units in convenient small piles along the site of the wall where they can be easily reached for use (the mortar may harden if you have to run with your wheelbarrow for another load of bricks to complete the wall). You may stack them loosely or merely pile them in a heap. Professional masons usually stand behind a wall when building and often stack the bricks there, but you may find it easier to reach over the wall to get your bricks until it rises high enough to make that impossible.

By placing a hose with its nozzle adjusted to a fine spray so that it will play on the brick pile for an hour or so before you use them, the bricks will be sufficiently moistened. Or, if it is more convenient, fill an old washtub with water and stack the bricks in it, letting them soak for at least a half hour before using them. Bricks should be removed and allowed to drain and dry a little, so that there is no free water on the units when laid, for proper bonding, as detailed above.

Concrete blocks should *never* be wet down before being laid, however, for wetting makes them expand and when laid wet, they may contract, when drying out, and crack the wall. Store them where they will stay dry and can be laid up dry; buying them wet is a hazard we must face, but if they are spread out to dry in the sun for several days, with covering of sheet plastic applied if there is rain or snow and also if they are covered at night to protect them from dew, they will be dried out thoroughly before being laid. Protect a wall in progress by covering it with polyethylene plastic or strips of tar paper used for roofing or some other adequate covering, thus keeping it dry and allowing it to cure properly and avoid cracks.

FOOTINGS AND FIRST COURSES

We shall assume that the footings have been constructed as detailed in Chapter 14, for footings and some sort of foundation will be necessary in all cases except those where a wall is only two or three bricks above ground. A thin footing of concrete on a 3-inch bed of tamped cinders or gravel will usually prove sufficient to support the wall. For warm climates, running two or three courses below ground level should suffice. The foundation and footing in cold climates should run sufficiently below ground level to prevent frost from heaving the wall, and also to provide a firm foundation. Use a footing, too, where the soil is light or unstable. This will keep the wall from tilting and cracking.

The primary courses are extremely important, for unless they are kept level (unevenness occurring in casting footings can be compensated for in the first few courses by varying the depth of the bed of mortar) all of the succeeding courses are likely to be out of plumb, and may even increase the error. Therefore spend time and care on the first courses, and the others will be kept in line more easily, with the corners square and plumb and the joint lines straight and even. Decide before you begin how high the wall is to be, and set a stake marked at the proper height to act as a guide. Also consider what the thickness should be: one brick, two bricks, or whatever size of block you will need to bear loads or withstand weights and pressures (as in the case of retaining walls). Low walls up to about a foot high may be built one brick thick—except for retaining walls—but any wall of seat height or higher should be two bricks thick, or else made up of backing-up blocks, hollow tile, or concrete, which may be veneered or faced with bricks. Veneer courses should be tied in with the back-up courses by means of corrugated metal called Tie Strips or Veneer Ties. Header courses can be used to tie the back-up and veneer together where brick back-up is used (page 245).

Walls of waist height or higher should be three bricks in thickness if they are to be retaining walls; or 10- to 12-inch blocks may be used if they are to be the finished face of the wall. The higher the wall the greater will be the need for strength. Iron rods may be used for bracing, either by inserting them through the holes in bricks (described under "Features found in some Bricks," this chapter) or by mortaring them into the hollow cores of blocks every couple of feet or so in high walls, thus lending strength to the wall, and assisting it to hold back the weight and pressures of soil when the wall is a retaining wall. A single course of brick veneer facing a cast concrete backing-up will be less likely to shift and crack if the concrete has been reinforced with horizontally placed rods, or by a combination of horizontal and vertical rods.

HOW TO LAY BRICKS

Whatever you may be going to use for building, the basic technique used in laying bricks will be your guide; for with blocks or hollow tile, or any other material, the procedure is very much the same.

First lay out a course of bricks along the foundation, which we will assume has been cast quite level and smooth. Lay them in a header course (see joint chart for explanation of header and stretcher courses) with about ½ inch left between bricks to allow space for mortar joints. This procedure is called "chasing out the bond" and will permit you to figure exactly the number of bricks you'll need per course. If you have a two-brick-thick wall, lay out a second course stretcher-style on top of this first one, with a back-up course behind it and centered on the facing course. This will let you figure how many bricks you will need for courses of this kind. If your courses can be adapted to even brick lengths you will save considerable time and energy by not having to cut bricks to odd lengths to fill out the courses.

If your wall is to be only a single brick in thickness, first lay a header course—bricks laid crosswise, not end to end. An exception is the SCR brick. Atop the header course and on succeeding courses, lay bricks end to end—stretcher courses—centered on the header bed course. No header courses will be necessary in succeeding courses in single-brick walls. In multiple-brick thicknesses, header courses are placed at intervals to tie the walls together and to give a certain decorative effect; or bricks may be inserted as headers in stretcher courses, singly or in multiples, for decorative, patterned effects. See the illustrations for English Bond, Flemish Bond, Diaper Pattern, and Rolok walls.

In single-brick walls you can achieve the same effect by cutting bricks in half and using them in the same patterns.

After you have noted the number of bricks needed for the various courses and transported a corresponding number to the vicinity of the wall where they can be easily reached, you are ready to start the final operations. Set a stake at each end of the wall and run a mason's line between them to give you the *exact* line of the face of the wall. You can butt every course of bricks against the line to insure straightness and plumbness in each succeeding course. The correctness of the succeeding courses will depend upon the correctness of this first course.

USING THE MORTAR PROPERLY

When your mortar is mixed and ready for use, place about a pailful on a piece of plywood 20 to 24 inches square and set it on a box or on the ground where you can easily reach it with your trowel. Scoop up a trowelful and begin at one end of the wall (removing the first course of the dry run of bricks as you go) to spread a layer of mortar about ½ inch or so thick on the foundation or footing for about the space of four or five bricks—from 20 to 24 inches along the foundation. This is called a bed joint and mortar should be spread so that it centers on the foundation or footing. Modern techniques call for a flattish mortar surface, perhaps tapering a little from side to side, to give a firm bedding. In succeeding courses, never spread more mortar than you can quickly cover with the bricks—20 to 30 inches, depending on your speed, skill, and weather conditions—for mortar begins to set very rapidly, especially in hot, dry weather. Bricks laid on a partially set mortar bed may make a weak bonding with it.

Then put the corner brick in place, setting it so that the face is laid squarely against the stretched line. The second brick to be laid, and all succeeding bricks in each course, will require one other operation. They must be "buttered" on one end before being set in place so that the vertical or "head" joint is filled with mortar. The method is simple: simply scoop up enough mortar on the tip of the trowel to cover the end of the brick with about ½ inch of mortar and wipe it off the trowel onto the brick end. After a few tries you will find out which is the best way to hold the brick, to fill the trowel tip with mortar, and to slide it off on the brick. Only experience can perfect your technique. Buttering will be used in all kinds of masonry work, and it should be perfected and made efficient so that you can swing into the work and complete it with dispatch and precision.

Once the brick is buttered, it should be slipped lightly into place on the mortar bed with a kind of swinging motion to butt it against the previously laid brick so that the mortar is squeezed out and the joints—both bed and head—are about ½ inch thick. Never place the brick on the bed mortar and then attempt to force in mortar for the head joint—it will never be easy or successful, although outwardly it may look so for a time. Also, if the brick is not successfully bedded the first time, do not remove it and then just replace it. Either add fresh mortar to the packed-down bed joint or scrape off the bedding mortar and replace it with a fresh lot, re-buttering the end of the brick, too. Not replacing the mortar or freshening it can only lead to failure of the joint—and any wall is only as strong as its weakest joint.

BRICK

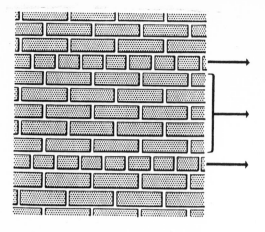

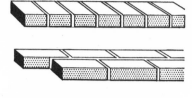

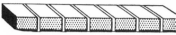

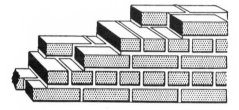

COMMON BOND

Placing header courses every fifth, sixth, or seventh course ties the double stretchers together for strength. Easiest, cheapest to build.

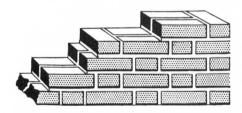

ENGLISH BOND

Header and stretcher courses alternate. Care must be exercised to keep vertical joints thin, to maintain proper spacing of headers.

FLEMISH BOND

Headers and stretchers alternate in same courses, each header being centered on a stretcher above and below. Easy to lay.

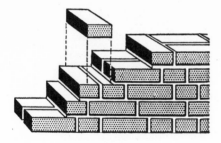

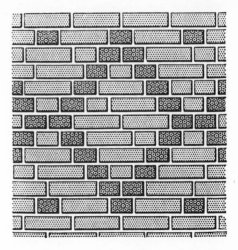

DIAPER PATTERN

Header courses are staggered to produce the diamond shape shown here, darker or lighter headers being used to emphasize the pattern.

ROLOK WALL

The header courses are laid flat with all of the stretcher courses laid on edge leaving a generous insulating air space between them.

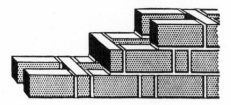

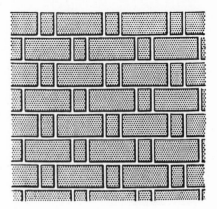

FLEMISH ROLOK

This variation of Flemish bond makes a good-looking wall, very strong and with the added value of insulating air between stretchers.

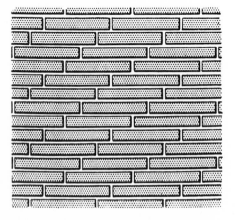

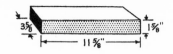

ROMAN BRICK
Increasingly popular is the use of Roman brick, a longer, shallower unit patterned on bricks used thousands of years ago in Rome. Although sizes vary with manufacturer and material, dimensions will be near those shown.

STACKED BOND
Bricks set above each other; although interesting in pattern, vertical joints need to be tied together frequently to obtain strength.

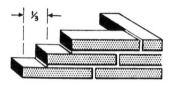

ONE-THIRD BOND
Bricks laid with each succeeding course one-third across from those in original course; joints make a pleasing diagonal line in wall.

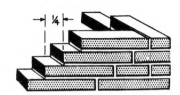

ONE-FOURTH BOND
Bricks laid with each succeeding course one-quarter across from that of original course, the joints forming a steeper diagonal line.

If, however, when the brick is swung into place and laid, it is slightly out of line or a little high at one end, it can be tapped lightly with the handle of the trowel to bring it into place. A little practice will give you the feel in your wrist of how hard to tap the bricks to bring them into place without dislodging them or squeezing out so much mortar that the brick is then too low.

When the wall is about half laid, move operations to the other end and begin there, laying the bricks in reverse order, working carefully and keeping the faces butted against the stretched line until only one more brick remains to be laid in the center. Butter both ends of that final brick and gently slide it into place, tapping it lightly down with the trowel handle until it is bedded properly. Be sure that the head joints are filled and tightly packed.

With the blade of the trowel, trim off the excess mortar squeezed out in bedding the bricks, so that the joints are flush with the bricks. Don't tool the joints immediately, for the mortar may be too wet and fresh to hold the shape given it by the tool. On the other hand, don't wait long or the mortar may have set too firmly to be successfully and easily tooled. This is a part of masonry work for which no exact time limit can be given, because of climate, weather, season, and other intangibles. Trial and experience will guide you to the exact time for tooling; but it may comfort the novice to know that there is usually plenty of leeway for tooling before the mortar is too set to allow it to be done. Tooling will not only shape the joint and remove excess mortar, but it will also compress the remaining mortar, making a good, weather-tight joint which will add to the textural beauty of the wall. (See Chart of Masonry Joints.)

Header courses and soldier courses necessitate a slight change of operation. Header courses are those courses in which the ends of the bricks face the finished face of the wall, you will remember, with the wide measure of the end laid *horizontally*. Soldier courses are those in which the wide measure of the end of the brick is laid *vertically*, or those in which the brick is stood up on its end when laid. In both of these instances the buttering will be done so that the side of the brick which butts against its neighbor is covered and makes a head joint.

HOW TO CUT BRICKS

Occasionally it may be necessary to cut a brick to fit a particular space in the wall: either to fill the final space between the last two bricks in a course, or to space out a course so that it ends properly for a window, a door, a gate, or some other opening. There is a good and a bad way to do everything. The bad way to cut a brick is to hit it with the mason's hammer and hope for the best. While some professionals become very adept at this, the amateur can waste a good many bricks and consume a good deal of time trying to learn the trick. Instead, we advocate the following method, which will cut bricks evenly and with little danger of their breaking and wasting good material.

Score the brick all the way around to a depth of about ⅛ inch at the juncture desired for the cut. Use a cold chisel for this and use a carpenter's try-square, if you wish, as a straightedge. Then put the brick on a flat, firm surface, place the broad mason's chisel vertically on the scoring, and rap it smartly with the hammer. This should cause the brick to break evenly and squarely along the scored groove. For rough work, such as for cutting a brick for a backing-up course, you may use the mason's hammer to strike the brick and chip off a portion of it, but it is seldom precise and is never as satisfactory as the scoring method for facing courses. Should there be any excessive irregularities on the broken-off brick, they can be chipped down by using the chisel-like blade of the mason's hammer.

SUCCEEDING COURSES

Always start at each end of the wall and build up four or five courses tapering up to the corners; then lay the courses in the intervening areas. When the wall is about level, build up the corners again for a few courses and repeat until the eventual height of the wall is reached. When a course is laid between corners, it should be kept level by using the mason's line as a guide. Always check corners with the mason's level as they are built up, or use a plumb bob to keep corners and wall vertical. To keep horizontal courses level, attach a line level to the tautly-stretched line in the center, thus obviating the need for frequent use of the mason's level. Always, when moving the line up to the next courses, be sure that it is securely fastened to the corner pieces and is tautly stretched and level, so that the succeeding courses will be straight and true. Check the face of the wall and the corners frequently to be sure they are kept in plumb, for slight leanings have a way of becoming exaggerated as the wall grows higher.

Should you be called away in the middle of the job or have to stop when the light fails in the evening, be sure to tool the joints and to clean off all mortar first so that you can make a good fresh start with the bed joints next day. Cover the wall with sacking or building paper, weighting it so that wind will not dislodge it. If you have to leave without tooling the joints, wet a burlap bag or other cloth with water and cover the wall to prevent the mortar's drying too quickly. But don't delay any longer than you must, for mortar sets quickly and you may not be able to tool the joints cleanly if you procrastinate. Keeping a bucket of water handy so that you can wash off the tool and keep it wet will help to give good, cleanly-tooled joints. Always try to finish a section of wall completely, having bricks bedded and set, joints tooled and finished before leaving the job. Then you will have a good, clean, well-built wall.

When you have finished the final course—usually a cap course or soldier course—finish the tooling of joints; then use the side of the trowel blade to scrape off any bits of mortar which may have dropped or splashed onto the face of the wall. Do this while they are still fresh and can be easily removed. Then let the wall cure for about two weeks before the final process.

CLEANING THE WALL

Clean excess mortar and splashes from faces of all walls as you go along. A soft bench brush will probably do it while mortar is fresh. Later, a wire brush will be required. Stubborn smears on block walls can be removed by rubbing with a piece of broken block. Brick walls may be completely cleared of mortar stains (and also of the white efflorescence that sometimes appears later on brick masonry) by applying a wash of hydrochloric acid—also called muriatic acid—in a *mild* solution. *This should be handled with care:* in its undiluted state it will dissolve and injure concrete and mortar and even in dilution will eat holes in clothing, damage painted surfaces and eat paint film, puncture galvanized pails and other metal receptacles. Therefore it should be mixed in a glass bowl or jar. Should the solution be spilled, immediately flush the area with plenty of clear water. Hosing it down will be best, if this is possible. If any splashes come in contact with the skin, they should immediately be rinsed off with lots of water and a saturate solution of bicarbonate of soda applied. To the cautious workman this would indicate the need for having a hose, a bucket of water, and some bicarbonate of soda handy while the acid is being used, just in case they might be needed.

The strength of solution recommended is 10 parts of water to 1 part of acid. Use a stiff old broom or a rag tied to a stick to apply the solution. Wear gloves while working. When the entire wall has been washed down with the acid, begin hosing it down with water, flushing it well from top to bottom. Should any blobs of mortar or splashes of cement still show, spot treatment with the acid will eventually remove them. Remember that mortar joints may also be affected by this acid solution if it is allowed to stay on them for very long, weakening them and leaving them open to weathering damage. Therefore it will be well to do any large walls in sections, hosing down before going on to the next section. If efflorescence reappears in a few months or a year, you can give the wall another bath of the acid solution, which should clear up the trouble and prevent its happening again.

FINAL TIPS ON LAYING BRICKS

Where walls of two or more bricks in thickness are to be laid, the back-up courses are always staggered so that the head joints do not coincide with those on the face course. They are usually centered on the face course so that at the end of the wall or on corners the bricks will fit neatly and require a minimum of cutting. Succeeding courses are staggered similarly, with no head joints, or as few as possible, running through the width of the wall. Headers, as well as header courses, are an exception to this, of course. Half bricks are used in the back-up courses where individual headers are called for by bonding pattern, being placed on either side of the header so as not to upset the staggering.

Back-up courses are buttered both on the end for the half-brick header and on the side which is adjacent to the facing course. Some professional masons butter only the end, and after the brick is laid and head joint set they throw in a little thin mortar to fill the intermediate joint between facing and back-up courses. We do not advocate this procedure because we feel it is not reliable; and in addition it requires the

NOMINAL SIZES OF BRICK

for use in figuring quantities of bricks needed for jobs.

Type of Brick	Width, inches	Height, inches	Length, inches
Conventional	4	$2\frac{2}{3}$	8
Roman	4	2	12
Norman	4	$2\frac{2}{3}$	12
Engineers'	4	$3\frac{1}{5}$	8
Economy or modular	4	4	8
Jumbo	4	4	12
Double	4	$5\frac{1}{3}$	8
Triple	4	$5\frac{1}{3}$	12
SCR*	6	$2\frac{2}{3}$	12
Firebrick	$4\frac{1}{2}$	$2\frac{1}{2}$	9

*Note that SCR brick is 6 inches wide and 12 inches long, a factor to be reckoned with in choosing brick for one-brick-thick walls. It is sufficiently strong to build walls for a one-story house, hence it is ideal for many kinds of garden walls.

mixing of a second, thinner batch of mortar. We feel that the method detailed above is better; any mortar which is squeezed out from the head joint or the intermediate joint can be left to become part of the next course's bed joint, provided that more mortar can be laid on to make the bed before the squeezed-out mortar sets.

You may not wish to buy the line holder given in the tool list (Chapter 10), and you may in this case use nails shoved into the bed joints near the corner on which to attach the mason's line for keeping courses level. This method is more trouble, but it can successfully be used.

When laying bricks in a curve or in a small circle with a short radius, it may be necessary to cut the ends on an angle to avoid wide, ugly head joints. Therefore it is a good plan not to use too small a radius unless it is absolutely necessary. It will be even worse when you reach the cap or soldier courses, for here the problem will be magnified if the entire length of each brick must be cut. About 3 feet would be the minimum radius recommended; a greater size would be preferable.

HOW TO LAY BLOCKS AND HOLLOW TILE

Even though the various blocks and hollow clay tile are much larger than bricks and differ in having hollow cores, the basic method of laying is really quite similar. Because they are larger units you may find them easier to lay in some ways, and the wall will grow much faster than is the case when bricks are used, due to their large unit size. They are usually set in single stretcher courses with a simple running bond—courses overlapping by half with the head joints centered on the block in the course below. Decorative patterns may be achieved by varying the laying or by using various-sized units, as we show in the illustrations which follow.

The main difference encountered in laying blocks is that the joints must be adapted and varied to suit their peculiar construction. Only

HOLLOW TILE—SOME TYPICAL SIZES

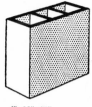

4″ x 12″ x 12″

4″ x 8″ x 12″

4″ x 5½″ x 12″

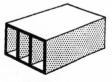

8″ x 5½″ x 12″

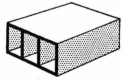

10″ x 5½″ x 12″

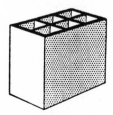

6″ x 12″ x 12″

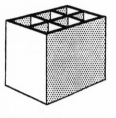

8″ x 12″ x 12″

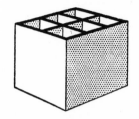

10″ x 12″ x 12″

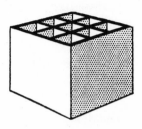

12″ x 12″ x 12″

THE BASIC SIZES AND KINDS OF BLOCKS

BASIC UNIT
8" x 8" x 16"

CORNER UNIT
8" x 8" x 16"

**PIER OR DOUBLE
CORNER UNIT**

JAMB OR JOIST UNIT
8" x 8" x 16" OVERALL

PARTITION UNIT
4" x 8" x 16"

WALL UNIT
6" x 8" x 16"

WALL UNIT
10" x 8" x 16"

FOUNDATION UNIT
12" x 8" x 16"

HALF BLOCK UNIT
8" x 8" x 8"

THREE QUARTER UNIT
8" x 8" x 12"

HALF FLUE BLOCK
8" x 8" x 16"

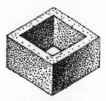

FLUE BLOCK
16" x 8" x 16"

the face, the back, and one end of most blocks are solid and smooth, so the mortar must be handled differently. The first course above the footing or foundation should be laid on a bed of mortar exactly as for bricks, but the flanges or "webs" which project on the end are buttered, so that they will butt against the smooth end of the preceding block. For succeeding courses, make the bed joint by buttering the front and back edges of the course below, ignoring the transverse webs or partitions of the blocks. Joints may be finished by merely scraping off excess mortar to be flush with the face of the wall or they may be lightly tooled,

OTHER USEFUL KINDS OF BLOCKS

SOLID UNIT
2" x 8" x 16"

PARTITION UNIT
3", 4", 5", 6" x 8" x 16"

¼ and ¾ SASH UNIT
4", 12" x 8" x 16"

FULL SASH UNIT
8" x 8" x 16"

CORNER BULLNOSE UNIT

DOUBLE BULLNOSE, END

DOUBLE BULLNOSE, SIDE

SINGLE BULLNOSE UNIT

ALL BULLNOSE BLOCKS ARE MADE IN VARIOUS WALL WIDTHS

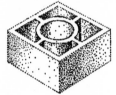

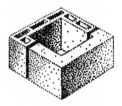

BEAM LINTEL UNITS
VARIOUS WIDTHS, HEIGHTS

VARIOUS TYPES OF CHIMNEY BLOCKS — CHECK LOCAL
SUPPLIER FOR TYPES AVAILABLE, FOR SIZES OF BLOCKS, OPEN OR SOLID

DOOR JAMB UNIT
FULL BLOCK 8" x 8" x 16"
ALSO HALF BLOCKS

SOFFIT FLOOR FILLER UNIT
8" x 21" or 24"
VARIOUS HEIGHTS

PILASTER AND HALF PILASTER
FOR ALTERNATE COURSES

DRAIN TILES, 8" LONG
4", 6" OPENINGS

COPING BLOCK
IN WALL WIDTHS

HALF-HIGH UNITS
4" x ¼", ½, ¾ LENGTHS

"TILE" UNITS, 12" LONG
8" WIDTH x 5" OR 4" HEIGHT

as with brick joints. The V-joint is most frequently preferred for both hollow tile and cinder and concrete blocks.

Build up the corners with three or four blocks, tapering downward in triangular fashion, checking carefully with the level to make sure they are plumb. Stretch the mason's line well and lay each course carefully. With their greater weight, blocks are more difficult to lift and set properly, and constant vigilance must be exercised lest the wall sag or belly outward. Because of their larger unit size the chances of error increase rapidly.

Finishing the top of the wall may pose a problem because of the hollow cores in the blocks. These should be covered with a coping course of thin solid blocks or with a soldier course of bricks. Coping blocks may also be used if they are obtainable; or it is possible to make a form of wood on top of the wall and cast your own coping of cement or concrete. All coping courses should overhang by 1 inch or more to carry drips out beyond the face of the wall and prevent their deteriorating or defacing it. With solid block, coping blocks, or cut stone there is no

HOW TO LAY BLOCKS

1. Mortar stiffened by moisture evaporation should be reworked; add water in small quantities, mixing and testing before adding more. Discard all mortar not used within 2½ hours of original mix at temperatures over 80 degrees F.; within 3½ hours when temperatures are below 80 degrees F.

2. Professional masons always make sure that mortar is sticky enough to adhere to concrete blocks being laid in walls. A quick vertical snap of the wrist makes mortar stick to trowel, keeps it from falling off the trowel as mortar is applied ("buttered") on edges of block.

3. Check the layout of wall by setting out blocks first for accuracy of layout and dimensions; then spread a full mortar bed, wider than wall, and furrow it with trowel as shown to assure plenty of mortar for blocks to rest on, complete coverage and a good tight joint.

4. Lay the corner block first, carefully positioning it; butter only face end of the next block (see photo 21E) and place it in position next to the corner block, pushing it well down into mortar bed, also against end of corner block. Butter several blocks at once for use.

5. It is efficient to set several blocks on end, convenient to where they will be laid and butter upright ends of four or five so that they can be laid quickly. Strive to keep vertical joints even, about ⅜" in width. Be sure mortar bed is laid ahead.

6. As soon as three or four blocks have been laid, the mason's level is utilized as a straightedge to assure correct block alignment. Tap blocks with trowel handle to bring into line, then use level to check the horizontal and vertical levels, again tapping into line.

7. After first course is laid, apply mortar only to back and front "shells" of blocks for bedding. Vertical joints may have mortar applied to blocks already laid or to new block being laid. Some masons butter both, thus assuring a tight well-filled joint.

8. Lay up corners first, four to five courses, checking each corner at each course for horizontal alignment, for horizontal level as above, and for being plumb. Work courses from the corners to make sure blocks are in same vertical plane for true walls.

9. Continual checking to make sure the wall is plumb is the secret of good masonry. Check corners at every course, plumbing both sides of wall. Between corners, use either a level or a straightedge as courses are laid to make certain blocks are kept in same plane.

10. In building corners, step each course back half a block—that is, the end of each course has next block laid crosswise. Check horizontal spacing by placing level diagonally across corners of blocks to be sure that they align.

11. To complete courses a final "closure" block must be laid between corners. Butter all four edges of block ends, also those of blocks on either side, to make doubly sure of a tight joint. Place closure block. If any mortar falls out, remove block, repeat operation.

12. To tie nonbearing block walls into intersecting walls, place strips of metal lath or ¼″ metal mesh across joint in alternate courses of wall; apply mortar over it. If one wall is to be built first, place strips in wall, then tie into second wall as courses are laid.

problem about finishing the top of the wall, for the front and back webs of the blocks are buttered as with any other course, and the coping course is buttered at the ends which butt against the previous block. But if brick is used, a different method may have to be employed to insure success.

Where the wall is constructed of 6-inch blocks, the soldier course of a single row of bricks will give sufficient overhang, but where walls are wider it may be wise to use a brick and a half or Roman brick for the coping course. Where a brick and a half are used, the joints will be staggered, naturally.

PAINTING MASONRY

Oftentimes the texture of masonry walls can be enhanced by painting them. New additions to old walls will be integrated with the old walls by covering up color differences with paint. Similarly, stone walls which have wildly unrelated colors can be pulled together and given dignity and greater effectiveness by being painted. A stone wall will lose that plum-pudding or raisin-bread appearance and become a coherent architectural mass, a thing of beauty and a joy for as long as the paint lasts. Old houses can be made to look fresh and modern when dismal, dark masonry walls are covered with light and appealing pastel colors, or with white or medium tones which will dispel the gloom and bring the house up to date. This is not a new idea, of course, for painted ma-

259

sonry and color-washed walls have been used in many parts of the world for centuries, and we have many examples of their use in our own country in the past.

Paint protects walls, helps to weatherproof them, and cuts down the need for repointing mortar joints and filling weather-induced pock-marks and holes. Some paints especially formulated for masonry use do not seal the pores in the ordinary sense, according to the claims of the manufacturers. They allow the masonry to "breathe," prevent condensation of moisture on the inside of the walls and also keep it from collecting behind the paint film, thus avoiding blistering with resultant peeling off of paint films. Possibly in retaining walls, garden walls and such, the condensation factor is less important than in house walls, but it should be considered for it will affect adherence of paint to the masonry and consequently call for more maintenance, repainting, and added work.

Oil paints are still available and used for masonry although other more recent introductions have cut into their use. We suggest that you read all the details of paints in this chapter, consider the various attributes of each kind, and then make your choice. In oil paints, either gloss or flat-finish paints may be used for brick, stone or block walls, as well as on stucco. Porous concrete or cinder blocks may need a little special treatment before the final coating. All surfaces must be thoroughly dry before being painted so that the paint will bond properly and not blister or flake off later on. It is always well to make a test of the wall surface by painting a square yard or so of wall with the primer coat to see if there are any alkaline places on it—what painters call "hot spots" because they seem to burn the paint. If blisters or discolorations appear, wire brush the wall over them and give two more primer coats to seal them off. If not, the wall can be safely painted after two or three days have elapsed without "hot spots" appearing. This alkaline test is necessary only on new walls; old walls which have weathered will presumably have dispersed the alkalinity. On most walls a single coat of primer, followed by a finish coat when the primer has thoroughly dried, will be all that is necessary. If you are using a colored finish coat it will be well to tint the primer to a hue approaching the color of the finish coat. This will prevent streakiness and thin spots in the final coat. If any appear, let the first finish coat dry well for several days and then follow it with a second coat to flatten out and give an all-over smooth and even coat.

Cement-base paints were formerly offered by a number of manufacturers and were much used. They come in dry form, to be mixed with water and applied with a brush. However, with the emergence of the newer synthetic-base paints, and the ease of using them as well as the greater range of color, the long life, and lack of maintenance required, we no longer feel that cement-base paints can be as fully recommended as in the past. The acrylic latex, alkyd, polyvinyl acetate, and latex bases are so versatile, so easy to use that we believe the home craftsman can choose them with assurance and use them for many kinds of projects. Most in their favor is that many types can be used on wood, both siding and trim, on shingles, shutters, and on primed metal as well as on masonry, something that cannot be said for cement-based paints.

If a cement-base paint had been used on an old wall that is now to be repainted, be sure that it is thoroughly cleaned off—powdery old paint wire-brushed and washed away—and the surface thoroughly prepared with a sealer coat. Consult your paint dealer for his recommendation on a sealer that will meet your needs. Do not expect any paint that is applied over peeling, powdering, or loosened coats of old paint to adhere and give a long-lasting, maintenance-free job. A little extra work in preparing the surface will save much labor later on of redoing an unsatisfactory job—unsatisfactory because the paint will not properly bond with the surface.

Synthetic-base paints are a fairly recent development and offer the home craftsman much that will be to his advantage. We have mentioned the wide range of colors available, from pale pastel to deep tones, also that many of them are recommended by the manufacturers for use on a variety of materials—wood siding and shingles, shutters and other trim, primed metal work (gutters and downspouts, for instance), as well as for all kinds of masonry, including stucco. Add to these virtues another solid advantage—easy application and quick drying—and the crowning virtue of all, quick and easy cleanup of applicators with water and soap, and you have something approaching the ideal masonry paint. Most of them can be applied with a brush, a roller, or sprayed on. However, we are old-fashioned in that we believe in the first coat being brushed on, so that all pores are well filled and the paint better bonded than is sometimes the case when a roller or other applicator is used. These paints dry within thirty minutes, for the most part, under ideal drying conditions, or a bit more than that depending on weather or climate conditions. But once on, they quickly harden and cure into a perfect coat. Should succeeding coats be needed—and new work frequently will need two to three coats because of the porous nature of the surface—they may be quickly applied without waiting hours and days as with other kinds of paint. These paints dry quickly to a splendid flat sheen and insure that dust pick-up and bug trapping in the paint surface will be practically eliminated.

The "breathing" action mentioned above is a distinct virtue in that it allows air to penetrate and condensation is virtually eliminated. Moisture is usually the cause of blistering and consequent peeling off of paint films from walls, and condensation on the inside of interior walls may cause trouble indoors, as well. A tough film of paint that will dry to a flat, velvety sheen, will resist cracking, blistering, peeling, will provide a water-resistant surface (even in driving rains, snow and ice), one that is also practically impervious to ordinary dirt and stains, chemical fumes, sea air, and smog. Such a painted surface is the aim of all painters and these paints come very close to providing it.

Most of these paints can be thinned with water, but be sure to *read the directions on the can carefully* and follow them to the letter. It is usually recommended that the first coat be used full strength in order to fill the pores, priming the surface and providing a proper base for succeeding coats. Masonry should be examined and repaired before painting. Look for breaks, cracks, holes, and voids in joints and fill them either with mortar mix or use a "block filler," purchasable at your paint shop and probably manufactured by the paint company whose paint you will use. Allow repairs to dry thoroughly before painting the masonry, and also be sure to remove all loose dirt, all splashes or drips of mortar, grease, oil, loose paint, and surface dust. Also make sure that the wall is dry.

Caution: *Do not use acid washes or zinc sulphate to neutralize the surfaces to be painted.* Instead, use water, either hosed or brushed on and if this does not clean the wall—possibly where grime and dirt are embedded in the pores—scrub with a household detergent dissolved in water to loosen and remove the dirt and oily substances. Then rinse down the surface thoroughly with plain water.

Where moss or mildew or moldy spots are present, remove them by a scrubdown with a solution of household bleach (such as Clorox) in the proportion of a gallon of water to a pint of bleach. Afterwards, rinse the surface thoroughly with plain water. *Never use acid washes* for removing moss or mildew.

The cleanup of rollers and brushes or sprayers is very easy with these paints. Merely wash them well with warm soapy water and then rinse them under running water from the tap and they will emerge as good as new within a short time.

These are the principal kinds of paints recommended for use on masonry and the home craftsman is advised to look into them and see the colors and kinds available before deciding against painting the walls. However, should a natural color of brick or block be desired—or where blocks have been chosen for their color quality, for the aggregates exposed, or for some other quality, look into the clear sealants or silicone-type sealants that will close the pores without changing the colors appreciably. In any case, whatever your final choice, you will be pleased that maintenance is cut down, that color has been added (if that is your choice) and that the beauty of the masonry is enhanced.

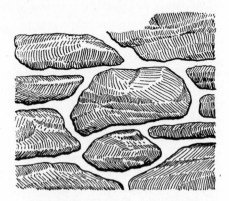

RUBBLE STONE MASONRY

Uncut stones, uneven in size and shape, are called rubble stones. Laid so that their longest dimension is roughly horizontal, they'll give a more natural effect.

COURSED ASHLAR MASONRY

Ashlar masonry is built with stratified stones or those cut and roughly dressed to a squarish shape. Shown here is a method to lay stones in informal courses.

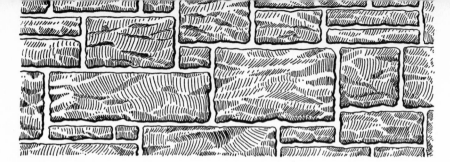

RANDOM ASHLAR MASONRY

In this type of work, the roughly squared stones are of various sizes and although the joints are fairly regular no attempt is made to form regular courses. Fit them together so that a pleasing sequence and variety of sizes occurs; get variety, too, with colored stones, but keep the colors close in tone to avoid a checkerboard look.

COMBINING MATERIALS ON WALLS

When there is an outcropping of rock to be connected to a wall of brick or other man-made material, knit them together with stones of related color interwoven and laid flush with the courses of brick. Don't, however, place stones at random in brick walls or allow them to overhang—too frequently a warty look will result.

FITTED NATURAL STONE MASONRY

Carefully chosen pieces of natural stone, cut to shape as necessary for fitting and mortared evenly will make a strong, long-lasting wall.

NATURAL STONE MASONRY

Stones selected to fit together and also to keep the face of the wall reasonably flat are pleasantly informal, interesting in texture.

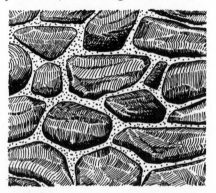

ERRORS TO AVOID IN STONE MASONRY

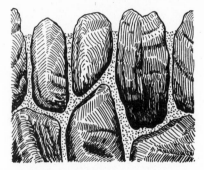

DON'T SET STONES UPRIGHT—YOU'LL NOT FIND THEM THAT WAY IN NATURE

DON'T SET STONES IN SLOBBERY MORTAR LIKE RAISINS IN A RICE PUDDING

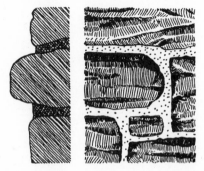

DON'T LET STONES PROJECT BEYOND THE WALL FACE UNLESS THERE'S A REASON.

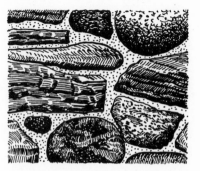

DON'T USE TOO MANY KINDS OF STONE OR YOU'LL GET A HAPHAZARD EFFECT.

A change in grade need not call for a retaining wall. Keep a low slope informal with large stones of varying sizes and shapes laid as a sort of pavement, to form an interesting rock garden with plants set between stones— making a triumph of what is a mundane necessity.

13

Choosing the Proper Joint

Perhaps the most important choice, next to selecting the masonry material you will use for the wall you are going to build, is that of the kind of joint you will use to bring out the inherent qualities of the wall materials and show them to their best advantage. Oftentimes masonry alone will have a beauty in itself quite aside from any architectural or plant embellishments, merely by an unusual choice of material and by thoughtful choice of the proper joint to show off the masonry material. The wall itself can be the showpiece and the planting merely the grace note.

Joints will create an interesting texture with sunlight, making shadow-effects that form a constantly changing and ever-fascinating pattern on the wall. Various textural effects can be obtained by the way in which the joints are shaped—by raking out the joints to get an indented shadow, or by squashing out the mortar in the joints and then either leaving them as an informal, irregular mass or shaping them so that they are extruded but formal. Even the color of the mortar can influence the final result. For instance, mortar which is dark in color will cut down the shadow effect on medium- to dark-colored walls. Conversely, white or light-colored walls with darkened mortar forming the joints will produce a far stronger pattern than if pale or white mortar is used. Dark masonry units with light-colored mortar will also have an impact because of the contrast of color. And all this is quite aside from the three-dimensional patterning caused by raking or otherwise shaping the mortar joints.

Where walls face the sun a good part of the day, more shadow texture will be apparent than on walls which receive only light reflected from the sky or from other buildings, as on north-facing walls. Hence, if it is desired to emphasize texture on northern walls the raking should be a little deeper, the shaping intensified, in order to obtain the proper joint shadows.

Sometimes, particularly where a very formal or refined sort of masonry work is used in modern or fine traditional architecture, it is attractive to have no joints showing at all or to subordinate the joints to the texture of the masonry units themselves. In that case, a flush joint is the answer. You may even tint the mortar to approximate the masonry unit color if you wish to pursue the flat effect to its final conclusion. Thus, by using flush joints with mortar in a related color, a very subtle textural effect with a subdued pattern is produced to make a unique architectural contribution to the general picture.

VARIOUS JOINTS AND HOW TO MAKE THEM

After the masonry units are laid and just before the mortar finally sets, the mortar joints are ready to be finished. This merely means that any loose mortar which was squeezed out when the masonry units were set in the mortar bed should now be removed. The trowel is usually employed to remove the major excess and final tooling is done with various tools which may be purchased or made in the home workshop.

A good, neat, well-made joint will finish the wall properly so that it has a professional look. A ragged, sloppy joint will make the wall look amateurish and bedraggled, reflecting no credit on your craftsmanship or on the home which your craftsmanship should beautify. Therefore you should exercise care and precision in striking the joints in walls you build, observing the basic rules and practicing them to achieve a well-built, good-looking, durable wall.

Hold the blade of the trowel at about a 30-degree angle from the wall, one edge of the blade resting flat against the wall so that it will scrape off the blobs of mortar that have squeezed out. Using a quick motion, holding the trowel so that it stays in approximately the same position in relation to the wall, scrape off the mortar and dispose of it by a quick jerking motion of the trowel or by tapping it lightly against some firm object. An occasional dipping of the blade in a bucket of water or a quick rub with a wet rag will keep it clean and prevent any mortar from adhering to it. Once the excess mortar is removed from the wall you are ready to proceed with the finishing work of tooling.

The head joints (vertical ones) are always tooled first, whatever method of tooling is employed. Then the bed joints (horizontal ones) are struck; and then comes the final removal of any blobs or other accidentally-left mortar which can be whisked off with the tip of the trowel blade. Certain of the joints may be finished with the point of the trowel, too. The flush joint, the struck joint, and the weather joint may be finished this way, and the trowel is also used for the final finishing of the tuck joint.

Flush joint.   The reason for all tooling is to compress the mortar somewhat and to remove any excess mortar from the finished wall. On the page of sketches, note how the edge of the trowel scrapes along the wall and removes the mortar until it is even with the surfaces of the masonry units. This joint is usually used with cement or cinder blocks; often it is employed on brick work where a smooth effect is desired.

Struck joint.   In this case the point of the trowel is inserted at about a 30-degree angle and run along evenly to compress the mortar and at the same time expose the upper edge of the lower course of brick. Vertical joints are either lightly raked (see raked joint) or kept flush for an even effect.

Weather joint.   The trowel point is again employed at about a 30-degree angle, this time to expose the *lower* edge of the *top* course of brick, the method being merely reversed from the preceding as detailed for the struck joint. Vertical joints are finished in the same way. This is

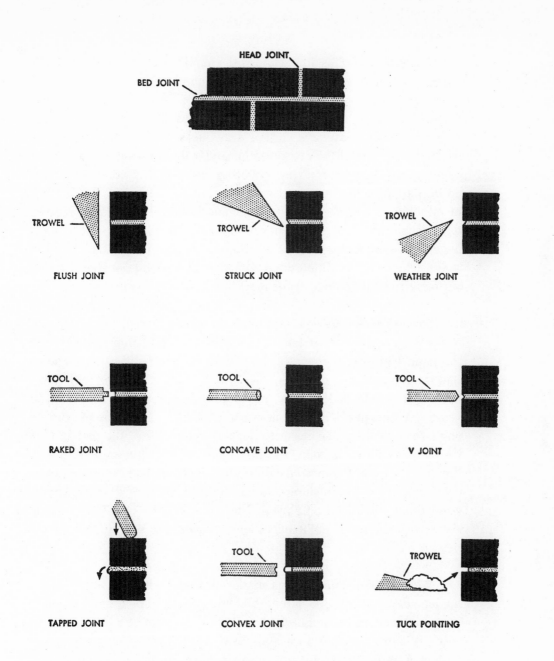

one of the best joints to give texture, especially in severe climates or those which are exceptionally damp, as this joint will shed water better than any of the others except possibly the flush joint. It will give nearly as much shadow effect as a raked joint without any of its drawbacks.

Raked joints. This makes one of the handsomest of masonry effects. If regular, even-sized modular units are employed, it will produce a good deep shadow line, emphasizing the geometric pattern pleasantly. It may be painted or left in its original finish, as you desire. However, in some climates it may have certain drawbacks. It will retain moisture, ice, and snow, and if the brick or other units are porous it may lay them open to cracking from freezing and other moisture deterioration. Where no insulation or moisture barrier is used on the walls of a house with raked joints, the inside walls may become rather damp. It may require more repointing than other types of joints (all joints may need repointing after a good many years of weathering); but even with these drawbacks, which do not obtain in all climates, let us point out that it is still our favorite joint and one of the most beautiful and most satisfying.

Concave joint. As weather-impervious as the flush joint, this one is made by using a short length of pipe or a round iron rod to shape the joint. Hold the pipe or rod level with the joint and at an angle so that the rounded side of the end fits into the joint. Draw the tool along, holding it always at the same angle. You should choose whether you want a deep concave or a shallow one and hold the tool at the proper angle to achieve it. The closer to the wall the shank of the tool is held, the deeper will be the joint, for the round end will fit more deeply into the area between the masonry units. A ¾-inch rod or pipe is about the right size to use for finishing a ½-inch joint. If joints are wider, larger-sized pipes should be employed. For concrete blocks, a jointing tool should be a minimum of 22 inches long, allowing for bridging irregularities.

V-joint. This joint can be made with a purchased tool; but, if a homemade tool is desired, a length of strap-iron can be ground down to produce the V, or the point of a trowel or the corner of a board may be used. Whatever tool is employed, however, the main danger to avoid is that of getting the joint uneven. A proper V joint centers the point of the V between courses of masonry units.

Tapped joint. A bit of practice will be necessary in order to produce the proper effect in making this joint. It depends upon the mortar bed's being just the right thickness so that the right amount of mortar is squeezed out when the masonry unit is tapped with the handle of the trowel. It should not be too thick, or too large a blob of mortar will be forced out, making the joint uneven and slobbery. A little practice will enable the craftsman to judge just how much mortar to use when laying the bed joint and when "buttering" the end of the unit to get the proper final effect. Correctly made, this joint gives a very interesting rough-cast appearance which is most attractive in the proper setting. It is not the most weatherproof of joints, but it will last about as long as any of the open joints.

Convex joint. For finely finished masonry work, such as fitted stone masonry, this is most useful. It is half-rounded and protrudes beyond the masonry line, giving a rather interesting line texture to a smooth wall. It is frequently used on semi-dressed or cut-stone walls, too. The tool may be used before the wall is scraped and cleaned, or the joint may be raked and refilled and the tooling done afterward. (See tuck pointing.) Due to the extra care necessary and the extra compression, this joint will probably be as long-lasting as any, particularly if tuck pointed.

Tuck pointing. The joints are raked as detailed above for Raked Joints, and then, by the use of a specially prepared mixture of mortar made with extra fine sand, the joint is refilled. This must be done, of course, before the original mortar has set, so that the two will bond together successfully. Use the point of the small pointing trowel with a small quantity of mortar and avoid dribbling it or splashing it on the masonry units. You may finally tool the joint flush or use the finishes for any other joints except the Raked Joint.

This method may also be used to repair a wall which has weathered to the point where mortar is softening and beginning to fall out. First clean out the old mortar well with an old chisel or sturdy screwdriver, then tuck point and tool to the desired finish.

Skintled brick work. Where a rough, informal, textured look is in order, this type of bricklaying may be desirable. While it is not strictly a joint, depending more on the use of warped bricks and the irregular placement and alignment of bricks than on joint finishing, it may interest the amateur if it goes with his style of home. He can get the hang of bricklaying without having any nervousness engendered by the need for perfect alignment of joints and courses. Don't attempt to use this with a formal kind of house or with the crisp clean lines of modern homes, but with the more informal and rustic type of home it will fit very well.

Quantities of Brick Needed for 100 Sq. Ft. of Wall Area

(Based on standard-sized brick—$2\frac{2}{3}$" x 4" x 8"—with a $\frac{1}{2}$" mortar joint. Overall measurement of $2\frac{3}{4}$" plus for brick and joint.)

Type of Wall	Quantity per 100 sq. ft.
1. Single Course Wall, brick set on edge	399
2. Single Course Wall, brick laid flat	616
3. Double Course Wall, solid 8″ wall	1232
4. Triple Course Wall, solid 12″ wall	1848

NOTE: Most brick veneers facing a backing-up course of concrete, cement or cinder blocks, or frame construction will be a single course, laid flat, as in line 2 above. It is possible to set the bricks on edge as in line 1 above. A variation of the 8-inch wall is the Rolok or the Flemish Rolok wall (see sketches), which will make a light but strong hollow wall.

14

How to Make and Use Concrete

Concrete is an excellent material for the home craftsman to use, when it is properly made. He can make it to suit his purpose, in quantities large or small; and it may be used in so many ways and for so many purposes that projects designed for concrete or adaptable to it would fill several volumes the size of this one. It is a very durable material. Because of our modern formulas with their exact measurements; because we have developed good methods of finishing concrete; because it can be adapted to very intricate casting forms; and because, finally, it can be reinforced so that it will bear heavy loads without cracking or danger of collapse, we have at our fingertips probably the most useful, permanent, and relatively low-cost plastic building material the world has ever known.

Not so long ago concrete was relegated to small areas or used for only workaday projects, but today it is used much more imaginatively. There are so many textural treatments possible, ranging from very rough pebbly surfaces—with the aggregates exposed and pebbles tamped into the wet surface of fresh concrete—to the very smoothest surfaces, which can be waxed and used as dance floors. With these possibilities and the use of coloring materials to give life and beauty to the concrete, we may well call it the most versatile of materials.

Concrete can be used in the old humdrum ways, or it can be employed in stimulating, creative ways to give piquancy to whatever is made of it and to heighten the effect of the garden picture. It has so much more to offer than just its commonplace and worthy qualities of permanence and durability that it should be considered for every possible project in the garden.

Before we take up ways and means, the formulas for mixing, and so on, let us look into the methods of handling it which may be employed by the home craftsman. Mixing and laying quantities of concrete is heavy work. Let us admit that immediately and proceed to see what can be done to lighten the tasks connected with its use. By planning your job so that it can be done in segments, none of them a large, backbreaking project, you can manage things so that it won't be too much for you to handle even though it may take you some time to achieve your goal. For instance, in building a terrace we can cast it in sections, rather than in one piece. It is better to do this for many reasons, one of which is that contraction strips are needed to allow for contraction and expansion due to heat and cold. If these expansion strips can be made with 2″ x 4″s, which may be left in the terrace or removed after the terrace blocks have been cast (the areas can be filled in with soil and planted to grass), you can do several squares at a time, and let them cure while you are going on to the next strip of squares. Your terrace will expand as rapidly as your time, your energy, and your enthusiasm will permit. Also it will present less strain on the pocketbook when taken in segments.

Walks and driveways may also be built in sections, and, although steps are usually better poured at one time in one piece, it is possible to cast them in sections if they are properly tied together with reinforcements which extend between the two units. It may be possible, of course, to dragoon the family or to hire a couple of strong-armed men for a day or two to help you to do large projects so that you can get them done with less strain, or if you feel they are too much to attempt alone. Concrete mixers can be hired by the day in most sections; they will cut down the labor of mixing large quantities of concrete such as will be needed for terraces or driveways, particularly if you have been able to hire a workman to do the shoveling and hauling so that you can do the finishing and lighter tasks.

TRANSIT-MIX CONCRETE

You should look into the possibility of buying concrete already mixed if you need a sizeable quantity. It is called transit-mix concrete. Many building materials dealers sell it by the truckload. Large tank trucks keep the mix spinning and churning as it is driven from the mixer to your home. You'll need a couple of wheelbarrows to haul it from the place the truck parks at your home, unless it is adjacent to where it is to be used, in which case it may be channeled from truck to project by metal conduits.

Transit-mix concrete offers many advantages; not least is the price, when compared with that of do-it-yourself mixes, particularly when the quantity desired is for more than a few yards of concrete. Very probably you will find that the ingredients alone will cost as much as, or more than, the price of the transit mix, and there will be the labor of shoveling, thoroughly mixing and final handling of the concrete yourself to add to that when you are considering the two methods. For home craftsmen with limited time, even if the price were a good bit more, it seems to make excellent sense to buy ready-mixed concrete because of the time saved as well as the labor avoided. You will find that transit-mix companies are, in general, quite used to supplying moderate or small quantities needed for average jobs such as a walk, a terrace, or a short run of retaining wall.

Or perhaps you will want to canvass the possibility of having concrete laid by contract, if you feel the project is beyond your abilities. Get a contractor to estimate on the work—be prepared to find that it will cost up to twice the amount you'll pay if you do the work yourself and buy only materials. It may well be worth it. However, you may still be able to pare the cost a bit if your contractor will allow you to do the digging out and grading which is necessary, and he may even let you build the forms if you can convince him you are competent. These possibilities will depend upon the good will and cooperativeness of the individual contractor, of course.

In the event that he will let you do something, it is a good plan to have a written contract specifying clearly what are the contractor's responsibilities and what you are to furnish. In this way neither of you will be expecting more than should be required, and unpleasantness can be avoided. Sample contracts of this sort may be available from branch offices of cement manufacturers, or you may write to the Portland Cement Association, Old Orchard Road, Skokie, Illinois 60076, for information on contracts and specifications for concrete work. This company also offers some helpful little booklets which give pictures of projects, tables of quantities needed for formulas for various projects, and other useful information on concrete and how backyard craftsmen can employ it to improve their homes and gardens.

"YOU-HAUL" CONCRETE

Another innovation to consider is the "you-haul" concrete, available in more and more places, particularly if your job is a medium-sized one. First offered on the West Coast and now spreading to other parts of the country, the you-haul is offered in three versions: trailer-haul, trailer-mixer, and pickup-mixer. The trailer-haul concrete is mixed to order while you wait, then poured into a box-type trailer hitched to your car's bumper. Never unhitch it while the trailer is filled or you'll regret it, because the weight is so great you'll never be able to reattach the trailer. The concrete should contain some elements that slow down setting and also keep the aggregates in suspension to avoid an uneven mixture. Probably your dealer will also supply, for a fee, wheelbarrow, tamp, shovels, and other finishing tools. He will also give you instruction on using the lifting device which pours the concrete out of the trailer.

The trailer-mixer looks a lot like a howitzer or cannon, and you can recognize it by this description. It also attaches to the car bumper and as you tow it the mixture is kept in suspension. It also has devices for pouring out the concrete on location. Both this and the trailer-haul method require some skill in driving with a trailer, maneuvering it, backing it into the exact spot where it is to be used as well as properly handling it on the road. Be sure you are capable of handling this phase of the job, too, before you choose these methods.

On the other hand, the pickup-mixer is easier to manage for it is merely a mixer similar to the latter type mounted on a pick-up truck chassis. You hire the whole thing for a fee, including the concrete, and drive it to where you are to use it, maneuver it into place and dispense the concrete. It may be the best answer for many craftsmen.

In all three cases, the cost of the concrete is a little less than that of the transit-mix, but not enough to make much difference unless you have a large job such as a driveway or a huge terrace to pave. You may feel it is not worth the extra trouble and labor involved in picking up the mix and transporting it yourself. In *all* cases, be *sure* that you have everything in readiness and the site completely prepared before the concrete arrives—forms in place, braced, ground leveled and filled, fill tamped—in short, be *completely ready*. There will be no time for adjustments and extra work to be done when a batch of concrete is waiting to be used and probably already beginning to set as it waits.

PORTLAND CEMENT

Portland cement is usually specified for general construction work. It is so named because, after it has been liquefied, cast, and cured, it is said to resemble a kind of limestone from the Island of Portland off the English Coast. It is a *type* of cement, *not a trade name*. Several manufacturers make it, using the fixed standards of the U. S. Government and the American Society of Testing Materials. It is usually packed in paper bags weighing 94 pounds, each equalling 1 cubic foot, dry measure. Although it is often obtainable also in barrels containing the equivalent of 4 sacks, these are seldom practical for the amateur's use.

Cement must never be stored in damp places, on concrete floors, or on the ground. It may absorb moisture through the bag even in comparatively dry basements. Opened bags will allow the cement to take up moisture even more quickly, absorbing it from the air, which will make the cement lumpy or solidify it altogether, thus rendering it unfit for use. Even unopened bags will become lumpy or solidify if not used within a reasonable length of time, particularly in humid climates; so it is a good plan to figure carefully and not order more than can be conveniently used at one time, and thus avoid the perils of storing it. On the other hand it is downright disconcerting, to say the least, to run short of cement if you are a weekend craftsman. It is better to have an *extra* bag in case it may be needed on a Saturday afternoon or Sunday if you are working then, for those are days when it is usually impossible to find a place from which it can be bought. If it is not used, you can always put it to good use the next weekend or the following one. Elsewhere in this text you will find explained the method of making stepping stones and paving blocks (Chapter 18), and it is recommended that any leftover concrete can be poured into the forms for these if they are kept in readiness. This will prevent waste by using every bit of excess concrete after any project, and you can utilize a partly-used or full bag of extra cement by putting it to use this way, too. If you have a partly-used bag which must be kept for a while, get a piece of polyethylene plastic and make an enclosure from it large enough to encase the cement sack. Using a hot electric iron over a folded piece of aluminum foil will seal a seam in the plastic. The top can be left open if you wish, merely by folding or rolling it several times to make a tight joint, and then putting two laths on and clamping them together with carpenter's C-clamps to hold it tightly closed. This will preserve the cement and keep moisture from it for several weeks.

AGGREGATES

Aggregates are any sand, pebbles, gravel, crushed rock, or mixtures of these materials which are added to cement to give concrete bulk and solidity. In mixing mortar only sand is used, but in making concrete the sizes and weights of aggregates are varied to produce the kind of concrete which will be best suited for the purpose for which it is to be used. For footings, for instance, sand, gravel, and crushed rock may be used. This mixture will also be found suitable for the first pouring or underlying mass for sidewalks, driveways, and paving, a finer aggregate being mixed with cement for use as the finish or top pouring. If a very fine, smooth finish is desired, extremely fine particles of sand must be employed to achieve it. Coarser sand can be used for average finishes; pea gravel or other aggregates of the desired sizes when brushed aggregate finishes will expose them. (See Brushed Aggregate Finishes later on in this chapter.) Finish coats may also be colored, the coloring matter not usually being incorporated in the entire mixture unless it is to be only 3 inches or so in thickness when finished. (See the section on Coloring Concrete.) For 6-inch slabs or even 4-inch ones, the colored layer need only be the top inch or so, thus saving money in buying expensive coloring matter.

Fine aggregates or sand varies in size from very fine beach sand (which must be well washed with fresh water to remove salt and organic matter from it) to ¼-inch particles. The varying sizes will fit together with the cement to make a stronger mixture than would those of one size only. All sand must be free from impurities—soil, plant or animal matter, and clay in any large proportion. If your hands become soiled when the dry sand is rubbed between them, it is probably not suitable for use.

Another test for the suitability of sand for concrete is to fill a quart jar half full of sand, and then add water until it is about ¾ full. Cover this jar and shake it well for about three minutes; then let it stand for an hour or more. Should there be more than ½ inch of silt on top of the sand after it has settled and the water is clear enough to see through, the sand is not suitable for use. It has too much clay or soil present to qualify as a good aggregate. Sand should be kept covered with a tarpaulin during rainy weather until it is used, for wet sand is difficult to measure and is both heavy and hard to mix.

Coarse aggregates are those larger particles which range upward in size from ¼ inch to as much as 2 to 3 inches in diameter. No aggregate should exceed about one-third the thickness of the wall or slab in which it is to be used. (1½-inch aggregates in 4-inch slabs, 2-inch in 6-inch slabs, etc.) Not all aggregates of this coarse type are round natural stones; crushed rock or stone may be used, and certain kinds of slag may also be employed. Reinforced concrete (such as is used for driveways or thin walls) should not contain aggregates exceeding 1 inch in diameter, but thicker slabs and concrete which is not reinforced may use aggregates up to the 3-inch limit. Broken or crushed stone should always be hard and clean, traprock, hard limestone, hard sandstone, and granites being the preferred kinds, although in some areas crushed fieldstone may be used. Old concrete smashed into suitable sizes may also be used, as may broken-up old bricks if they are the hard-burned kind, and particles of broken terra cotta and clay tiles may also be incorporated if they are available. Shells were used in Georgia and Florida and elsewhere along the sea coast as aggregates and have proved very durable there, as witness the structures of early times which are still in good condition.

Gravel usually refers to roundish pebbles of various sizes. It comes in different sizes and grades, sorted by sifting through various-sized screens at the gravel pit. If you are using gravel for a brushed aggregate finish, you may want to see if colored aggregates are available—pink, dark brown, yellow, and even black (actually a dark grey) sometimes being available at a slightly higher price. When exposed in the finish, color will add to the beauty of the effect and insure its being unusual and different.

Soil and clay should not be present in gravel, nor should the pebbles be coated with mud or dust, for that will adulterate the cement and prevent adherence of the cement to the pebbles. You can wash gravel by flattening the pile to about a foot or so and letting a hose spray on it for several hours; then when it is dry, turn it and wash it again.

WATER

The water used in mixing concrete and cement is exceedingly important, for if it is unsuitable it may have an adverse chemical reaction and weaken the mixture. It must not have alkali or oils in it—don't use old gasoline or oil cans for measuring it unless they have been *thoroughly* washed out with a detergent—nor should it be very acid, since all of these will react adversely against the cement, preventing it from achieving the proper strength. Nearly any water which is suitable for drinking can be used in mixing concrete, provided it is clean and has no silt or vegetable matter in it. Always wash out pails or other receptacles for measuring water before using them to prepare concrete.

CONCRETE FORMULAS

It is hard to specify a single formula for use in every project for concrete, but according to the latest information available, good concrete is good concrete. One that uses little water and an adequate amount of cement is desirable and probably will suffice for most jobs. The various kinds are described by maximum-size aggregate used.

One factor not always taken into account—and of vital importance to the strength of the finished concrete—is the presence of moisture in the sand and other aggregates used. These are seldom truly dry, for when they are dug from the gravel pit or if they have been stored outdoors, there is usually some water present either from rain or other natural moisture or from the place it came from. Note the cautions below for adjusting the amount of water added to the mixture.

Two Basic Formulas for Concrete Mixtures

KIND OF WORK	Parts cement	Parts small aggregates	Parts large aggregates
Footings, foundations and non-waterproof walls, etc.	1	3	4
Walks, driveways, curbs, floors, steps, basement walls, etc.	1	2—2¼	3

Note: Water Required. The amount of water in proportion to the dry ingredients determines the strength of the concrete, its water-tightness, and its ability to last a long time. In general, the less water is used, the better and more durable will be the concrete, but there must always be sufficient present so the smaller aggregates will fall into solution and fit between the larger ones, when mixed well with the cement. A drier mix is more difficult to manage but even the rank amateur can manage it.

Average work: Footings, foundations, and non-waterproof walls. If sand is slightly damp, add up to 6¼ gallons water to each sack of cement (1 cubic foot). If it is wet, decrease water to 5½ gallons and if it is very wet, decrease to as little as 4¾ gallons.

Finer work: For walks, driveways, curbs, floors, steps, and basement walls with a greater proportion of cement, for moderately damp sand use about 5¾ gallons of water; 5 gallons, if it is wet; and 4¼ gallons if really wet.

In all cases, add water *little by little,* mixing it in well with dry ingredients until the exact consistency has been arrived at.

HOW TO MIX CONCRETE

Never mix too much at one time. How much *is* too much? Probably a good rule to follow is to mix a sack of cement or less at a time, depending upon the measurement of the concrete you need for the job. It can be measured out by the bucketful to achieve the proper proportions. Where, for instance, the formula for paving concrete is given as

        1 sack of cement (which is 1 cubic foot)

        2½ cu. ft. of sand

        3½ cu. ft. of gravel

the proportions will remain the same if you use single measures or multiples of

> 1 bucket of cement
> 2½ buckets of sand
> 3½ buckets of gravel or large aggregates.

A metal wheelbarrow is convenient for mixing up a concrete batch, being waterproof and impervious to scraping with the hoe or shovel used in mixing. It has the additional advantage of being transportable so that you can wheel it to the sand and gravel piles, and measure them and the cement directly into it; after the water has been measured and mixed into these, the wheelbarrow can be wheeled back to the job. Or it can be filled with the dry ingredients and the water mixed in at the site where the concrete will be used.

You will need a shovel and a hoe. A square-ended shovel or garden spade is the best to use, but any shovel can be used if you haven't either of these. A garden hoe can be used, although a mason's hoe, which has holes in its blade, will more quickly mix ingredients.

The procedure for mixing concrete is as follows:

Measure out the sand and place it in a ring in the wheelbarrow, or on a large piece of outdoor plywood if you are not using a wheelbarrow. Then measure out the cement and place it in the center of the ring. Turn the sand and cement together with the shovel to mix it until the whole mass is a uniform brown-grey with no streaks of either sand or cement visible in it. Next, measure out the gravel or other aggregate you are using and spread it on top of the mixture; then turn it again with the shovel to distribute the gravel evenly throughout. When this has been done, make a hollow in the dry ingredients and pour in a little water, mixing it and turning the dry ingredients into it with the shovel or the hoe. Add water, a little at a time, mix, and then add more water, until all of *the proper amount of water* has been used. You may use less than is prescribed in the chart if the mix seems to be of the proper consistency, or just a little more if it is too stiff, but be careful in adding water lest you use too much.

Keep mixing the concrete until it is thoroughly blended, scraping up every bit of the dry ingredients which cling to the bottom of the wheelbarrow or mixing platform and working it in. Practice will enable you to use less muscle without any less thorough mixing, perhaps, but never skimp on the mixing. Better sore muscles than a weak batch of concrete, a mixture which may fail to do its job properly. Use the concrete within a half hour of the time you finish mixing it.

Always mix the ingredients in the order given above: sand and cement, then gravel, and finally the water, whether you are using a mechanical mixer or mixing by hand; and you will be assured of proper blending.

HOW MUCH TO MIX

That will depend upon what you are doing with the concrete, on the depth of the paving, on the width, length, or height of the wall, and on other factors. Paving for terraces and for walks should be a minimum of 3 inches thick, or, even better, $3\frac{1}{2}$ to $4\frac{1}{2}$ inches thick. Driveways should be not less than 4 inches thick, according to many authorities. We submit that 6 inches should be the minimum depth for, even though you do not plan on having heavy oil trucks or other heavy equipment riding on the concrete when you are building it, future necessities may arise for weighty vehicles to traverse it. Thinner slabs may be reinforced with welded-wire mesh, made especially for this purpose or with $\frac{1}{2}$-inch steel reinforcing rods. In frost areas, reinforcing will help to keep slabs intact as will contraction joints. These latter should be placed at 9- to 10-foot intervals, closer in narrower slabs such as walks, terraces, and so on.

To determine how much concrete will be needed and what materials to buy for the job, compute the overall square footage and figure on the depth of the slab. Then determine how much of each material you will need. The following table may be helpful as a guide:

For Covering 100 Square Feet to a Depth Of:*

	3″	4″	6″
Sand	15 cu. ft.	$19\frac{1}{2}$ cu. ft.	30 cu. in.
Cement	6	7.8	12
Gravel	21	27.3	42

*All ingredients figured on the basis of the 1: $2\frac{1}{2}$: $3\frac{1}{2}$ formula recommended for paving concrete.

COLORING CONCRETE

Coloring concrete will give added interest to paving, a floor, steps and stepping stones, and you can even simulate natural stone colors that will allow concrete to harmonize with natural stone used for walls and garden features. The main criticism (and all too often justified) one hears about colored concrete is that it is too bright, that the end result is often garish, dominating the scene and making it difficult to create a harmonious color scheme that will include the house, flowers, and other garden features. Although concrete colors may fade a bit and tone down with aging, there is no reason why colored concrete cannot be harmonious and natural-looking from the beginning. Care and taste must be exercised to achieve a natural, subtle effect.

There are three methods by which one can color concrete: 1. Mix coloring agents in the concrete itself. 2. Spread the coloring agent on top of fresh concrete and trowel it into the finish. 3. Color the surface of finished concrete with a dye, a stain, or a paint.

The colored concrete can be mixed by the home craftsman, or it may be ordered from a transit-mix company, if you are willing to accept the limited colors and the uncontrolled effect you may get. If you mix it at home, you can mix up some preliminary samples and cast some stepping stones or other paving and get some idea of what result you will obtain, adjusting the final colors accordingly. One cannot tell from wet concrete mix or even uncured concrete what the final color will be. Only when it is thoroughly dry is the result truly visible. But, with some preliminary experimentation, you will have a fair idea of what the final color may be.

The coloring agents most frequently employed are usually mineral oxides, especially prepared for this use, and available through your cement supplier. The usual colors are: red, black, green, blue, and yellow. By mixing two or more colors in varying quantities you can achieve some wonderful hues. For instance, by mixing a bit of black with the red you will get brownish and chocolate tones, with deeper tones appearing as you add more of the colors. Blue and green will give a deeper blue-green color which can be toned down and made more harmonious with the surroundings by killing the brightness slightly with a bit of black added. Yellow added to red will produce orange-red tones, and blue added to red will give a cooler, more purplish tone.

Again, by adding a bit of black you can tone down the brightness.

It is generally believed more tasteful to use only a few or closely related colors in paving—that is to say, don't use red, blue, green, and orange blocks in a single area. Rather, use blues and greens together; reds and oranges, reds and blacks with greys, chocolate and brown tones. Observe nature and see how successfully colors are blended together in rocks and how they seem to occur naturally in related hues with many subtle variations.

Some authorities recommend using white portland cement rather than the normally grey-colored cement in order to get "clearer, brighter colors." If you want brilliant, circus-toned paving, by all means achieve it this way, but if you want a pavement that does not draw attention to itself, that lies flat on the ground, unobtrusive as a tasteful rug, we suggest using the grey-colored cement and toning even that down a bit with a little black coloring matter.

MIXING COLOR IN CONCRETE ITSELF

Color mixes must be blended more thoroughly and mixing should take longer than for ordinary concrete mixtures. Don't skimp on mixing or you may have streaky uneven colors in your paving. For thin pavement—stepping-stones and the like—you may want to color the entire mix but, because of the cost of the coloring ingredients, for larger areas and for thicker pavings particularly, you may want to consider two layers of concrete—a base layer with a color layer laid on top. Mix regular concrete with a variety of aggregates and pour it into the forms to within an inch or even a half-inch of the top, or final surface level, and level it fairly well. Allow it to set till surface water has disappeared, then pour in the colored concrete mixture.

This mixture should be made *without* course aggregates, though the proportion of cement to sand and water should be approximately the same. Level to the tops of the forms, then complete the process of finishing, floating, and troweling in the same manner as for other concrete. (See the photographs demonstrating this.) Cure the slabs carefully, protecting them from spotting and staining during the process.

TROWELING IN COLOR

This method is about as reliable when properly done as the above method and certainly easier than mixing the concrete in two coats or layers. Mix and pour regular concrete level with top of forms and strike off; level with a bull or hand float, after the free water has evaporated from the surface. The dry color agents are then sprinkled or shaken over the surface, care being taken to spread it as evenly as possible. "Dry-shake" colors are offered ready for use, packaged by the manufacturer, and usually consist of white portland cement blended with specially graded sand and the mineral oxide color.

Allow the dry-shake to absorb moisture from the surface of the concrete for a few minutes, then with a hand float, float it into the concrete surface, blending it in thoroughly. Repeat the process of spreading the dry-shake but this time use only about half the quantity, and then float it into the surface again in the same manner. Trowel the surface at once, then, after it has set up sufficiently, a fine steel-troweling should be given (followed by a final steel-troweling if a very smooth surface is desired). For most floors and pavings, a light brushing of the surface with a push-broom will give a little surface "tooth" or texture, giving traction for wet weather use and not unpleasant in dry weather.

PAINTING, DYEING, STAINING

These require a different approach and are usually done only on old or finished concrete. If the floor or paving has previously been painted, it must be thoroughly cleaned and then repainted. Dyeing and staining will not work satisfactorily on previously painted surfaces, no matter how well you think you have removed all paint. Paint may be applied by brush, roller, or spray, and formulas specifically made for use on masonry should be used. A recent formula which incorporates epoxy is long-wearing, and has proved satisfactory for most uses, we hear.

For a concrete slab, unpainted, unfinished, whether new or old, dyeing and staining will be possible and give satisfactory results for a number of years. It can be redyed or restained and brought back to life again, if you wish. These two methods wear longer than paint, require less upkeep, and give pleasing effects because the colors are absorbed into the pores of the concrete, creating a more stone-like natural effect. The floor or paving should be thoroughly cleaned, scrubbing it with a solution of 1 pound of trisodium phosphate to 1 gallon of water. This should be flushed or washed off thoroughly with clear water. Grease spots or oil stains (particularly on pavement that has had cars standing on it or where cooking has been done) should also be removed to insure even coverage of the colors. A solution of lye— 1 pound to a gallon of water—poured on sawdust or shavings and allowed to stand overnight or at least for several hours will usually lift the spot. *Handle the lye solution with care.* Lye is strong and dangerous skin burns can result from splashes or careless handling. If one application of lye does not entirely remove the grease stain, repeat the application one or two more times. Scrape the stained spot with a putty knife after the sawdust is carefully removed and disposed of, then flush with clear water.

Next apply an acid etch to open surface pores so that the dye can penetrate deeply, and to remove any further impediment to absorption. One gallon of muriatic acid combined with 2 to 3 gallons of water in an enamel pail—note this carefully, for the acid will eat unprotected metal—will give the proper solution. Brush it on the slab generously and scrub it in, letting it stand till all bubbling ceases, then work it into the slab brushing it up and down and then across till there are no slick spots left on the surface. Flush the slab with clear water to remove all acid; then let it dry *thoroughly* before applying color.

Stains may be applied with a roller or a brush and you have a choice of greens, greys, reds, and browns in a variety of tones. For porous surfaces, two coats may be necessary, with a day elapsing between coats to allow drying. Concrete floor wax toned with the color you used may be used to finish the surface, if you wish, but this is usually only done on interior work.

Dyes are applied with a stiff brush. Sometimes a second coat may be required for a final dressing. Colors come in the same range. If some spots are lighter, or glossier, scrub them more vigorously till the tone seems the same as elsewhere on the slab.

One final caution: Always read manufacturer's directions on the packages and follow them completely to insure good results. The above are the best general directions we can give; specific ones must come from the maker of the products.

ALLOWING FOR SUBSURFACE DRAINAGE

All grading, leveling, and construction of forms must be done before concrete is mixed or poured. Where there is any likelihood of subsurface seepage or excessive moisture, it is well to excavate 2 to 3 inches extra to accommodate a layer of tamped gravel or cinders. This will insure subsurface drainage and prevent any sinking or heaving of the slabs from frost or bogginess. In bad cases of subsurface moisture, drainage tiles laid under the paving to carry away the excess moisture will suffice, provided that there is some place to which the water can drain from the tiles. If any filling needs to be done, it should be well tamped, wet down, and, when its surface has dried, tamped again. The process should be repeated several times until a good firm surface is obtained. Allowing several months or a year of settling before casting a slab on fill will also do the trick. If your slab must cross a recent excavation for a gas or water line, reinforce the slab above this fill, for cracks and sinking are sure to occur, particularly if your project is a driveway, which must carry heavy weights. Such trenches should be soaked and exceptionally well-tamped before the slab is cast over them.

Note: In digging out for wall footings, make sure that the footing is below the frost line in cold areas. The usual recommendation for measurements of wall footings is that they should be the same depth as the measurement of the wall width, and twice as wide as the width of the wall. Thus, if your wall is 12 inches, make the footing depth 12 inches and the width 24 inches, with the wall centered on it.

BUILDING FORMS

If your project is the paving of a terrace, a walk, or a driveway, it will probably require only the simplest of wooden forms. Sometimes they may even be incorporated into the paving by using redwood, cypress, cedar, or some decay-resistant wood which has been well treated with wood preservative. (See Chapter 4.) For some patterns which are obtainable by using this method and from which you can adapt your own design to fit the space you are to fill, see the illustrations for designs and also the photographs elsewhere in this book.

On the other hand, you may want to cast your paving and remove the forms, using them elsewhere for further casting and saving the cost of further forms, as in the case of building a long walk, or in casting a paved terrace in rows of units.

Note: For pre-casting paving blocks, as well as for casting slabs in place, a stiff mix of concrete with a minimum of water will allow the forms to be taken off very soon after casting for reuse. For paving

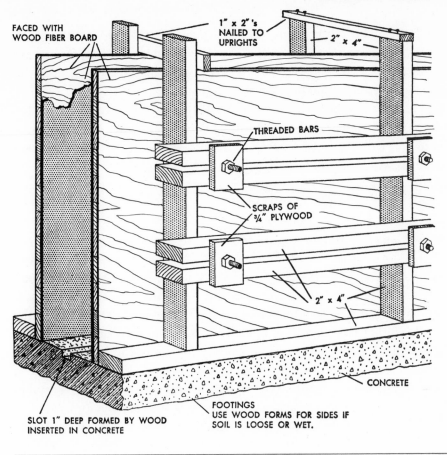

FACED WITH
WOOD FIBER BOARD

1" x 2"'s
NAILED TO
UPRIGHTS

2" x 4"

THREADED BARS

SCRAPS OF
¾" PLYWOOD

2" x 4"

CONCRETE

SLOT 1" DEEP FORMED BY WOOD
INSERTED IN CONCRETE

FOOTINGS
USE WOOD FORMS FOR SIDES IF
SOIL IS LOOSE OR WET.

blocks, cast them on a plywood pallet leaving them on it to cure and harden; or they may be cast as specified in Chapter 18.

It will be well, in fact, to plan the terrace paving or walk so that it can be paved in units; then you can do only as many units as you may be able to handle in a day. About all that the home craftsman can be expected to accomplish in one day is about 50 square feet of 4-inch-thick paving. If you can break this up into 4- or 3-foot squares (or oblongs of the equivalent square footage) you will find that you can do about three 4-foot squares or five to six-3-foot squares per day. In the beginning, have as many forms built for the squares as you think you may be able to accomplish in a half-day's work—let us say two 4-foot or three 3-foot. Use selected 2" x 4" stock which is straight and clear grained for the forms. The forms should be well brushed with old crankcase oil from your car to prevent the concrete from sticking to them if they are to be removed. If they are to remain as part of the paving, don't oil them; merely use wood preservative on them.

Stakes set outside the forms will hold them solidly; crosspieces in the forms should be securely nailed into place. Start with one end of the form and nail the stake securely to it so that the top of the board will be level with the soil, or with whatever level you wish the concrete to attain. Using the mason's level or a long carpenter's level, place the other end of the form on a stake and secure it.

Let us digress to point out that terraces, driveways, and walks all need a slight pitch to enable moisture to drain away. Terraces are pitched away from the buildings they adjoin; driveways are crowned so that they drain from the center to both sides, unless there is some good reason for draining to only one side; and walks also are usually crowned. On occasion it will be advantageous to pitch driveways to drain in the center so that the water will be led down to an outlet or sewer. About ½ to ¾ inch per 5 feet of length is usually sufficient for drainage on terraces, and a little more can be used for driveways and walks.

To return to the building of forms, be sure to use good sturdy stakes that are long enough to anchor firmly in the soil. Brace the forms against other objects; or use several heavy building blocks to hold them if you are not certain the stakes will hold, as may be the case in sandy soil. Use as many intermediate stakes as may be indicated by the length of the forms. Wet concrete is very heavy and may burst out of the forms or push them aside if they are not securely anchored. If you wish to cast concrete with a long curving edge—as on a terrace, a driveway, or a walk—thin plywood may be used for the form, or outdoor hardboard, cut to proper width. Double or triple the boards and you'll have fewer stakes to set and the form will be strong and pliable. Or use 1-inch boards of suitable width, making saw cuts a third to halfway through to allow them to be bent. Cuts are made at regular intervals, a few inches apart, with cuts farther apart for arcs with a larger radius. Whatever the form material used, make sure that the curve is a true arc with no bumps or jogs or "corners" occurring anywhere. Stake well so that the forms will hold the concrete and not be pushed out of place, thus assuring a long, flowing curve.

1. Excavate to required depth—base plus thickness of concrete slab—roughly leveling soil. Remove all sod and vegetable matter, digging out soft and muddy-moist spots and refilling with sand or gravel. Also loosen hard, compacted spots; tamp well to give uniform support to slab.

2. Build forms of straight kiln-dried or seasoned wood, staking on outside of forms at intervals to brace and hold firm. Drive stakes in straight to ensure that forms are plumb and true. Level forms, using a carpenter's or mason's level both lengthwise and crosswise.

3. Fill form with sand or gravel base to proper level, taking care not to put forms out of plumb. Base should be compacted by tamping, filling in any hollows that result, and retamping until the base is reasonably level and even, made ready to receive concrete when it is poured on.

4. If forms are to be reused, wrap them with plastic or kitchen waxed paper, holding in place with masking tape. Should the forms be permanent, remaining as dividers between concrete slabs, protect tops with masking tape for removing excess concrete.

5. Concrete should be relatively stiff in consistency, containing only enough water to produce readily workable mix that will integrate well. Some experts dampen base thoroughly—but with no puddles—before pouring concrete, thus preventing water being pulled out of mix by base.

6. To avoid air spaces or voids at edges of forms, work concrete with a spade or hand floater to compress it. To even and smooth surface, use a straightedge screed, resting it on forms, working with a sawlike motion, with a little concrete ahead to fill low spots.

7. After slab is well leveled, use a long-handled bull float (it can be homemade) to smooth or level surfaces, and to take off excess water which bleeds to the top and must be removed to retain strength of concrete. Work float lightly back and forth to smooth and fill depressions.

8. Where a rounded, finished edge is desirable on walks and edges of driveways, terraces, walk joints and elsewhere, use an edger after bled water and water sheen on surface have gone and concrete has started to stiffen. Run edger lightly back and forth alongside form.

9. Next step is to "float" the surface. This removes edger marks, helps to embed large aggregates just below surface and also consolidates mortar at surface in preparation for final finishing. Metal floaters slide easily, make smoother surfaces than wooden ones.

10. Immediately use steel trowel to produce a smooth, hard surface. Keep blade flat against surface but take care to prevent edges cutting into concrete. More trowelings with successively smaller-sized trowels and a certain amount of pressure produce an exceptionally smooth surface.

11. Steel-trowelled surfaces may be slippery in wet weather. Slightly roughened surface may be accomplished by brooming surface with a long-handled, soft-bristled push broom. Protect slab while it is curing with a covering of burlap, kept wet by spraying with a hose.

12. After concrete has hardened sufficiently not to be marred, remove coverings and replace with a layer of sand or straw for at least three days, keeping it moist by sprinkling with water regularly. Or build earth dikes around edges and fill with 1" of water.

POURING THE CONCRETE

When your forms are ready, water the area inside them, soaking it well on the evening before you are to pour the concrete. This will prevent the soil (or dry gravel if you are using a drainage layer) from pulling water from the concrete and weakening it. Early morning is the best time to pour concrete in hot weather, because heat makes it set quickly, sometimes preventing proper finishing and curing. The ideal time for concrete work is cool or moderately warm weather, so it is best if you can schedule your work to take advantage of this. If the temperature is likely to go below freezing, it is not wise to attempt a concrete job because freezing will weaken and destroy it.

Place a plank over soft ground so that you can wheel up your barrowload of concrete and pour it into the forms, dumping it with ease. Concrete should be used within a half hour of mixing for best results. After it is in the forms pat it into place with the back of a flat shovel. Pay particular attention to the corners, ramming the concrete in to be sure no air holes remain. Roughly smooth it by patting with the shovel; but don't spread it too much, for the next pouring will be placed adjacent to it, not on top of it (unless it is the finish-coat pouring). Then quickly mix the next batch before the first sets. If you are unable to finish the entire slab for any reason, scratch the entire surface just before it sets and roughen the sides of the last pouring so that the next day's work will "take." Then, before pouring the following day, hose it off and brush on a mixture of cement and water about the consistency of heavy cream, using an old broom or a whisk broom just before pouring the next batch of concrete. Reinforcing is installed on top of the first layer when poured, too. We shall cover that subject later in this chapter.

PRELIMINARY FINISHING

Fill the forms to the top and then, using a "strike board" or "screed" (a 1″ x 4″ board long enough to ride along the tops of the forms), scrape off the excess concrete and roughly level and smooth the concrete inside the forms. The finish coat or final pouring is usually made with more sand and fewer aggregates in the mixture, unless a special aggregate surface is desired. The finish coat may be colored if desired, but the finishing will be the same in any case. After the "screed" has been used, use the level once more to check whether or not it conforms to the level or pitch you desire; then smooth away any further irregularities with the float (see Chapter 10), using a swinging motion. A little practice will enable you to do this easily and quickly.

FINAL FINISHING OF THE SURFACE

You may choose from a number of ways of finishing the surface, according to your tastes; the use to which the concrete is to be put; and how much effort you can put into finishing and maintaining it. A smooth finish is good to look at, but it may be dangerous in wet weather if it must be walked on very much. On driveways little traction will be obtained for climbing grades or for braking to a stop. Smooth surfaces are good for dancing and are the easiest to maintain, however. They are perhaps best used inside garden structures or on an enclosed porch or loggia.

Medium-rough finish. This is achieved by brooming the surface of concrete, and it is achieved with the same wooden float, any depressions being filled in by floating concrete off high spots into them so that the surface is leveled. Don't overdo it and float the surface too much, or you may draw too much of the water to the top, causing the surface to be weak and watery. Experience is the best teacher in this respect; but we may say that it is better to let well enough alone at first, for you can return just before the surface has set and go over it again. You can make the float marks all in the same direction, or you can make them in swirled circular motion so that they go all over and not in a particular direction, according to your wishes for the finish.

Combed rough finish. This is achieved by brooming the surface of the green or nearly-set concrete before it is too hard to take the marks of the straws of a broom or stiff bristles of a brush, yet is hard enough to retain them. A push-broom with bristles that are fairly soft, one with stiff, coarse, and heavy bristles, or even an old kitchen broom may be used, according to how deeply combed you wish the texture to be. Note that these textures are likely to be dirt-catchers and will need scrubbing and hosing down, as well as sweeping, to remove dirt and dust. Their advantage is that they provide surface interest, good traction, and also cut down sun glare. Eventually, of course, they will weather and wear and become more smooth. Some craftsmen brush one square of a terrace one way, and the next at right angles to it in a kind of checkerboard effect; or the brushing may be done at an angle.

Smooth finish. If, after considering the hazards outlined earlier, you still want a smooth finish, we must warn you that it is the hardest to achieve. It will take from two to four trowelings after the rough float-finishing to get it smooth enough. The first troweling must be done just after the float-finishing, but the metal trowel (see tool illustrations Chapter 10) should be lightly handled this time to prevent water and fine particles from being drawn to the surface where they will wear and flake off in the future. Use the steel finishing trowel just heavily enough to smooth out the float marks; then allow the concrete to set for half- to three-quarters of an hour. By then the watery or wet, glossy look of fresh concrete will have disappeared and the second troweling can take place. It will take strength and really tax your muscles to do it, but the result will be worth it. A third and even a fourth troweling may be necessary if a very hard, extremely fine and glossy surface is to be achieved, and each time you must wait for the surface to harden a bit before re-troweling. After the first troweling the trowel must be employed with enough pressure to make it ring as it is drawn across the concrete surface.

If a concrete floor is desired indoors, this will be the surface to use. After being finished and well cured, it can be waxed for dancing. It can be very easily cared for, since no dirt or dust will find harbor in its fine texture. Because of its close, even surface it will reflect sun and cause glare, but if it is used under shelters or shade this need not be a factor for concern.

Exposed aggregates. This extremely rough, but durable, finish is obtained by using a stiff coarse-bristled brush and water to scrub away the surface concrete, removing the small aggregates and cement to expose the larger aggregates. The final surface layer should contain only sand and uniform aggregates of the size which it is desired to expose, probably chosen for color as well as for uniformity of size. This layer is finished as for the medium-rough finish and allowed to harden

FINISHING EXPOSED AGGREGATES

1. Exposed-aggregate finishes are colorful, agreeable in texture, give a rustic effect that is most decorative. Choose aggregates uniform in size and compatible in color, not less than ½" to ¾" in diameter. Distribute them evenly on top of concrete immediately after screeding.

2. Embed aggregates initially by patting them in gently but firmly with a flat board—a 2" x 4" is a good size and weight—resting it on both sides of form so that surface remains flat and even. Be sure to work in well until aggregates are thoroughly embedded.

3. As soon as concrete is firm enough to support a man on a board, hand-float the surface using a metal floater. Make sure that all aggregates are embedded just beneath the surface so that grout completely surrounds them, with no holes or openings left in the surface of the concrete.

4. Allow surface to set for an hour or so, and test; then expose aggregates by brushing with a stiff-bristled pushbroom or a straw broom, hosing surface as the work is done. Large slabs may require use of a retarder, sprayed or brushed on, to slow setting and allow brooming work to be done.

for about four to five hours in average weather, or longer in cool weather, less time in hot weather. A surface which has hardened too quickly due to hot weather may be scrubbed with a bristle brush (not a wire brush), and hydrochloric (muriatic) acid, in a solution of 1 part of acid to 5 or 6 parts of water. Rinse and hose down with water after the aggregates have been sufficiently exposed, to keep the acid from continuing to eat away the concrete.

This rough surface, attractive though it is, has some disadvantages. It is not a good play surface for children, being extremely abrasive, and it may also prove inconvenient to use with some kinds of furniture. It will hold dust and dirt and is not easy to sweep, so that hosing and an occasional scrubbing may be necessary to keep it clean. However, its attractiveness with informal structures and even fairly formal ones (it mates particularly well with modern architecture), and its compatibility with rough-sawn boards facing houses and fences will recommend it to many. It provides excellent traction in all weather and looks attractive even in very large areas of paving. It is best used for paving terraces, walks, and steps.

THE CURING PROCESS

After the surface has been finished and has hardened for five or six hours, it should be wet down with a hose, the nozzle set to a fine spray. *Don't do this too soon* or you may wash out the cement and fine aggregates on the surface. Wet it down again the next day and then protect it from sun and winds for about a week to two weeks, ten days being the average time. You may cover it with building paper, 6 inches of straw, a canvas tarpaulin, or 2 to 3 inches of garden soil. In very hot summer weather it should be sprinkled with the hose every morning and afternoon (building paper must be removed, as must canvas) and, if you are using the soil covering method, you may make a little dam around the edge of the slab and fill it with water once a day to a depth of 1 inch.

New chemical aids to curing are already available through the manufacturer and will soon enter the do-it-yourself retail market, so we shall mention them here. One such is Sealtite Cure-Hard. These are either rolled on or sprayed on the surface to assist in giving an even, well-cured, hard slab.

At the end of the curing period coverings are removed and the surface well washed with water. It will be ready for use as soon as it has dried out completely hard, and will be perfectly durable. It will be well, however, not to use it too much or too vigorously the first week or so after it has been cured.

REINFORCING THE CONCRETE SLAB

Probably in most terrace paving and sidewalks little or no reinforcing will be necessary; but in places where the soil has been filled or where it is unstable, where there is a wide fluctuation of temperatures, or where stress and strain from weight occurs, as on driveways, it will be well to use reinforcement. Various aids may be employed, including steel reinforcing rods of ½-inch diameter, set 6 to 12 inches apart, and crossed by other rods at right angles. Or welded-wire mesh reinforcing, specially made for this purpose, may be used. However, a thicker slab may make the concrete fully as strong as a reinforced one (and at no greater cost) unless heavy weights may cross the concrete—such as oil delivery trucks —in which case reinforcing may be wise.

Reinforcements are usually laid on top of the first pouring, which is 3 to 4 inches in thickness; then the final pouring is made on top of it. Occasionally, where the aggregates are not large, reinforcing is placed on brickbats or low stakes which will not protrude through the concrete, and then the pouring is done through the mesh, being well tamped afterward to be sure that no air spaces remain where the concrete has not worked through the wire mesh completely. The first procedure is more to be recommended for amateur use.

MAKING FOOTINGS

Concrete footings for brick or block walls as well as for cast concrete walls are usually specified for double the width of the finished wall, to provide an adequate base for the wall. A good rule, as we have previously stated, is to make the footing twice as wide as the wall centered on it, and the depth of the footing the measurement of the width. Thus, an 8-inch wall would have a footing 8 inches thick and 16 inches wide. No forms need be built unless the soil is very sandy and loose, trenches usually being sufficiently squared off to be used for forms. Footings should be reasonably level on the top though they may be rough, because block or brick walls need a base that is level enough so they can start the ground courses at true level and maintain it to the top of the wall. It is usually recommended that footings be placed below the frost line in soil, and this is a good rule to follow in severe climates.

Where foundations are poured as well as footings, consider the pouring of an extra 4 to 6 inches of foundation in front of the wall, to be used as a mowing strip if lawn is to be brought up to the wall. This will eliminate hand cutting and trimming, and still give a neat, easily maintained edge to the lawn.

POURED CONCRETE WALLS

These walls give the real craftsman great leeway in design, for any thing which can be built in wood as a form can generally be cast in concrete. Concrete is a plastic, fluid material admirably suited to construction work in the garden.

If poured concrete walls are high—unless they are retaining walls—they may prove very heavy work for the amateur craftsman; but if they are of seat height or less they are not too ambitious a project. High walls require not only more elaborate forms and better bracing and reinforcement, but also ramps or some system to transport barrow-loads of mixed concrete up to where they can be poured into the forms. This is less of a problem where a retaining wall is being built, because the barrow can be wheeled up the hill, and planks set out across the excavation to the forms for pouring the barrow-load of concrete. Low walls require less elaborate forms and can be poured by tilting the wheelbarrow directly from the ground, or by setting a plank on a concrete block or two if a higher vantage point is needed.

Build all forms and do all excavating necessary, piling the soil nearby for refilling once the project is completed, before beginning to mix the concrete. Do not rush things. Be certain everything necessary is completed and ready before going on to the final step of mixing and pouring.

Forms may be faced with outdoor plywood or outdoor hardboard to give above-ground walls a smooth face. Where that is not desired, any lumber 1 inch or more in thickness can be used for the forms. The main thing is to use strong lumber with no knots of a size to weaken it; straight, unwarped boards are preferred. For making curved wall forms, 1" x 6"s may be used, with saw cuts every few inches halfway through the boards. For large curves, every 6 to 8 inches will do, and for small curves, every 2 to 3 inches. Wet the board thoroughly before bending and securing it to the framework. It is also possible to use a framework of 2-inch lumber cut to the curve desired and face it with plywood or hardboard, being sure that the frame is well braced to withstand the weight of the poured concrete. Outdoor plywood or hardboard may be used, the ¼-inch weight being preferred.

While pouring the wall, pour each layer over as much length as possible, making layers 6 to 8 inches deep, and tamping them into place with a board to force out air holes and compact the concrete. A smooth-surfaced wall is assured if a flat shovel is forced down between the form and the concrete to push large stones and aggregates away from the surface and into the middle of the wall. Build up the wall with these 6- to 8-inch layers and, when you reach the top, level it off and finish it with the screed board; then float and finish-trowel it.

Forms should be left in place on poured walls for at least a week after the walls are completed. Place a canvas or building paper cover over the wall so that it is protected from too-fast drying and curing. Remove it and sprinkle the wall daily with water to keep it damp through the curing period.

Expansion joints may be necessary if there is considerable variation in temperature in your climate. These will be placed vertically every ten to twelve feet. So that strength is not lost, the wall may be keyed to preserve unity, being cast with a kind of tongue-and-groove effect vertically; or it may be cast with a piece of plywood secured to the form where the joint is desired, with reinforcing rods extending from section to section through holes bored in the plywood. Rods should be placed every 3 inches in the wall's height at this point.

Reinforcing rods. Walls over 2 feet in height and less than 8 inches in thickness may need to have steel reinforcing rods inserted every 6 inches horizontally, and overlapped so that no part of the wall is weakened by having all the rods end together at one point.

If for any reason the wall cannot be finished at one pouring, shove short reinforcing rods down into the fresh concrete every foot or so, letting them project into the area of the next pourings. Brushing all rough surfaces of the previous pouring with a creamy paste of cement and water just before pouring again will also aid in making the next pouring adhere to the partly-set one beneath it. If, however, your wall is as large as such an unfinished job might indicate, perhaps you will want to look into hiring a concrete mixer or buying ready-mix concrete.

Capping.   Finished walls should be capped to prevent them from weathering and to give them a finished look. Bricks in a soldier course, thin concrete blocks, coping blocks, tiles, or wooden seats may be used to top the wall and protect it. If the wall is to have a wooden seat on it, be sure to insert the bolts in the still-wet concrete so that you can later attach the seat with ease. Bolts should be sunk at least 4 inches into the concrete and should project high enough to allow a nut to be placed on the threaded shank without its showing above the seat's surface. Holes are countersunk to accept the nuts when the seat is being made.

PLANNING FOR MOISTURE

Where a poured wall is used as a retaining wall, remember that moisture from the hillside will collect behind it and cause pressures, unless you insert weep holes every six feet or so to allow the water to drain out. Or you may insert a line of drain tiles behind the lower part of the wall to carry away excess water, arranging for it to exit at the end of the wall where it can drain away. A gutter is sometimes provided at the foot of the wall face to carry away water which may come out of weep holes or flow down from the hill over the face of the wall, preventing its running across walks, lawns, and terraces below.

Where walls may have to resist a good deal of dampness or moisture, refer to our formula chart, where you will note that a $1 : 2\frac{1}{2} : 3\frac{1}{2}$ ratio is recommended. The richer cement mixture will make a more water-resistant concrete.

Walls may be extended by casting an overlapping pier on the back side, or by drilling into the existing wall and inserting reinforcing rods to tie the two together. By painting, stuccoing, or plastering them, the extension can be made almost unnoticeable.

Speaking of plastering reminds us to cover the subject of finishing the face of the wall. Cast concrete walls may be painted with masonry paint, stuccoed, plastered, or left with the grain of the wood forms showing, if you fancy that and if they are worthy of being exposed. Sometimes striated or sandblasted plywood is used for the facing of the form to give the striated or "driftwood" texture to the finished wall. The wall may be painted if you want a colored finish. (See Chapter 11 for information about masonry paints.)

WALLS TO RETAIN TREES

With hillside sites being used more extensively every year, we find that we must grade to make level living areas, and this means that we must either destroy trees or find some way to retain them and their roots. The intelligent thing to do is to keep the trees for shade in the living area, building a retaining wall out near the perimeter of their roots. Such a wall is best kept low, about seat level or a little higher, so that less damage is done to the roots and so that the wall need not be made exceptionally strong, merely strong enough to hold the hill and to withstand pressure from roots in future years. It should be fairly waterproof, because the roots of the tree will have to be watered for several years to assist them in getting over the shock of being cut for the building of the wall, and to enable them to reëstablish themselves and send their new roots downward.

Where it is necessary to fill in around existing trees, *always be very careful.* Hundreds of trees are killed each year by people who foolishly believe that they can fill two, three, or four feet over roots of trees and still have them live. Trees *won't* accept this treatment because their feeding roots near the surface need air and moisture; when this is shut off they have no recourse: they die. If you must fill in over a tree's roots, put an 8-inch layer of heavy crushed rock down, out to the perimeter of the outermost branches; put 4 inches of gravel or sand on top of this; and then fill in with soil, leaving a well around the tree with openings into the stone and gravel layer to permit water to go out to it. Make such a well at least 4'0" in diameter (larger if possible), and try to avoid having the fill be more than 2 feet in any place.

Young trees adjust better to having roots amputated than do older ones. Therefore, if there is a choice to be made between saving a young tree or an old one for shade, choose the younger. If you must use an old tree, leave as much root anchorage as possible. Sometimes when you are digging footings you will encounter a large root below ground. While a small one may be safely cut, any large root should be saved. Dig the footings about six inches away from the root on each side to allow for expansion, running them 12 inches deeper, for about 12 to 18 inches outward, than the rest of the footing. After footings are cast and cured, lay a pre-cast window or door lintel across the root to carry the wall's weight. Allow for 2 to 6 inches of expansion room above the root, too, or it may exert pressures as it grows which will crack and upset the wall.

It is possible to resist the pressure of a hill on a retaining wall by methods of construction, too. Use vertical reinforcing rods bent to an L-shape, letting the lower angle extend out into the footing, and fastening the uprights to horizontal rods with wires. Extend the footing on the face of the wall to 8 to 12 inches beyond the face so that the "foot" will help to resist outward pressures. The face of the wall should slant back or be "battered" from the base to the top, making the wall in profile slightly wedge-shaped. About 1 to 2 inches per 2′0″ of height is a good ration for slanting the walls.

REPAIRING CONCRETE

Should concrete become damaged by winter freezes or develop cracks and chips from any other source, patch it promptly when the weather permits. This can now be done with much less work and far less trouble than was once required. New compounds with additives in them make the job easier than using conventional portland cement and sand mixtures. The quick-drying latex, vinyl, or epoxy-fortified compounds will make a stronger patch than ordinary cement-and-sand, performing in ways that it cannot, and on all but very large patching jobs will certainly be the answer to the home craftsman's prayer. They are, of course, more expensive, but their advantages far outweigh the price.

The latex type comes packaged in two components—a powdered cement and a liquid latex. Mix the two, blending them into a smooth and workable consistency according to package directions. This will harden in a half hour to an hour, so do not mix more than is needed for the minimum time span of drying. Although package mixing directions can be used in general, for various purposes you can vary it to suit the job at hand. For building out a broken corner, for instance, keep the mixture thicker so that it can be molded and will hold its form. For small chips and fine cracks, make it thinner so that it will flow easily. Latex cement can be spread as thinly as $\frac{1}{16}$ inch, may be "feathered out" on the edges to meet existing surfaces. Conventional concrete mixes must be used at least a half-inch thick and this usually means chipping and cutting out old concrete to make room for it. The thin latex mixtures are also excellent for patching hairline cracks and splendid for smoothing pitted or pockmarked concrete.

Epoxy patching cement is tough and, because epoxy is the same adhesive used in the powerful epoxy glues, it has the strongest possible bonding strength. Not only can it be used for patching, it may also be employed to set brick, tile, flagstones, and slate; for bonding glass, steel, and ceramic tile, too. It comes in kits with two bottles—one is emulsion, the other a hardener—and a bag of dry cement. The liquids are mixed according to proportions detailed on the package, then stirred into the dry cement. It may also be "feathered out" in thin layers so that there is no need to chop out concrete or to prepare it in any way other than to clean it well with a stiff brush of all dirt, and remove grease or oil from the surfaces before beginning to patch. Be sure to fill all cracks and depressions well and tightly pack them to insure a good bonding.

Vinyl patching cement comes already mixed in a dry compound. All that is necessary is to add water and mix well, then use it. Like latex cement, a layer as thin as ⅛ inch may be applied, making it ideal for pockmarked or rough concrete. It has greater adhesive strength than conventional sand-and-cement mixtures and it can be used to set brick, tile, stone, and even glass. It is not affected by winter freezes and thaws in cold climates, and it resists chipping and cracking.

Remember that all of these compounds set very quickly, so that no more than the amount that can be used in something under a half hour should be mixed at one time. Divide larger jobs and do one part with the first batch, then mix another and patch the next part.

For large patch jobs, particularly where a section must be replaced entirely, the conventional concrete and sand mixes may be used, and probably should be, because of the price difference and also because of the quick-drying qualities of the additives in the patching cements.

These, then, are the essentials of concrete work. We have tried to delineate as well as we could for the amateur craftsman the procedures and the methods to employ; and we have tried also to show him what a wide and exciting range of possibilities there are for the craftsman to explore. We hope that you will enjoy your concrete constructions; that you will integrate them well into the garden scheme; and that they will make your outdoor living more practical and more enjoyable.

15

Building Garden Steps

A large part of the charm of the old Italian and French gardens lies in their use of steps. The change of level in Italy is made mandatory by the landscape in which hills abound; but, more frequently than not, in France the use of steps seems to stem from the aesthetic theory which is so beautifully displayed in her great formal gardens.

In our country today we are using more and more hillside sites for our homes each year, carving out terraces from the slopes with bulldozers and leveling our living areas into terraces which step down from the house, from the street, and from terrace to terrace as we need more land for outdoor living. This comes from a practical need, but there is an aesthetic theory which can be applied, too, for the two are not incompatible. For instance, in the new trend in American gardens we see many more hillside plots left as Nature made them, with wild trees and shrubs kept and others planted to complement them, and with fewer hillside lawns to create problems of maintenance. It is only the living areas around the houses which have been leveled for terraces and living lawn areas. Paths thread downward through the natural parts of a hillside, with a few unobtrusive steps inserted here and there when the slope becomes too steep for a path to negotiate easily. Sometimes the paths are a series of very wide steps, a kind of ramp broken here and there by risers.

In more formal treatments, steps are used to climb up to terraces raised above the level of the house or the garden, or to descend to those placed below it. Little retaining walls are built to divide the garden into two levels, both charming the eye and giving the illusion of more space, because the division will lend a visual interest to the whole garden picture.

Steps, steps, and still more steps . . . that seems to be the order of the day in American garden design, and the choice of kinds of steps to use in your own garden has never been wider. Ramp steps for climbing long slopes gently, with the minimum effort; circular steps; free-form steps with flowing curves; wooden steps to wooden-decked terraces; water-washed stone steps laid to simulate a natural outcropping of rocks; brick-and-plank steps; cast concrete steps; concrete-and-plank or concrete-and-brick steps; cut-stone steps; steps made of old railroad ties . . . the list is practically endless. Every day more new and exciting ways to use building materials are found, ringing new changes on the classic combinations and making wonderful new combinations of materials to suit our changing styles of houses and gardens.

RELATIONSHIP OF RISERS TO TREADS

Risers	Treads
2″–3″	20″–24″
4″	19″–20″
4½″	18″–19″
5″	17″–18″
5½″	16″–17″
6″	15″–16″
6½″	13½″–14″
7″	12″ (11″ min.)

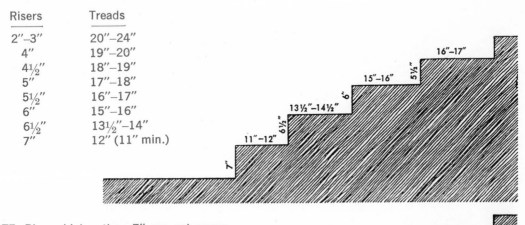

NOTE: Risers higher than 7″ are not recommended. Treads wider than 24″ may use risers from 3″–7″. Lawn planted treads should be at least 20″ wide.

Exposed-aggregate concrete, boxed by broad planks of treated wood, create a formal-informal atmosphere that suits the modern house. Plantings in soil spaces beside steps and terrace help to soften still further this geometric effect, making it homelike and human.

Heavy steps lead outward from a wooden deck whose floorboards lead diagonally to the stairs. Note that the wide redwood planks on steps are separated by narrow redwood boards, same width as those used on the deck. The deck is edged with heavy wood, of same dimensions as edges of the steps.

Even a stairway can be dramatic, making its emphatic statement in the general garden scheme. These cast-in-place steps have planting space left at the end of every other step, making a place for a fine collection of sun-loving plants needing little water.

PLANNING YOUR STEPS

Granting the need for steps, the next question to be asked is, what kind of steps? How many, and where shall they be placed?

The shortest distance between two points is a straight line, but when we build steps we may find that this is not the safest way, the most comfortable, or the most satisfying in design. Steep runs of steps are frequently dangerous. They are always less comfortable to ascend and are perilous to come down. There are certain primary proportions between the width of the tread (that part on which you walk) and the riser (the height of the front part of the step). Let us look for a moment at the chart giving measurements that will give you the best proportions to use in planning your steps, assuring you of a good, easy climb whether the slope is a long gradual one or shorter and steeper.

There have been tests made which show that broad treads and high risers are just as tiring and frequently as hazardous to use as the commonly-acknowledged dangerous combination of high risers and narrow treads. In general we might say the rule is that the closer to a ladder steps become, the more effort must be put forth in climbing them, and that the closer they come to the level or like a ramp, the less effort will be expended in negotiating the slope.

But the width of steps is important, too, though the "elbow room" factor is often ignored in building steps. Where they must pierce a retaining wall or be built beside a wall, thus necessitating a bannister, the width from side to side becomes important. We have prepared a chart which shows the minimum widths for steps to permit the passage of one or two persons, and you can judge proportions for wider sets of steps.

Don't forget that garden steps must be at least as wide as stairs in a house, for gardeners are constantly toting flats of plants and other garden equipment up and down them. Lawn mowers, garden carts, and other wheeled equipment must frequently negotiate garden steps in order to be used on the various levels. Perhaps you might solve this traffic problem by building ramps of concrete, brick, or stone (whichever material faces the steps) alongside the steps for the gardener himself. Or if expense or lack of space rules this out, wooden ramps built to fit the steps can be put in place when needed and taken up and stored in the garage or tool shed in between uses.

RECOMMENDED DIMENSIONS FOR GARDEN STEPS

Minimum width of steps for one person is 2'6" but 3'6" is better, allowing for carrying things or on a long flight of enclosed steps allowing use of a handrail placed on wall. Minimum step width for two persons is 4'6" but 5'0" is better, particularly on stairways enclosed on both sides. More than four steps may require handrail, fewer usually don't.

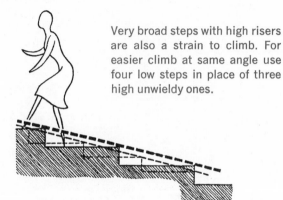

Very broad steps with high risers are also a strain to climb. For easier climb at same angle use four low steps in place of three high unwieldy ones.

High risers and narrow treads make climbing more difficult and dangerous as well.

MATERIALS

The choice of material will be governed by many factors. First of all the style of house and garden where the steps are to be built must be considered. The materials must be compatible with those used in the retaining walls, house walls, terrace paving, walks, and other adjacent features. If the general effect desired in the garden is natural and informal, the steps may be built of rough stone brought in from fields and hills nearby or bought from stone suppliers. In this case, do not use too many different kinds of stone or too many varieties of texture, or you will end up with a little bit of masonry which looks like an exhibit in the Natural History Museum's geology section, rather than functional steps which keep their place because they are simple, quiet, and *useful*.

Stone steps may be set either in the soil—with gravel or cinders providing a drainage layer beneath them so that they will remain where they are set—or they may be laid in informally formal patterns such as those seen in "Steps With Opposing Circles" in the sketches we present. These are usually best laid in concrete and mortared in place. You may leave little pockets of soil here and there in which to plant small flowering plants or living material of other kinds. In Wales, at the estate of Lord Aberconway, "Bodnant," we saw some circular steps in which a few inches of soil had been left at the rear of each step except the top one. All of these were planted with rings of tiny violas and the riser of each step formed the background for a mass of lavender bloom. This charming planting idea may be adapted to your own garden, using annuals, perennials, or succulent plants.

Where ramp steps or short runs of steps are inserted in a long walk up a slope having sharp breaks, it is best to use the same material for the treads as that found in the path in which they occur. Blacktop, gravel, crushed bluestone, marble chips, rough stone or concrete, brick or lawn—whatever the walk material may be, it can be adapted to the steps. Risers should also be compatible: wood, stone, brick, or any of the materials which are shown on the pages of drawings and plans.

Occasionally where long walks—particularly straight ones—must be used, with steps occurring only occasionally, a change of pace and texture will add interest to an otherwise commonplace and rather tiresome walk. Our circular steps which are entitled "Break Straight Lines with Circles" are a case in point. If still more interest is needed or desirable, bricks or stones may be set into the concrete steps in radiating patterns as shown.

A dash of spice or a pleasing bit of individuality may be given to a garden which is otherwise rather usual by angling the steps, turning the flight at a slight angle or even making a 90° turn. Look at the sketch called "Steps With Variety and Unity," which demonstrates how low steps can take a turn for the better in a small garden. In "Steps Can Be a Feature of the Garden" we show how to avoid making a too-long flight of steps where a considerable distance must be covered due to a sharp change of level. By inserting two landings, varying the color of the cast concrete with insertions of brick or tile paving on the landings, and making the steps still more individual by extending the risers and planting the soil boxes which end each tread with flowering or other colorful plant material, one can make a real feature out of the necessity of steps. Also, by breaking the retaining walls into sections, we avoid a high, blank, rather ugly expanse of masonry. The trick of extending the risers and planting the treads may be adapted to most of the other steps we show to give them a fresh, unusual, modern look. Or you may want to plant ivy alongside the steps, training it across the risers to keep them green throughout the year. An occasional nipping back of stray shoots will keep the ivy within bounds with very little work.

PRACTICAL CONSIDERATIONS

The practical side of building steps must not be ignored. We advocate the use of adequate foundations and footings to insure the permanence of your work. A deep footing beneath the bottom step will materially help to prevent the weight of the concrete and masonry mass from sliding down the hill, as may be the case if concrete is poured into an excavation which merely parallels the land's slope without the use of this "slide stopper." Where steps are laid on soil, without concrete, use gravel or cinders as advocated above in this text, and be sure that drainage away from this layer is also provided, if possible. Such steps should be slightly overlapped so that the lower step assists in supporting the one above.

Where steps are planned to be finished and to tie in at the specific level of a wall or to match the level of an existing walk, be sure when you are planning the concrete foundations to allow for the depth of the paving or tread-finishing material plus the mortar joint. Then when the steps are completed you will have no hiatus between levels of step and adjoining wall or walk.

There are so many places in small gardens in which a few steps will make all the difference in the world, not only in comfort and convenience but also in visual effect. We urge you to look over your garden and see where steps might fit into your landscape plan. If you decide that there is a valid use for them, study the examples we offer in our sketches and plans and choose the one most suited to your purpose. But don't rush into construction until you've thoroughly studied all aspects of the problem. Make sure that your choice is right and that you know just what to do. Be sure to make the excavations for drainage and footings adequate, put the drainage layer in place, build the forms for the concrete, and then begin to move fast, once everything is in readiness.

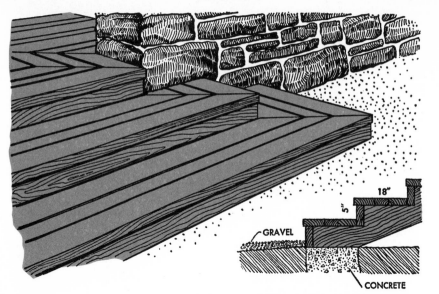

WOOD AND STONE COMPLEMENT EACH OTHER

The crisp lines of planed wood against the natural informal lines of rubble stone laid in a low wall above a gravel terrace will give a visual contrast to the garden composition. Steps should be set on a concrete footing to prevent sagging; a bolt anchors them against downhill thrust. Leave 1/2″ between boards, slope steps to insure drainage.

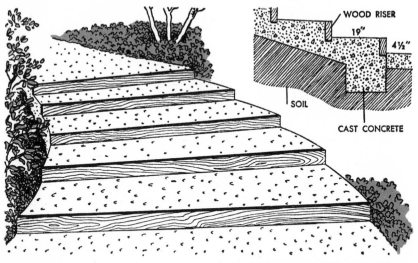

WITH STEPS, EASY DOES IT

Let the slope of the land govern the kind of steps you choose. Use steeper, narrower ones where slope breaks abruptly, curving broader steps with low risers for gentle slopes. Set treated redwood risers in concrete steps for permanence. Always use a deep footing at bottom step (see sketch) to prevent weight from thrusting steps downhill.

314

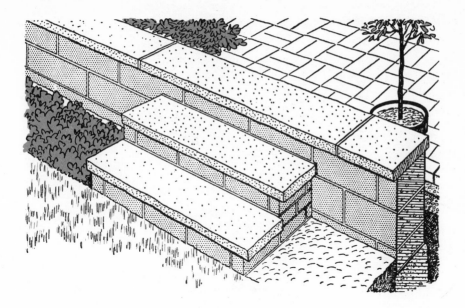

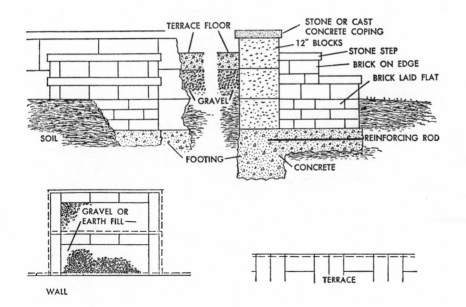

TERRACE FLOOR

STONE OR CAST
CONCRETE COPING

12" BLOCKS

STONE STEP

BRICK ON EDGE

BRICK LAID FLAT

GRAVEL

SOIL

REINFORCING ROD

FOOTING

CONCRETE

GRAVEL OR
EARTH FILL—

WALL

TERRACE

A SECRET EXIT FROM A TERRACE

By using a wall one step high around a terrace which is only slightly above the lower level, a "secret" stair is possible. On the terrace side it presents the look of an unbroken wall, the location of the steps being defined by a potted plant or a flower bed let into the paving, while on the garden side steps ascend between flower or shrub beds.

BREAK STRAIGHT LINES WITH CIRCLES

Relieve the monotony of a long, straight walk by inserting circular steps. Used in those shown here are risers of cinder or concrete blocks mortared to a concrete base before being filled in with brushed aggregate concrete, which is also the material used in the brick-edged walk. Bricks set in a radiating pattern in steps lend interest.

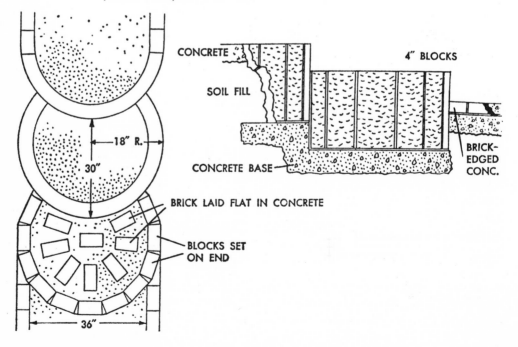

18" R.

30"

CONCRETE

SOIL FILL

4" BLOCKS

CONCRETE BASE

BRICK-EDGED CONC.

BRICK LAID FLAT IN CONCRETE

BLOCKS SET ON END

36"

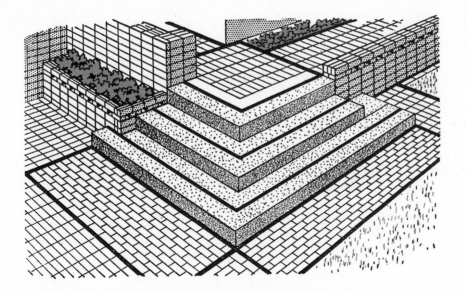

CORNER STEPS ARE INTERESTING

Descending from a walk which goes around the corner of this house, corner steps give easy access to the lower terrace and its adjacent lawn. A raised plant bed in the angle of the terrace echoes the pierced brick retaining wall—note stacked bond—and brick paving areas are defined by preservative-treated wood also set in concrete steps.

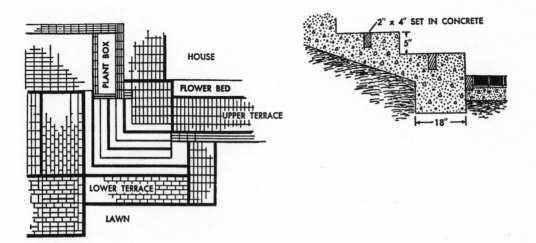

STEPS WITH OPPOSING CIRCLES

Circular steps go particularly well with our formal or informal traditional architecture. Any material may be used to construct them, but field stone laid dry as shown here, is informal and small plants can be grown in the cracks between stones. For the best visual effect use fewer steps above the circular landing than those descending from it.

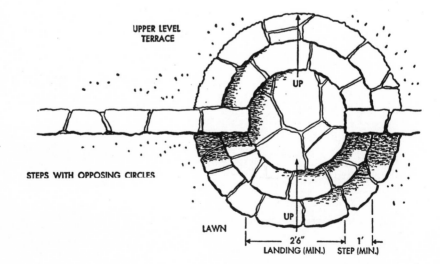

UPPER LEVEL
TERRACE

UP

STEPS WITH OPPOSING CIRCLES

UP

LAWN

2'6"
LANDING (MIN.)

1'
STEP (MIN.)

STEPS WITH VARIETY AND UNITY

In sections where ledge rock is plentiful or quantities of other masonry materials are easily available, the garden will gain in interest from changing levels of plant beds. Vary the widths of the beds and the heights of walls but keep unity by using only one kind of material. A flat platform by the steps encourages the use of colorful potted plants.

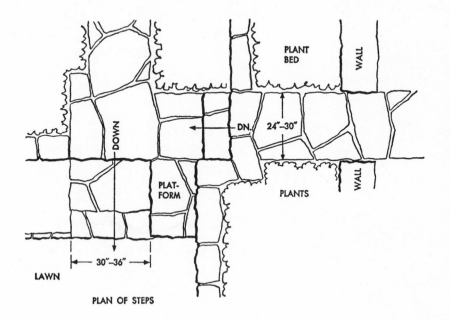

PLAN OF STEPS

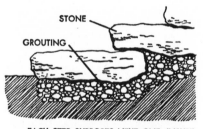

**EACH STEP SUPPORTS NEXT ONE. INSURE
DRAINAGE BY USING GROUTING BELOW**

STONE

GROUTING

MORE OPPOSING CIRCLES

Here opposing circles of stone
are used in a wild garden with
a stretch of sloping path for a
landing. Top steps bow toward
us, the wider, shallower bottom
ones bow away from us. Because
these steps are informal, they
may be adapted to either traditional cottages in a natural
setting or modern houses which
fit well in natural landscapes.

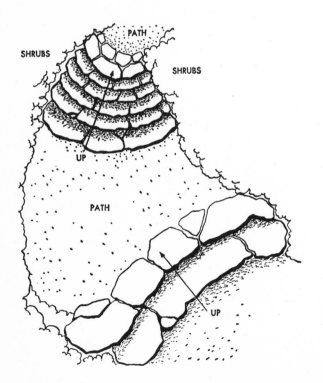

SHRUBS

PATH

SHRUBS

UP

PATH

UP

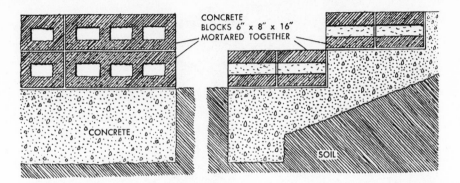

CONCRETE BLOCKS 6" x 8" x 16" MORTARED TOGETHER

CONCRETE

SOIL

WIDE STEPS WITH CHARACTER

Concrete blocks are long-wearing and make steps that are interesting in texture and as unusual and unique as can be. Laid on a concrete base with a deeper footing in front (to prevent down-thrust of gravity from dislodging them) they lead from level to level with dignity and character. The holes in the front may be left open, or if you wish, fill with pebble patterns (see Pebble Mosaic).

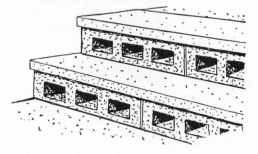

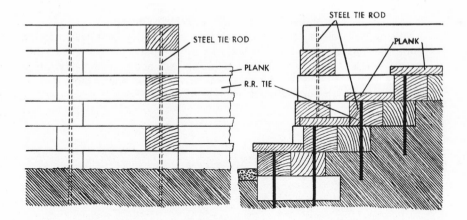

STEEL TIE ROD

PLANK

R.R. TIE

STEEL TIE ROD

PLANK

TIES THAT BIND WALLS AND STEPS

Old railroad ties can frequently be had for the hauling. Originally impregnated with a preservative, they usually will last several more years in your garden. To make a wall, lay them up tier after tier, stepping back slightly as shown, or keep perpendicular. To prevent frost's thrusting them out of line, bore holes and insert steel tie rods. Use 2" planks to surface steps and prevent wear.

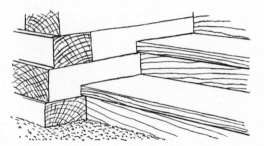

WOODEN PLANKS TEAM WITH BRICKS

Brick steps which echo the design of the walk are held in place by 2″ x 6″ planks of redwood or some other preservative-treated wood that is stained or painted, unless a weathered wood effect is desired. Bricks may be mortared on concrete base (see sketch) or laid without mortar on bed of tamped cinders or gravel. Distance A equals brick height.

BALANCED BUT NOT BISYMMETRICAL

Retaining walls beside steps need not have exactly the same treatment on both sides. As shown here, one wall can curve around to hold a plant bed on the uphill side, lower side has retaining wall above plant beds. Steps may be placed at the end of retaining wall or along it wherever there is some logical architectural or garden division to follow.

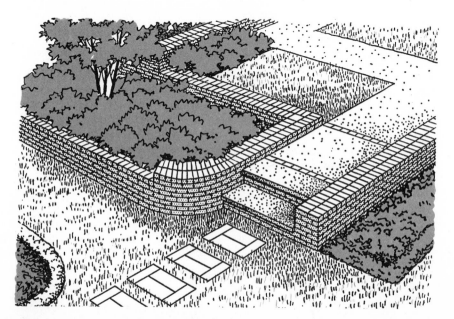

USE RAMP STEPS FOR LONG SLOPES

Broad steps with low risers may be the answer on long slopes where paths
are too steep for an easy climb. Break long runs of steps with a level landing
or perhaps a rustic seat to rest on. Make steps 2'6" wide with 4"–7" risers;
floor them with sand, blacktop, concrete, bricks, gravel or wood, according
to use, form and cost.

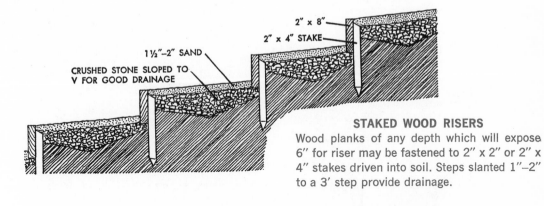

2" x 8"

2" x 4" STAKE

1½"–2" SAND

CRUSHED STONE SLOPED TO
V FOR GOOD DRAINAGE

STAKED WOOD RISERS

Wood planks of any depth which will expose
6" for riser may be fastened to 2" x 2" or 2" x
4" stakes driven into soil. Steps slanted 1"–2"
to a 3' step provide drainage.

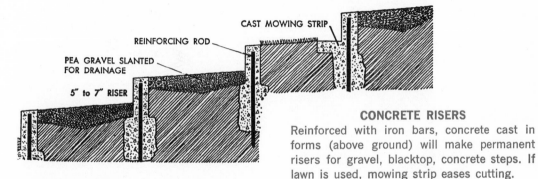

CAST MOWING STRIP

REINFORCING ROD

PEA GRAVEL SLANTED
FOR DRAINAGE

5" to 7" RISER

CONCRETE RISERS

Reinforced with iron bars, concrete cast in
forms (above ground) will make permanent
risers for gravel, blacktop, concrete steps. If
lawn is used, mowing strip eases cutting.

STEPS

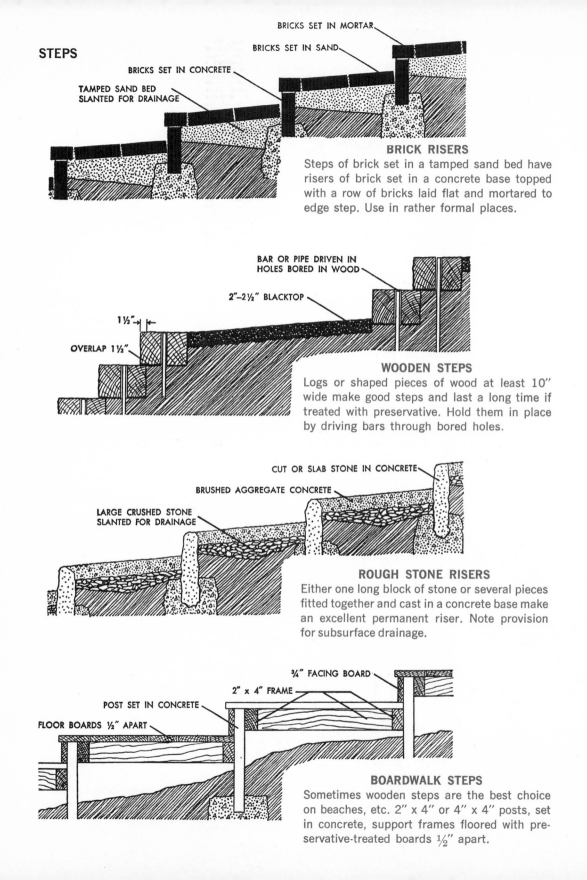

BRICKS SET IN MORTAR

BRICKS SET IN SAND

BRICKS SET IN CONCRETE

TAMPED SAND BED
SLANTED FOR DRAINAGE

BRICK RISERS

Steps of brick set in a tamped sand bed have risers of brick set in a concrete base topped with a row of bricks laid flat and mortared to edge step. Use in rather formal places.

BAR OR PIPE DRIVEN IN
HOLES BORED IN WOOD

2"–2½" BLACKTOP

1½"

OVERLAP 1½"

WOODEN STEPS

Logs or shaped pieces of wood at least 10" wide make good steps and last a long time if treated with preservative. Hold them in place by driving bars through bored holes.

CUT OR SLAB STONE IN CONCRETE

BRUSHED AGGREGATE CONCRETE

LARGE CRUSHED STONE
SLANTED FOR DRAINAGE

ROUGH STONE RISERS

Either one long block of stone or several pieces fitted together and cast in a concrete base make an excellent permanent riser. Note provision for subsurface drainage.

¾" FACING BOARD

2" x 4" FRAME

POST SET IN CONCRETE

FLOOR BOARDS ½" APART

BOARDWALK STEPS

Sometimes wooden steps are the best choice on beaches, etc. 2" x 4" or 4" x 4" posts, set in concrete, support frames floored with preservative-treated boards ½" apart.

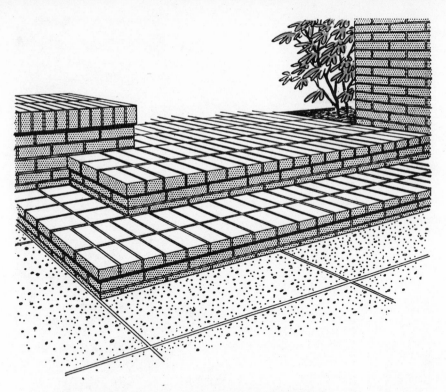

LOW, BROAD STEPS OFFER A WELCOME

Today, houses often are built on sloping land, the backyard being higher than the rear terrace by the house; or possibly the front door is placed a little below street level so that the rear terrain may be used to better advantage. In most cases, broad, low steps are the most inviting, making the transition between levels agreeable, easy.

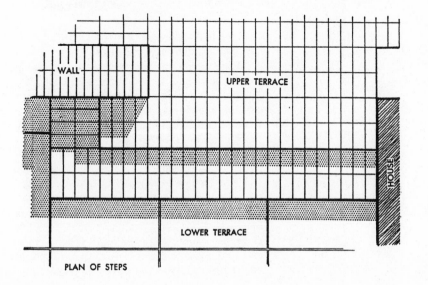

WALL

UPPER TERRACE

HOUSE

LOWER TERRACE

PLAN OF STEPS

STEPS

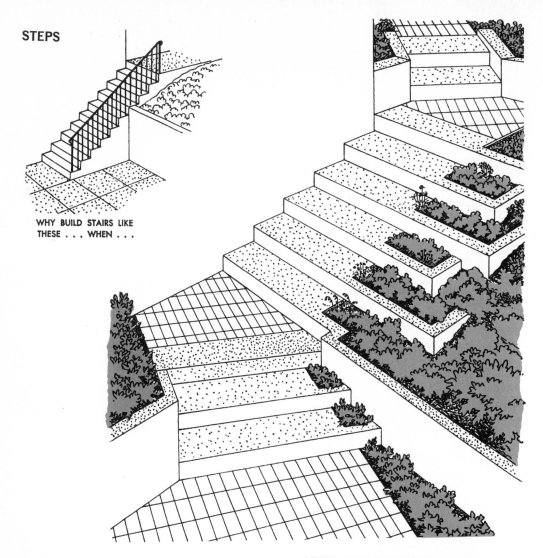

WHY BUILD STAIRS LIKE
THESE . . . WHEN . . .

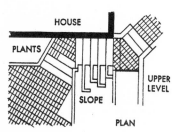

HOUSE

PLANTS

SLOPE

UPPER
LEVEL

PLAN

326

STEPS CAN BE THE FEATURE
OF THE GARDEN

A straight line is the shortest distance between
two points—and sometimes the dullest, too.
Steps needn't always follow the stiff, straight
side of a building; instead, give them more
visual charm and make them easier to climb,
too, by turning them at landings placed where
they'll give a rest space for climbers. Widen the
treads a little, reduce riser height a bit, and
you can dispense with a handrail. Two steps
placed above the corner of the house will gain
space for the lower terrace. Cut down the
height of the retaining wall so it won't be grim,
and plant the resulting bank with ground cover
and add some planting boxes to the end of
each step. You'll have a local showpiece!

16

Planning a Terrace

A terrace, according to the traditional definition, is a paved, usually unroofed area, either adjacent to the house or placed elsewhere in the grounds wherever its use would dictate a suitable location. Many terraces are raised above the level of the surrounding garden, but not all are, or need be. Of recent years such paved areas have been called "patios" but properly speaking (according to its Spanish or Spanish-American origin) a patio is either a paved courtyard, entirely enclosed on all four sides by the house, or a partial courtyard, enclosed on two or three sides with the remaining side or sides bounded by a high garden wall, thus ensuring privacy and protection during use. In this chapter, as elsewhere in the book, the outdoor living area will be termed a terrace, in the traditional meaning of the word.

There are many people who choose their terraces merely by thumbing through glossy magazines. When such a person finds a glamorous picture, perhaps something planned for quite a different kind of climate and different sort of house, he thinks, "That's pretty! That's for me," and then proceeds to have some version of that terrace built in his garden.

But a terrace is far too expensive and permanent to be chosen so lightly or so frivolously. There are a good many practical considerations which enter into the choice; they should be taken into account and enter into the planning *before* the terrace is built, and not afterward when regrets set in.

Among the factors which should come under scrutiny are:

- What is your climate?
- What size is your family?
- What are their interests and inclinations? Their ages?
- What do you expect as to durability, function, future use?
- What can you spend for a terrace?

This latter factor is a vital one for it is the kingpin, oftentimes, of the whole matter. The pocketbook usually settles a good many things about our way of life, and this is no exception. A terrace should not be too elaborate for its surroundings. It should not overpower its setting, but it should, on the other hand, be adequate. It is an outdoor room and as such it is "capital equipment" which will add to the value of your property if you ever wish to sell it, provided that it is properly planned and durably built. Also, sometimes it is possible to start out with a small terrace and a big plan, adding pieces each year until it is the size you really want it to be. Perhaps it can start out as a gravelled terrace and then, when it is possible, concrete, stone, or brick can be laid on the gravel. If you want to do it or have it done all at once, you may want to take out a loan and spread the payments over several years. If you do it yourself, adding to it year by year, you'll have the use of the terrace as it expands; you will spread the work over several years and the cost, too, thus saving the interest charges on the loan.

First of all, a terrace is for use. What do you want from your terrace? Is it to be an outdoor room where you will entertain, eat, sit, and where the children can play in clement weather? What time of day will you be using it most? If it is to be used mostly, or frequently, at night, can it be lighted? If it is to be used a good deal for eating, how close to the kitchen is it; or do you want to put cooking facilities beside it?

If you have children, the terrace is a good play-place for them. It may be that it should be placed adjacent to the part of the house where the mother is most likely to be at work, so that she can supervise and watch their play. Perhaps you can satisfy two requirements, that of eating outdoors and that of supervised play, by placing the terrace adjacent to the kitchen, where food can be passed outdoors from a window or from a convenient door to be served before it cools.

Should you wish to use the terrace for evening entertainment or for sitting outdoors to watch television after the children and older members of the family have retired, the terrace is best placed away from the

WAYS TO ACHIEVE PRIVACY ON SMALL PROPERTIES

In many neighborhoods it is difficult to get privacy because houses are close together and because upstairs windows overlook living areas of neighborhood gardens. By careful planning, by taking into account all of the factors involved, it is possible to assure complete protection from all angles. In sketch below are shown several methods. The fence which surrounds the front garden has a baffle entrance rather than a gate, thus assuring protection of the living terrace from observation by passersby as well as from adjacent properties. The terrace shelter is partly roofed over with slats to give summer shade as well as protection from eyes on upper floor of house next door. Black arrows show sight lines.

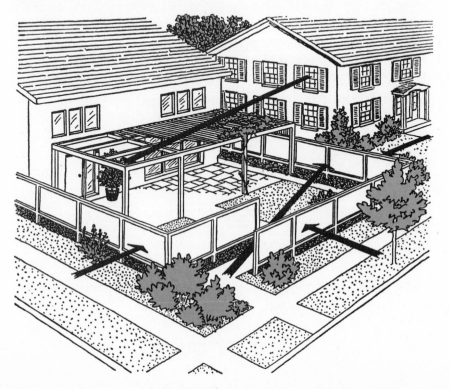

sleeping rooms, so that the noise of conversation or entertainment and the lighting of the terrace will not disturb the sleepers. Alongside a detached garage, or perhaps an attached one which extends away from the sleeping rooms, would be a good location. This would give a chance for plantings near the house, as well. Some people prefer to divorce the terrace from the house and place it in a corner of the property, using high fences to give it privacy; while others prefer to have the terrace attached to the house so that there is an easy flow of living into and out of the dwelling.

The weather and the sun must also be taken into consideration when you are planning your location for the terrace. Remember that the afternoon sun is hottest about three o'clock, and that a terrace on the southwest or west of the house is likely to be unpleasantly hot during the afternoon—even into the early evening if it is paved—and it retains the heat unless it is very well shaded. Even if it is shaded, it is usually hotter during those hours of the late afternoon. Hence, many people today are placing their terraces on the east or southeast side of their houses, even though that side may face the street or be close to a neighbor's house. Devices such as privacy fences and baffles will protect them from passersby and from the neighbors.

On the other hand, if the only logical place for your terrace is the southwest or west, making use of a large existing tree to shade it or planting a good-sized one to grow up and give you shade during the hottest hours may be the answer. Until trees get large enough, and even afterward, annual vines on a framework covered with trellis slats will give enough shade to make the terrace pleasant. When the trees get large enough to shade properly, the trellis slats may be removed and the vines dispensed with. You may leave the trellis framework, or take it down, as you wish. It is also possible to use roll-up materials for temporary shading. Snow-fencing, bamboo or basswood porch shades, and other materials can be rolled up and stored indoors during the winter, when any light and heat will be welcomed indoors and on the terrace. They can be lashed to the framework during summer so that winds and storms will not dislodge them.

Take into consideration the prevailing winds, too. In many places a cold wind, a strong wind, or a hot, dry wind may make the vulnerable side of the house less desirable than the side which is sheltered. It is sometimes the case that this windy side is the only possible location for the terrace, due to the layout of the grounds. In this case the terrace can be placed on this side and wind barriers erected in the form of baffle fences; louvred fences; fences with plate glass inserted where there is a view to be preserved but a wind problem to be overcome; or fences with wind-directing tops, which will cause the wind to leap over the terrace before sweeping to the ground again. Plantings, too, may be installed, with evergreens preferred, but with quicker-growing deciduous material also a good possibility. This living wind-break will be very effective when it has grown to proportions large enough to prevent the wind from sweeping over the terrace.

Is your climate variable? If sudden showers are likely to come up without warning, perhaps a covered terrace or a partially-covered one may be the answer to this problem. Terrace furniture can be quickly wheeled or carried under cover and, when the shower is over, brought out again for use. If your climate is cold, it is possible to tuck a terrace in the angle of two walls which face the sun, so that you will have a protected warm spot in which to sit early in the spring and late in the fall, when other parts of the property may be found to be too cold for comfort. This brings us to another point which you may wish to consider.

MULTIPLE TERRACES

A trend of the past few years is toward more than one terrace on the home grounds; in view of the points brought out in the discussion above you can see why this has come about. If the property is large enough or if the pocketbook can stand it, several terraces would seem to be the best possible answer for all the questions brought up.

A terrace on the eastern side of the house is pleasant for breakfasts before the sun gets up toward its zenith and becomes too hot for comfort, and the same terrace can be used for cocktails in the afternoon and for evening entertainment. Terraces on the west or northwest can be used for morning, noon, and evening entertaining. They are also pleasant in the off-seasons when a bit of warmth in the afternoon is welcome, provided that wind does not interfere with the pleasure of sitting out. A small terrace outside a bedroom is an agreeable place in which to sit in dishabille or even to sleep, privacy being maintained by a fence. In the case of an older member of the family, such a terrace outside his room would provide a pleasant refuge when the younger members of the family might be monopolizing the main terrace for their own entertaining. It would also give him a place in which to entertain his own friends in peace and privacy. Parents of teen-agers who allow their children to entertain extensively at home will appreciate such a refuge during the green season when the youngsters are in possession of the main terrace.

Multiple terraces extend the pleasures of living and give a richer expression to healthful outdoor life. Having the childrens' sand box next to the terrace makes it a play-place for them during the day; the terrace can be taken over by the parents at night. When the children grow up, the sand box can become a plant box or a pool, or the area be turned into a flower bed. One terrace may be a covered one or have an enclosed porch or a garden house placed alongside it; while another terrace may be left open to the sun to give you a choice of sitting-out places for all kinds of weather, as well as all times of day.

All of these considerations and any others which occur to you should be carefully evaluated and worked out so that you can make the most intelligent choice possible. Probably you will have to compromise somewhere along the line, for practical considerations must always shape the ideal and will influence the placement, the construction, and the embellishment of the terrace.

TERRACES ON PROBLEM SITES

Not so long ago terraces were never built on anything but level, flat land, and, if a house had to be built on sloping land, terraces might be considered but they would probably not be built, due to the expense of excavation. Today, however, terraces can be built anywhere and everywhere, even on very sharply sloping hillsides where houses have been perched to take advantage of the view. In such extreme cases the terrace may become a wooden deck built on posts and joists and fitted around existing trees whose branches have been trimmed to the proper level, thus taking advantage of them for shade while opening up enough space for use around them. If the trees are surrounded by the deck on a very high level, sufficient space must be left open around the trunks so that they can sway with the wind without endangering the terrace.

On less violent slopes, a terrace can be made by excavating with a bulldozer or by hand (depending upon the extent of the work and the energy of the builder), the excavated soil being used as fill on the downhill side, probably behind a retaining wall. The fill should be well soaked with water, tamped, and allowed to settle for some time before it is finally paved, to avoid cracks and settling. If the uphill side is more than a foot or so high, it may also require a retaining wall where the excavation has taken place. If there are existing trees on the site the retaining wall may be curved, square, or triangular shaped to save the trees and contain the important roots. It is dangerous to fill in around a tree more than a few inches, for roots need air as well as water. The fill may cut these off and smother and kill the tree. Thus, if your tree is on the downhill side, it would be well to curve the retaining wall around it to leave four or five feet of clear soil at the original level about the trunk.

Very frequently such retaining walls can be the real feature of a terrace rather than just a regrettable necessity. If the wall is a low one, it can be utilized as extra seating space for parties and picnic luncheons for large groups. A pool or a raised flower bed can also add to the interest and beauty of a terrace retaining wall when incorporated in its structure. Often two retaining walls—one on the edge of the terrace kept low for seating purposes and to contain a flower bed, and another one higher up on the hill to hold back the rest of the bank—will be a better solution than one high wall would have been. For one thing, the pressures will be distributed over two walls instead of concentrated on one; and for another, the wall will look less forbidding and be less of a problem to plant and make attractive.

The vogue for houses of the split-level, tri-level, or whatever they may be called, variety seen everywhere nowadays brings in new problems of terrace placement, but offers excellent opportunities for using a terrace to tie the house to its site and keep it from looking peculiar, as so many houses of this type seem to look. The terrace can be built to give living space outside the living room on a raised, intermediate plane, tying in the wing and its two stories with the other section of the house; bringing them into scale with each other; thus making them more interesting architecturally, a function which was performed to a certain degree by the porch on Victorian houses.

Speaking of old houses and porches, many an old house can be made to look up-to-date and to perform today's functions well if the porch is removed and a terrace built to replace it. Most old porches are too narrow and skimpy for today's uses and they are seldom well planted or screened for privacy. Sometimes the porch can be left, if it is in good condition and looks well on the house, with the terrace built out in front of it and at its level, the floor being of wood if the porch is constructed of that, the joists being set on a masonry foundation. This will not only make it possible to enjoy your home more, but it will add to its value and enhance its saleability should you ever wish to dispose of it.

Frequently a terrace with its fence, its trellis, or view-breaking structures can integrate an unrelated pair of structures, such as a house and a detached garage, or a house and a tool shed or other building. The breezeways, so popular in recent times, can also be brought into use by adding terraces to them. Too small to be used as a sitting place and too wide to be considered merely a passageway from house to garage, the breezeway becomes a real feature when it includes a terrace, and has a function beyond that of connecting the house and garage.

HOW TO LAY OUT YOUR TERRACE

Once you have determined what your problem is, you are on the way to solving it. Pace off and measure the area you have decided to use for the terrace. Perhaps if you drive stakes on its perimeter and run a cord around them you can visualize better what the area is; assess its qualities and its drawbacks. Outdoor furniture can be placed in position within this area, or boxes and other makeshifts if you don't yet own the furniture you'll be using on the terrace. Test it to see if the furniture is out of the way of traffic from the house across the terrace. See if it is large enough for the uses you plan to make of it, and if its shape and outlines are pleasant to view from the house as well as from the terrace itself. Then, if you want to make revisions in size, shape, and even in placement, you can do so without any more trouble than taking up the stakes and trying them elsewhere.

When finally, you have come to the conclusion that you have the ideal solution to the problem, it will then be possible to begin the excavation, the building of retaining walls, the laying of the drainage beds, and the other preliminary operations, and eventually to finish the terrace. These preliminary skirmishes with the problem may sound a little unnecessary; but if you have any confusion of mind, they are the ideal way of resolving and clarifying anything which has bothered you.

WHAT MATERIALS TO USE

The determining factor of what materials to use in paving the terrace may be their expense, but other factors should not be ignored. The suitability of the materials, their color, texture, and durability should also enter into the choice. What you use for retaining walls, what kind of fencing and shelter or trellis are also factors to take into account.

Paving.   There is such a wide choice of materials here in such an array of colors and textures that all you must determine is which is best and what you can afford to pay. If you feel that you do not want to go into permanent paving, you can employ gravel, bluestone, marble chips, brick dust, granite chips, pine needles, tanbark, or choose from many other substances. All have their merits and most of them are

Even poured-concrete terraces need not be dull. In fact, amateurs may find it expedient both in time and for the budget to cast a block or two at a time, lifting the forms to use for the next squares or oblongs of varying sizes. Grass between makes a pleasing pattern.

A carport-storage project offers a terrace partly covered, partly open to sun. The quickly cleanable cast-concrete carport floor is balanced by a concrete terrace paving with 2" x 4" cedar screeds set in 4' x 4' squares. Access walk encloses flowers.

Set inside a large area of brushed aggregate concrete paving, a smooth-finished concrete "rug" may be used for dining out or for a pleasant place to sit under shelter in the evening. If finish is really smooth, it can be waxed and used for dancing.

Contrasting lightly-textured terrace paving and steps with raked-joint retaining wall is an effective means of using textures in masonry. Broad, easy steps lead to landing and carry down to the lower level from there. Note "weep-hole" drain set in the wall.

relatively cheap and widely available. A well-kept lawn may be the answer in some places, although it will require constant repair and upkeep due to the wear it will receive, Also, chairs and tables will punch holes in turf after wet weather, and it will be less possible to use a lawn during rainy weather than would be the case with quick draining, quick drying stone materials.

In modern use, these loose materials are usually contained within areas defined by wood boards, masonry, concrete edgings, or metal lawn edgings. They have many advantages; but the main disadvantage is that they are not permanent. Weeds root among the particles and require constant vigilance; children love to play with gravel and other stone chips, throwing it at each other or the house, even taking bucketfuls of it to dump on the lawn, where the gardener will find it does his mower little good. In wet weather particles will adhere to the soles of shoes and be tracked into the house, where they will injure polished floors and linoleum. But for all these disadvantages there are many assets: advantages which will probably far outweigh all the drawbacks. The textures are very pleasant, and the variety of colors available in the materials makes it possible to complement any setting. Sometimes these loose materials are used in conjunction with other paving: gravel or crushed bluestone may be used to surround stone or concrete stepping stones, or a rectangular terrace may have some squares paved with permanent materials and other squares with loose stone chips or brick dust. Brick dust or colored stone chips make a very attractive border around a terrace, or a definition line across it, one part being paved with permanent materials and the other floored with gravel.

Tanbark and pine needles are good natural materials to use for flooring a terrace with a very wild, woodsy surrounding. They keep a soil floor from becoming muddy and unpleasant to use in wet weather. They key in well with surroundings where most paving would be out of place; but the fire hazard should be kept in mind where people may be smoking and drop cigarettes and matches to the ground. Tanbark in wet weather may have a rather pungent smell for a year or two, and it may make the soil acid.

Permanent paving is, of course, ideal. Its expense will range from a moderate price for bricks laid on sand to sizeable sums for tile or mosaic laid on concrete. But, at any price, permanency is desirable.

Bricks laid on sand or set permanently on concrete with mortar give a chance for a variety of patterns and colors and have a pleasing texture. They are among the most versatile of paving materials, lending themselves to many uses because of their small modular size, which allows them to be used in sweeping curves, to be laid flat or on edge, to be set into other materials to form paths and thus direct the flow of traffic across a terrace, as well as to lend spice and variety in a number of other ways. Note the patterns we show herewith and consider them for the terrace you are planning, for brick may be the best choice for you. If so, be sure to order hard-burned brick, because it will wear better than common brick, which is softer and will disintegrate after a few years of hard use.

Concrete blocks either those especially made for paving, which come in a variety of sizes and colors, or the type used for building walls, are adaptable for use. Either the 2-inch or the 3-inch thickness will serve equally well. In many parts of the country a large 24-inch square block about 3 inches thick is now available, and in some parts various other sizes of this same kind of block may be had. You may also cast your own blocks (see Chapter 18), either in their permanent location or elsewhere, to be assembled and laid when you have the requisite number for the job.

Blacktop can be used for paving and has the advantage of being inexpensive, reasonably permanent, and not bad looking. It has an odor for a time after being laid, particularly in hot weather, and at that time it is also likely to dent from chair or table legs. Because it is dark it will absorb heat more than light-colored paving does, and will hold it into the evening on really hot days. It is best installed by professionals, although some amateurs find it not too difficult to do. Our own feeling is that it needs the small but heavy steam or powered rollers which professionals employ to flatten it and make it level.

Concrete is one of the best and most adaptable of materials for paving terraces. It can be used as a base for brick, slate, or other stone, and when used by itself can be cast in any shape, combined with other materials, and colored in a variety of pleasing shades. (See Chapter 18.) By varying the finish of the surface, anything from a finish as smooth as slate which can be waxed for dancing to the nubbly texture of brushed aggregate can be obtained. Properly made, concrete is absolutely permanent, and will last and wear well for a lifetime.

Many effects are possible with concrete. We refer you to Chapter 14, How To Make and Use Concrete, for the various methods of achieving these effects. The most recommended surface is the medium-rough finish because it has sufficient "tooth" to make it safe for walking in wet weather, yet is smooth enough to have a refined texture, if that is what you feel is best for your terrace. The combed rough finish and the exposed aggregate finish are both interesting textures and give considerable traction, being quite rough. They are especially suitable to modern homes and to traditional homes in rustic or picturesque styles and will give contrast to smoothly finished walls.

Concrete with redwood, cedar, or other durable wood between

areas of concrete used as forms and left there for division lines is a modern interpretation which makes a terrace very interesting. Sometimes the areas are squares; sometimes oblongs; and sometimes bricks are laid in crosswise or lengthwise patterns between the cross-strips of wood, making the paving into a very interesting and unusual feature. You will see examples of all of these in our sketches, and they will give you many ideas from which to select. (Chapter 17.)

Stone has much to recommend it. Its texture and color and its natural beauty are only enhanced by time and weather, which makes it an ideal paving material. It is likely to be a most expensive material, particularly if it is in the cut-stone category. However, in some regions stone occurs naturally in shelving strata. If so, it may be possible to use a crowbar to dislodge pieces large enough to use for paving and to carry them home from the country.

Slate is available in most places, too, and makes a handsome pavement. It is a little fragile and is likely to chip if anything hard is dropped on it, so that it is best used mortared to a concrete base with the joints well finished. It may come in squarish shapes or in irregular shapes of triangles and oddments of all kinds which can be fitted together in what is called "crazy paving" style. Many people like to mix all the colors of slate—black, red, green, grey—in one pavement, but it is usually in better taste to limit it to one or two hues, because the colors of slate are vivid, especially when wet. Slate, being a very fine-textured stone, is likely to be a little slippery in wet weather. This may be a factor to consider if the terrace must be crossed a good deal in wet weather.

Tile is used extensively in many sections of the country where it is available and where it is priced within the means of the average home owner. It may be either colored or plain, combined with brick or other kinds of paving, and is good for outlining areas of concrete. The patterning of the tile makes a pleasant relief from the usual dull all-over effects, giving the terrace distinction and originality. Amateur ceramists can embellish their own tiles with fired and glazed designs to line the terrace pool or decorate a masonry seat on the terrace. Spanish and

Portuguese courtyards have used tile with great effect; there is no reason why modern uses cannot be found which will express our own day.

Wood blocks are used in many sections where they are available at a reasonable price. This is treated at some length in Chapter 17. But wood used for decks is something to consider everywhere. 2″ x 4″s or 2″ x 6″s on a good sturdy framework set on concrete piers firmly embedded in the ground make very handsome floors. Either countersunk screws or large nails fasten the boards to the framework. Wood should be well treated with wood preservatives (Chapter 4) before being laid; then a coating each year of boiled linseed oil should keep them in good shape. The wood can be stained a deep, dark color, or finished in its natural shade. The floor boards can be laid in many interesting patterns: in large squares with the direction varied in the laying so that alternate squares or oblongs are laid at right angles to each other, or perhaps diagonally across the terrace or in chevron fashion. About ¼ to ½ inch of space is left between the boards, these cracks allowing the deck to drain well and simplifying the sweeping off of dust and dirt.

Cut flagstone can be laid on sand, on concrete, or directly on the soil itself, although the latter is not highly recommended. Random squares—squares and oblongs of various sizes fitted together so that they give a pleasing variety of sizes but maintain a certain regularity at the same time—may be purchased and fitted together in a subtle and beautiful grouping, making a good pavement which does not draw attention to itself but is interesting to contemplate. Flagstone (usually sandstone) is never overly-colorful or overly-patterned and therefore stays a part of the background, unless it is somehow featured. But by itself it is never obtrusive.

Pebble mosaic, too, will add interest to paving. Used in strategic places, such as in front of a door as a kind of "welcome mat," to border a flower bed or a pool, or to break up a large area of flat concrete paving, it can be very piquant and add considerable interest to the terrace. (See Chapter 19.) If you are very ambitious, and have available large quantities of the proper kind of pebbles, you can make a great feature of pebble mosaic, but it is not recommended to impatient people or to the dilettante.

OTHER FEATURES FOR THE TERRACE

The terrace can be made more livable by the additions of various other appurtenances. Trellises and shelters will make it more beautiful and provide shade and shelter when needed; privacy fences and baffle fences will give seclusion and protection from passersby and neighbors who are too near. Pools will give a pleasant, cooling tinkle of water and provide a very attractive grace note with their plantings and living sound of water. Planters of various kinds, whether built up alongside the terrace or into the paving itself, will bring color and beauty to the outdoor room that most terraces become; and seats, whether of masonry or wood, whether permanent or demountable and transportable, will add to the attractiveness of the terrace and extend its use. We suggest that you look through the chapters which deal with all of these and see if you wish to incorporate them in your plans when you are building your terrace.

Terraces not only augment the living space of your home, but they add a new kind of life to your property. We cannot recommend them too heartily, and you will take double pleasure in one if you build it yourself.

A narrow side yard can provide a paved area for family use with a tall fence (this one is made of Western red cedar) on a sturdy frame to assure privacy. Cedar 2" x 4"s with spacers make the seat, set on pipes in concrete, and additional cedar 2" x 4"s are laid in the concrete.

On slopes the best way to get sitting-out space is to build a deck. Floor it with spaced 2″ x 6″s laid flat, frame the guard rail with 2″ x 4″s with a 1″ x 8″ board let into the front and make the seat of alternated 2″ x 6″s and 2″ x 4″s set on concrete blocks.

An entertaining deck has an original curving-front floor made of 2″ x 4″s set ½″ apart. It may be precut to curve, or roughcut, nailed in place, and the curve cut with a saber saw. The shelter has a center roof of opaque plastic for diffused light.

A too-small paved terrace need not daunt the craftsman. Extend it by adding raised decks of Douglas fir 2" x 4"s, also seen in the seat set on 4" x 4"s. Extend a retaining wall with a wooden one and top both with a low fence of evenly spaced 1" x 1"s in a 1" x 3" frame.

Ideal for a beach house but adaptable anywhere, a redwood deck features planks set on the bias in alternate directions; plant beds and barbecue pit may be let into the deck, raised seats on the side might have hinged tops, making equipment storage boxes.

A wood deck need not be square and blocky. A curving free-form edge is most effective—even steps may be rounded. Redwood floor is echoed by step, both edged with ½" boards kerf-sawed at intervals of ½" to ¾" and bent in shape.

The strong directional lines of the floor of this deck give it a crisp contrast to the natural curves of the plants. Douglas fir 2" x 4"s set on edge ½" apart, are framed by mitered 2" x 4"s. A step or two similarly styled leads down to the sloping ground.

345

A strong curve is saved from harshness by the material—redwood—and the linear effect of the floorboards. Beams supporting it rest on small concrete piers sunk in the ground. The fence is also framed in redwood, and bamboo (which is replaceable) gives a light, airy texture.

A built-in seat utilizes the guard rail of the deck for attaching its slanting back members, giving stability. The deck is made of Douglas fir 2″ x 4″s set on edge, while the seat has them laid flat, both on seat and back, spaced ½″ to assure drainage.

An oriental simplicity suffuses this garden scene, with its wood deck of Douglas fir 2″ x 3″s, its wooden access ramp and its use of container-grown plants. Silhouetted foliage casts alluring shadows on the semi-opaque plastic covering the high privacy fence.

Any sitting-out space is welcome in warm months, but one that is for both sun and shade is doubly valuable. A sturdy redwood trellis above a pool and a 2″ x 4″ floored area can be vinecovered in summer. Wood-floored decks in two ascending levels behind provide sunny sitting.

17

Paving Walks and Terraces

The question of what to use for paving garden walks and terraces seems to give the novice a bit of trouble. Perhaps it is because he has such a wide choice of material. However, once he has chosen the material to be used to surface the desired area, he must then give his attention to what underlies this paving. No matter how attractive the paving materials may be or how unusual the texture or the pattern obtained in laying down the paving, no terrace or walk is more durable or better than the underpinnings on which it is laid. Therefore let us take up the question of how to lay paving materials permanently so that this knotty problem will be settled.

Paving, as much as any other element in the garden, can give the outdoor picture distinction. Therefore the choice should be made with care and discretion. Similarly, once you have chosen what is to be laid, it is well to be sure that it will be permanent, easy to maintain, and safe to use in all weather throughout the year. You can cut the cost of building a terrace or a walk about in half if you do the work yourself. It is not the easiest kind of project, but perfectly possible to do it yourself.

Precast concrete blocks in various shapes of standard modular sizes are widely available for use in paving walks and terraces. Some are colored, some the usual gray of concrete. Preparation of a bed of well-tamped leveled sand is a necessity for complete success.

A play walk for the children circles the easy-to-tend lawn area, joining the paved terrace in the foreground and widening at the right to meet the fence with its attached seat. More lawn and a plant bed are seen in front of the fence at the left of the picture.

The first move is to measure out the area your paving is to cover (we will assume you have read Chapter 16 if it is to be a terrace), and to stake it out with stakes every 3 feet or so; and it may help to run a mason's line around the stakes so that you can see exactly how it shapes up and make revisions before it has cost you any labor which may have to be put in again. When it is laid out to your satisfaction, figure out the approximate square footage to be covered; then visit your building materials dealer. Select several of the materials you like and wish to consider. If your dealer can be persuaded to lend them to you, you might take samples of the paving materials home with you to try on the spot to see how colors look with your house or any other architectural features. If not, then you will have to exercise your imagination. You will be able, once you have chosen two or three likely paving materials, to figure out comparative costs for the materials from the square footage to be covered. Don't forget to figure in the costs of the underpinnings— sand, gravel, cinders, concrete, or whatever the paving material will rest on.

Whatever surfacing you finally choose, there will be only one of two basic methods used in laying the paving: mortaring it to a concrete base (this is the most permanent and satisfactory way, and naturally the most expensive), or laying it on a tamped base of sand, cinders, or some other porous material, or sometimes on a combination of two of them. Cinders or crushed rock may be used where the soil is exceptionally wet, or where it is so heavy with clay that drainage is poor when heavy rains occur. To make a good flat base for the paving material, a layer of sand on top of this coarse drainage layer will be necessary. The sand should be tamped down well to level it off.

There is a third paving material which we shall consider later on— concrete cast in place—but for the moment we are concerned with loose units of paving materials to be laid for walks or terraces. Of all the infinite number of paving patterns possible in a great variety of materials and sizes of units, all will be laid in one or another of the above methods.

SAND BASE METHOD

Because it is the simplest and cheapest method, this will appeal most strongly to the amateur. But considerable care and skill must be exercised to obtain really good results, even though it would seem that basically all one must do is to grade the soil to the proper depth, lay a bed of sand to the proper thickness, and then place on top of that the paving material chosen. If any of these basic operations is not done well, the result will be far from pleasing and will require much work of replacement and maintenance.

Let us first remember that sand will settle even after it has been wet down and tamped. The weight of the paving material will pack the sand somewhat, and the weight of use and traffic will pack it still further. This will make the level of the paving uneven, and it will be lower than you had thought it would be in the beginning. Therefore allow ½ to ¾ inch of space for settling and, when paving walks, crown the center a little so that, where the general traffic will be, repeated use will gradually pack down and level the walk. Should any particular units sink out of line, it is comparatively easy to pry them up and fill under them with a bit of sand and then replace them.

With both walks and terraces it is wise to start with the boundary courses. While the soil may be allowed to remain more or less roughly level, these edges must be precise and carefully placed. Usually boards are used for edge guides, being nailed to stakes and set with the aid of a level, if the walk is to be level and not slant to conform with a hillside. But a walk should be level on both its edges, no matter how it may slope, so the guide boards must be leveled.

Place the boards so that the top is exactly at the edge where you wish the paving to begin, with the stakes on the outside. After the first board is set, do a trial run with the units (we are thinking particularly of bricks, but the principle holds good for any modular unit paving material) to see how many you can place across the width of the walk with good tight joints. Then place the other board on the other side of the walk and level it, stake it, and make ready to start work. With terraces or pavings broader than garden paths, you may have to set some intermediate boards to assist you in leveling; or put in stakes at intervals driven to the exact depth desired, so that the top of a stake can be the guide for a good straight board or a mason's line, which will help to keep things going well and to keep the paving level. Note that all terraces should have a slight pitch for drainage in wet weather, even though the paving units when laid on sand will allow rain to drain into the joints and disperse in the sand and drainage layer below.

A minimum of 2 inches of sand should be used for the bed, placed either on the soil or on the aforementioned 2 to 4 inches of well-tamped cinders, gravel, or well-crushed stone. Level the sand roughly with a board and then soak it thoroughly with a hose, the nozzle adjusted to a medium fine spray so that it will not dislodge the sand too much. After the soaking it may be necessary to refill any low spots and soak it again. The sand should then be leveled with a screed board, which may be cut to rest on top of the side form boards and shaped to crown the walk, or used as a square leveling board if a terrace is being laid. A 2″ x 4″ is frequently used, but a 1″ x 4″ or a 1″ x 6″ may be used if desired. The edges which project over the board will guide it and permit fairly accurate leveling. The board will be cut to scrape out the sand to the depth of the paving units that are to be used. Every now and then remove the sand which the screed board has pushed up in front of it.

LAYING THE PAVING

Now that the sand bed is ready for use, you are ready to lay the bricks or other units. The secret of success in laying paving is to be found in the constant tamping and careful fitting of the units, and in the frequent checks of the levelness of the paving. Keep your level, a short board about 1″ x 4″ x 13″, and your heaviest hammer always at hand. It is possible to use the board and hammer to tamp an entire row of 3-inch wide bricks, to set them. *Never hit the paving material directly with the hammer* or you may damage it by cracking, chipping, or gouging it. Use the board and hammer also in front of a row of units to tamp it back firmly against the previous row. If you are laying paving in hot weather, don't just leave it when you must stop in the middle for the day. Always dampen the sand base with a hose, again using a fine spray to prevent dislodging the sand which has been leveled. This will keep the sand in place and moist, so that it can be slightly dampened the following day and the next lot of units laid on it.

FINISHING TOUCHES

After all the units have been laid, place a shovelful of sand here and there on the surface of the paving and then sweep it across so that the sand fills in the cracks of the joints between units. You will find that dry sand sweeps better and will fill more compactly; and, if the cracks are well filled, there will be less shifting and settling in the future. After all the cracks are filled, use a hose with a gentle spray to wash more sand into the cracks and to settle what is already there. Then, when it has dried, sweep up the excess sand and remove it. Occasionally, for a year or so, you may want to put on some more sand and repeat this process to prevent soil from settling and filling up the cracks, which would permit weeds and grass to take root.

The wooden edging may be left in place if desired—redwood, cedar, or cypress well brushed and soaked with wood preservative being best for this purpose—or the edging may be removed. In either case, fill in with soil next to the paving or edging and sow grass seed, if the paving abuts on a lawn; or fill with good topsoil if there is a flower border next to the paved area.

It may be that you will want to use a brick edging set in mortar or concrete, in place of the boards, to contain the sand and the paving material, thus making a permanent edging to the walk or terrace. (See the sketch in this chapter called "Methods of Laying Brick Paths and Various Ways of Edging Them.") The brick may be set on end in the concrete base or laid flat, with mortar between bricks in both cases. Paving bricks may be laid on a sand bed between these permanent edgings; or they may be laid on a 1-inch mortar bed on top of a well-tamped sand bed; the sand bed in both cases being wet down and allowed to settle, and then screeded to level it. These permanent edges will form the edge on which the screed rests as it is pulled along to level the sand bed. Note that paving should be set from ¼ to ½ inch higher than is eventually desired, to allow for settling.

PAVING SET IN MORTAR ON A CONCRETE BASE

The most permanent bedding for pavement and the one requiring the least maintenance is, of course, concrete. The paving material is usually mortared in place on top of this concrete underlayer. Occasionally paving units are laid on a 1-inch sand bed placed on top of concrete to obtain the softer texture of the unmortared pavement. This may be desirable in many places.

In cold climates big concrete slabs should be reinforced with wire fencing or steel rods (see Chapter 14), and everywhere it is a good plan to place expansion joints between slabs every 6 to 8 feet to

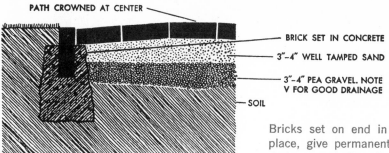

PATH CROWNED AT CENTER

BRICK SET IN CONCRETE

3"-4" WELL TAMPED SAND

3"-4" PEA GRAVEL. NOTE
V FOR GOOD DRAINAGE

SOIL

Bricks set on end in concrete hold path in place, give permanent edge. To insure good drainage and allow for settling, raise the center about 1½" to every 3' of path width.

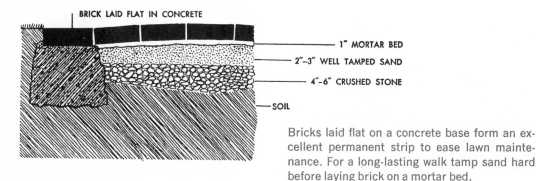

BRICK LAID FLAT IN CONCRETE

1" MORTAR BED

2"-3" WELL TAMPED SAND

4"-6" CRUSHED STONE

SOIL

Bricks laid flat on a concrete base form an excellent permanent strip to ease lawn maintenance. For a long-lasting walk tamp sand hard before laying brick on a mortar bed.

allow for contraction and expansion during temperature and weather changes.

Good drainage under the concrete is essential; 3 to 4 inches of crushed rock and tamped cinders will insure this. On top, pour the concrete in the manner described in Chapter 14, where the formula for concrete is also given, a 3- to 4-inch slab sufficing for this underlayer. In the case of a very wide walk or a terrace, it will be well to divide it into sections and build forms for them so that they can be conveniently cast. You will have to allow for the depth of the paving unit plus mortar joint when casting, so that your paving will come out even with the soil line or whatever you have determined as the final height of the pavement. Be sure that these forms are at the height you wish to have the concrete end, so that they can be used for support for the screed. After one section is cast and set you can remove the form, and the resulting

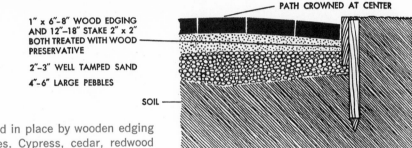

PATH CROWNED AT CENTER

1" x 6"–8" WOOD EDGING
AND 12"–18" STAKE 2" x 2"
BOTH TREATED WITH WOOD
PRESERVATIVE

2"–3" WELL TAMPED SAND

4"–6" LARGE PEBBLES

SOIL

Paths may be held in place by wooden edging fastened to stakes. Cypress, cedar, redwood are best to use, but treat all woods with preservative to kill decay bacteria.

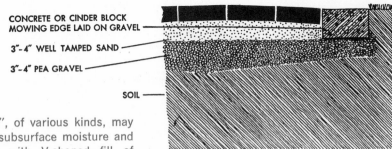

CONCRETE OR CINDER BLOCK
MOWING EDGE LAID ON GRAVEL

3"–4" WELL TAMPED SAND

3"–4" PEA GRAVEL

SOIL

Blocks 4" x 8" x 16", of various kinds, may also be used. Drain subsurface moisture and avoid frost damage with V-shaped fill of crushed stone or gravel or large pebbles.

joint, when the next section is cast, can serve as an expansion joint for the slab.

You need not wait for the concrete to cure completely before laying your paving material. As soon as the slab will support your weight and is dry enough to work on in comfort you can start laying the paving units. Spread a mortar bed for several units just as for brick laying (see Chapter 11) and butter the paving unit so that, as it is swung into place, it butts against its previously-laid neighbor and leaves a joint of mortar about ½ inch in width. The mortar bed should also be about that depth when the unit is laid. Check with the mason's level frequently to maintain the units in good relationship to each other, with no dips and no hillocks in the finished surface. Be sure that no air pockets are left under the units where frost and moisture can enter and cause havoc later on. Bed the units well and avoid this.

The joints may be tooled so that they are flush, or if you want a more interesting texture tool them lightly with a 1-inch pipe (see Chapter 13). This will emphasize the joints slightly, without raking them so deeply that they will collect dirt and débris and make the paving hard to keep clean.

We recommend that the paving units be soaked with water just as you would soak them if they were being built into a wall, so that they will not extract the moisture too rapidly from the mortar. After the paving has been laid for half a day or so, sprinkle it lightly with a hose set to a fine spray, or with a sprinkling can, to prevent its setting and curing too rapidly. Spray it again once or twice a day (more frequently if the weather or the climate is hot) for about a week, after which time it will be properly cured and you may cease.

OTHER PAVING IDEAS

Occasionally today one finds wooden paving blocks being used where wood is available and cheap. Waste blocks from sawmills, limbs or large tree trunks sawed through and used with the rounds of the wood facing upward, redwood blocks, and many other variations are possible according to the availability and the price of the wood in your region. All of these mentioned should be set on sand on top of 3 to 4 inches of crushed rock, because good drainage is mandatory if the wood is to last very long. Sand is filled into the spaces between square blocks, and smaller rounds or triangles of split wood are forced into the spaces left between the round slices of limb or trunk.

Wood paving is never so permanent as brick, stone, and others of the more impervious materials; but it has a very rustic, pleasant look which complements certain types of homes, and it will last for some years if properly laid with good drainage below it. Various hardwoods and redwood stand up better under wear than do softwoods, but all will last longer if they are well soaked with wood preservatives before being laid. (See Chapter 4.)

Wooden decks are also coming into greater use, particularly where the terrace must be built out from a house on a hillside, and where it is not desirable to bulldoze into the hill to level a spot for the terrace. On other hillside sites, part of the terrace is on level ground and a wooden deck extends the living space out from it, wood being used in part of the leveled ground paving to integrate it with the deck part of the terrace. Occasionally in some places it is possible to get wood such as old railroad ties just for the hauling. These can be used for paving; if they are sawed up and set on end with the crossgrain part taking the wear, they will last several years, for they are well creosoted before being laid as ties for the railroad. This type of paving is well suited to tying in with a wooden deck, the sympathy of materials being evident to all observers.

On the following pages you will find a number of suggested patterns for paving which should give you a good start in choosing the design you'll feel best able to create, or the one most suited to your particular paving problem. Study them well—perhaps even lay out some of the patterns as a test to see which may suit you best, using borrowed bricks or other paving material if you are not yet ready to make the final investment. You may come up with a fresh new pattern of your own which will be more to your liking than any we offer here. If so, by all means use it.

But whatever pattern you finally choose, don't forget to do the job of putting it down *properly*. There is no substitute for care and good craftsmanship.

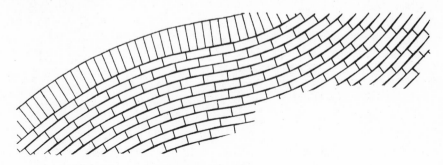

FLOWING CURVES

Today garden borders use sweeping curved lines. Terraces can echo them informally. If curves are kept shallow, bricks set on edge will leave only small spaces between them.

TERRACE PAVING

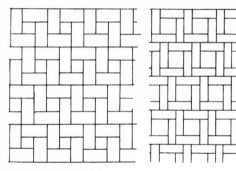

WHIRLING SQUARES
Bricks laid in this pattern give an interesting offbeat effect. Centers can be filled with broken bricks cut to fit space or with cement tinted to harmonize with color of the bricks.

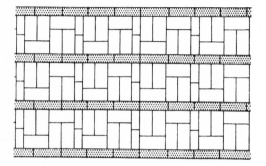

DIRECTIONAL LINES
Long lines of brick laid on edge, parallel to the house, can make a boxy square terrace seem longer and narrower, especially if bricks are darker than those laid flat.

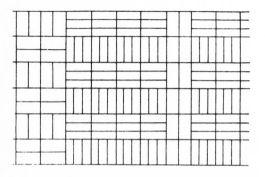

DIRECTIONAL PATTERN
Adaptable to walk or terrace, long lines are interrupted occasionally with cross courses laid flat. Further emphasis for length can be achieved by alternating the colors of courses.

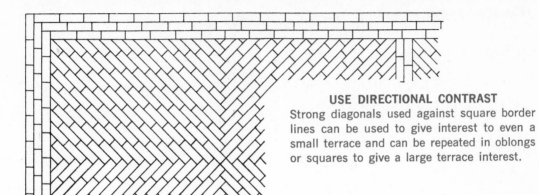

USE DIRECTIONAL CONTRAST

Strong diagonals used against square border lines can be used to give interest to even a small terrace and can be repeated in oblongs or squares to give a large terrace interest.

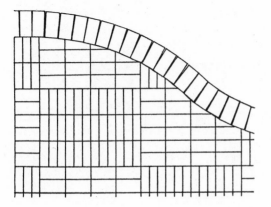

USE LINEAR CONTRAST

On small terraces using only one material, obtain contrast with line—curves against straight lines; extra linear effect of brick laid on edge against that of the brick laid flat.

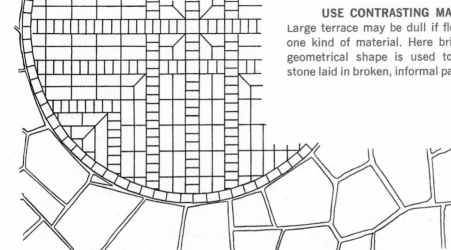

USE CONTRASTING MATERIALS

Large terrace may be dull if floored with only one kind of material. Here brick paving in a geometrical shape is used to contrast with stone laid in broken, informal pattern.

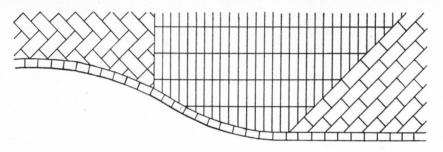

TRAFFIC GUIDES

Vary the patterns and direction of paving and guide traffic across the terrace as if you had built paths to contain it. Keep the areas of pattern unequal to avoid "jumpiness."

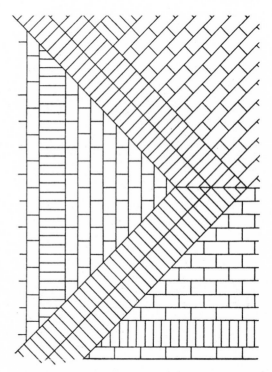

VARIETY SPICES THE TERRACE

Add interest with angular patterns of brick paving, using strongly defined header courses to divide areas. Particularly good with modern, it fits well with traditional houses, too.

USE WOOD WITH BRICK

Brick paving laid flat, on edge, at an angle or herringbone style in large squares (3'–4') is given unity by 2" x 4"s of cypress, redwood, or other wood treated with a preservative.

SQUARE OF SOIL FOR PLANTING

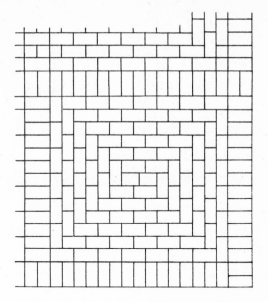

CONCENTRIC SQUARES
Build rows of brick around a pair of ¾ bricks until the desired area is covered. Lay the border rows at right angles, either flat or on edge, and repeat oblongs to cover the terrace.

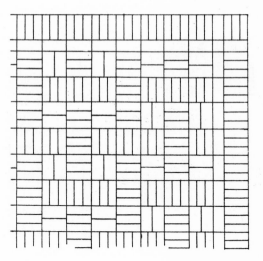

LARGE SCALE BASKETWEAVE
Bricks laid on edge contrast with bricks laid flat in this paving design from Mexico. Further interest can be obtained by using two tones of brick or by filling some spaces with cement.

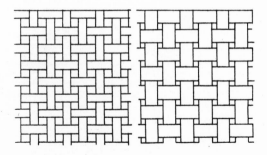

SMALL SCALE BASKETWEAVE
Laid on edge (left) or flat (right) and centered on the long side of the cross brick, this is a most useful pattern. Fill the holes with broken brick cut to fit or with mortar cement.

WHY NOT ENTER ON THE BIAS?

Where the entry is set between two uneven projections of the house, don't think you must square off the entry platform. Make it on the bias, leaving a triangular bed for permanent plants, dress it up with pots of flowers and houseplants through summer. Pave adjoining walks with brick, too, and if you wish, extend the walk into a sitting terrace.

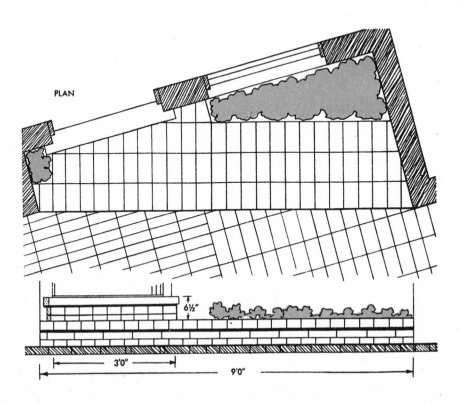

PLAN

6½"

3'0"

9'0"

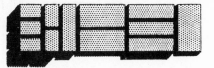

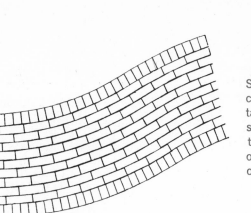

CURVING WALK

HOW TO COMBINE BRICKS

Shown here is a graphic chart of how bricks can be used in combination. (In all cases mortar joints are included in figuring the modular sizes.) Three bricks laid on end horizontally or two laid on end vertically equal the length of one brick. Two bricks laid flat or three laid on edge equal one brick length.

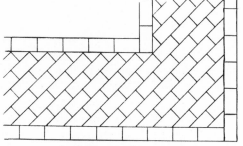

DIAGONAL AROUND A CORNER

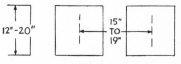

STEPPING STONES

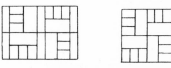

VARIATIONS OF PINWHEEL

BASKETWEAVE

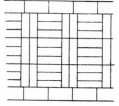

BASKETWEAVE VARIATION

WHIRLING SQUARE VARIATIONS

BASKETWEAVE

WHIRLING SQUARE

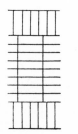 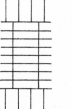

VARIATIONS OF LINE EFFECTS

LINE EFFECTS

MORE LINE EFFECTS

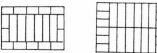

STEPPING STONES

Where the traffic across the lawn or through plantings is heavy enough to cause wear, set in stepping stones. They may be made of bricks, taking their cue for pattern from that of the walk or the terrace with which they connect. Several of the designs shown on these pages utilize brickbats or broken bricks to advantage, saving money by using waste materials. Note the variety of patterns possible, particularly noting variations of bordering the paths. If you do not find a design here which you like, work out one of your own—brick combinations possibilities are infinite.

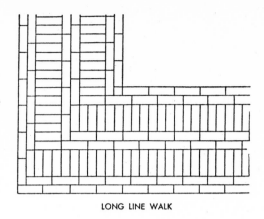

LONG LINE WALK

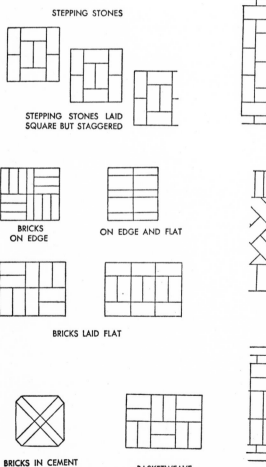

STEPPING STONES

STEPPING STONES LAID
SQUARE BUT STAGGERED

BRICKS
ON EDGE

ON EDGE AND FLAT

BRICKS LAID FLAT

BRICKS IN CEMENT

BASKETWEAVE

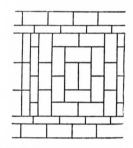

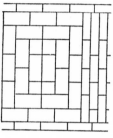

CONCENTRIC OBLONGS

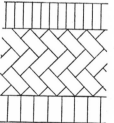

VARIATIONS OF HERRINGBONE

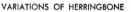

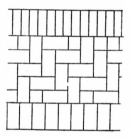

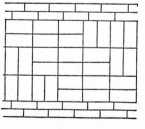

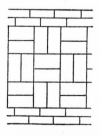

VARIATIONS OF BASKETWEAVE

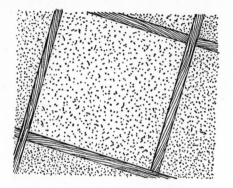

CONCRETE WITH WOOD
Make forms of redwood or other preservative-treated wood and fill evenly with natural or tinted concrete for variety.

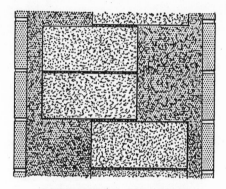

COMBINATION FOR WALKS
Concrete blocks paired and set in brickdust or gravel bordered with bricks laid on edge make a good, visually attractive path.

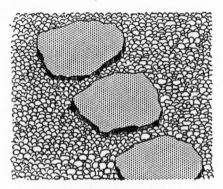

PEBBLES OR GRAVEL
Informal terraces or path areas can use gravel or larger pebbles with irregular stepping stones to make a walk.

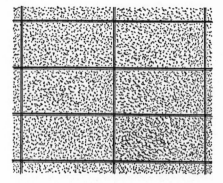

BLOCKS OF VARIOUS KINDS
Inexpensive blocks—cement, cinders, pumice, adobe, etc.—when set in sand, are easily replaceable if broken.

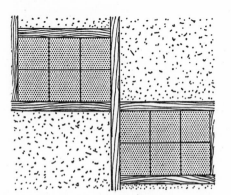

TILE AND CONCRETE
Terrace tile, fairly expensive to use on big terraces, may be used in selected sections with wood and concrete.

USE WOOD FOR DECKS
Where soil fill would be expensive and terrace must be elevated, build a deck of wood and floor it with wood as well.

RANDOM SQUARES

Stones of varying sizes and shapes, roughly squared off, can be fitted together to give a serene, stable effect.

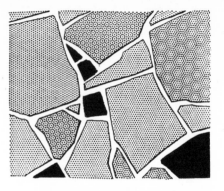

CRAZY PAVING

Vari-colored stones of many angular shapes can be used to advantage where paving needs informal touches.

MOSAIC OF PEBBLES

Small water-smoothed pebbles of same size, color make decorative patterns set in cement.

SMALL BOULDERS

Trees or plants set into paving need water. Set similar sizes of stones around plant in soil, sand or peatmoss.

TREE TRUNK BLOCKS

Wood blocks treated with preservative make good paving; fit rounds together, then fill holes with sand or soil.

SQUARE WOOD BLOCKS

Pave with redwood, cypress or other hardwood blocks or use sections of railroad ties bounded by ties laid flat to form blocks.

18

Cast Your Own Paving Blocks

You can make your own paving blocks, if you wish, by casting them in any size, any shape, or any color that you desire. It is not really a hard job, although it will entail making your own concrete, finishing the surface of the blocks, and making the form in which they are to be cast. If only a few are made at any one time, however, it should not be too difficult a project. If you make the forms and hold them in readiness, you may be able to save money by using leftover concrete from other projects, getting a dividend with no expenditure of money or effort. You may even want to mix a bit extra for the other projects each time and save the labor of a special mixing, which would be required for making the blocks alone.

All that you'll need is some lumber for the forms, some building paper or felt, some polyethylene plastic, some sand, cement, gravel, and a place to mix the concrete. Perhaps a metal wheelbarrow is as good as anything for the purpose. You'll also need your garden hoe for mixing the concrete, a straightedged board to use as a screed, and a trowel and float to give a final surface to the blocks.

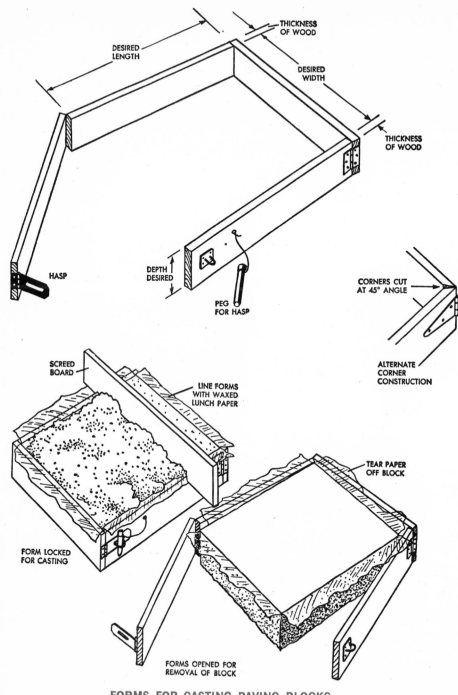

THICKNESS
OF WOOD

DESIRED
LENGTH

DESIRED
WIDTH

THICKNESS
OF WOOD

HASP

DEPTH
DESIRED

PEG
FOR HASP

CORNERS CUT
AT 45° ANGLE

ALTERNATE
CORNER
CONSTRUCTION

SCREED
BOARD

LINE FORMS
WITH WAXED
LUNCH PAPER

TEAR PAPER
OFF BLOCK

FORM LOCKED
FOR CASTING

FORMS OPENED FOR
REMOVAL OF BLOCK

FORMS FOR CASTING PAVING BLOCKS

MAKING THE FORMS

The only requisite for the forms is that the wood be straight and true—not warped out of shape—and be reasonably smooth so that the concrete will not stick to it. Oddments of 1-inch boards can be ripped down to an even width, the height you have chosen for the thickness of the block, cut to the proper length, and any roughnesses smoothed down. It is a good plan to cut several sets of sides and ends simultaneously, clamping the boards together while they are being cut so as to insure that all will be exactly the same dimensions and that all the forms will be even and squared up. Hinges are installed on the outside edges of three corners, and a hasp on the fourth with a wooden peg to keep it firmly closed. Thus, when the block is cast and has set sufficiently, the hasp is loosened and the hinged form can be peeled off easily. Make at least two forms; half a dozen or more will be none too many if you plan to make many blocks for paving. Fewer than that will make it hardly worth while to mix up a batch of concrete for the casting. By using a rather stiff concrete mixture—one made with less water—the cast blocks will hold their shape and allow the forms to be removed almost immediately and re-used.

Before casting the blocks, brush the forms with old crankcase oil saved from an oil change in your car. Any sort of oil or grease can be used, so long as the forms are thoroughly brushed with it before each casting to prevent the cement's adhering to the wood. Or, if you prefer, you can wrap the forms with waxed lunch paper held in place by a strip of masking tape. To remove it, cut the paper with a razor blade and lift off the form.

PRELIMINARY OPERATIONS

On a *level* surface place a piece of building paper or felt somewhat larger than the form. On bare soil this will prevent the moisture from the concrete being lost in the soil, or the block picking up grass and stray bits of twigs and soil which may have to be removed if they are to be mortared onto a concrete base later. If you are using a garage floor, a walk, or a driveway for the casting, this will keep the blocks from adhering to the floor surface. On these surfaces use at least 18 inches extra paper all the way around the forms, so that the splashes, blobs or screed scrapings will fall on the building paper and not on the cement or concrete floor. A piece of plywood or hardboard also makes a good casting base. Cover it with polyethylene plastic so that the concrete will not adhere to it. Such pallets can even be used on rather uneven ground, when properly underpinned to make them level.

Either a trowel or a large-sized vegetable- or fruit-juice can may be used to scoop up the mixed concrete from the wheelbarrow so that it can be poured into the forms. Be sure to fill each corner well. Use a block of wood to ram the concrete gently down into each corner and to joggle the rest of it gently to release any air bubbles and pockets. Do this when the form is half full and again when it is completely filled. When it is half full, too, is the time to insert any reinforcements of wire, etc., you may wish to use. Then fill the form to the top, pack it down well, and add a bit extra so that it bulges a little on top. Using the screed board— any good straight board a little longer than the width of the forms can be utilized—gently scrape the concrete forward with a zig-zag motion, a kind of sawing motion. Let the board ride on the tops of the form boards as it removes the excess concrete, and as it scrapes off whatever water rises to the surface, too.

Finishing can be done with the float and trowel (for method see Chapter 14) for a smooth surface; but if you want a textured surface or a patterned one, you will want to use other implements once you have floated the surface smooth. A good stiff broom or an old whisk broom may be used to make swirls or straight striations in the surface when the concrete is just beginning to set. It can also be used to scrape regular patterns into the surface or to brush out the nearly set cement to expose the aggregates for an exposed aggregate finish. (This is detailed and also shown in photos in Chapter 14.) Various other implements may be used to imprint patterns in cast blocks: small tin cans either round or square, thin wood blocks or strips of tin soldered together in crosses, diamonds, stars, triangles, or other geometric or freeform shapes, or cookie cutters—all of these may be utilized to print designs at regular intervals or as your fancy dictates.

CASTING BLOCKS IN OTHER WAYS

If you wish to cast terrace or stepping stone blocks without the trouble of building wooden forms, you can sometimes use the soil itself for the form, casting the blocks in the exact location in which they are to remain.

Be sure, however, that the soil is reasonably firm and not too sandy, so that it will hold its shape and not crumble, destroying the edge you want to maintain for the block. Water the soil well a day or two in advance of digging it out so that it will be moist and hold together. Cut the stepping stone or terrace block spaces in the lawn or the soil with a sharp, square-ended spade, then dig it out carefully with a trowel, using a putty knife to remove the corner earth and keep the corners square, if that is the effect for which you are striving. Stepping stones should be at least 2½ inches thick, with 3 to 3½ inches being preferred, and they should be cast on two or more inches of pea gravel or sand or cinders to provide drainage and prevent frost from heaving them. The concrete is poured in on top of the drainage layer and lightly rammed with a wooden block when the form is half full; then reinforcements are set, and the rest of the hole is filled and lightly tamped to release air bubbles. The screed board or the trowel and the float alone may be used to level off the concrete with the surface of the soil. You may prefer to use freeform or simulated "crazy-paving" forms rather than squared forms for your stepping stone or paving. They will look more natural if you striate or lightly brush and roughen the surface of the concrete, and will be less dangerous underfoot in wet weather than if they had smooth surfaces.

If you want your stones colored, you may mix two batches of concrete. With the natural batch fill the forms to within about ¾ inch of the top. Then pour the colored concrete layer and finish it as described for ordinary concrete. See Chapter 14 for details on colored concrete.

WHAT SIZE FOR THE STONES?

If blocks are to be used for stepping stones, the size is largely a matter of individual preference, but will also be governed by the distance from center to center of the stones. As they are finally laid they should be comfortable for the average adult. Fourteen to 18 inches center to center is a good average for the casual garden stroller. If, on the other hand, the blocks are to be used in a pattern (see paving section Chapter 17) on a terrace or a walk, then they must be worked out carefully to an exact modular size so that they can be easily fitted together when they are laid. In this case the usual ratio is 2 to 1—that is, the length equals twice the width. Make the length about ¼ to ½ inch longer than the exact double measurement of the width to allow for joint spaces between blocks when they are finally laid.

Blocks larger than 16″ x 32″ are practically impossible to transport, and are unwieldy and heavy even if they must be moved into place nearby. The maximum size we recommend is 12″ x 24″ for a 3-inch thickness. These may be placed on rollers and guided into the place where they are to be permanently set; then shifted a little with a crowbar for final placing. Blocks smaller than the above sizes are easily handled.

Consider the need for half blocks, two-third, and one-third sizes, if the blocks are being laid in a staggered block pattern; and make the forms to cast them exactly as for the full-sized blocks, but use the proper dimensions. Hexagonal, triangular, and octagonal forms may also be worked out by the adept amateur; when carefully measured and well cast they will fit together exactly and with beautiful effect. Diamond patterns also may be used, being made fat or thin according to the space and the effect wanted. Long thin diamonds will make a narrow space look longer and narrower if the diamonds are used with the long measure parallel to the length of the narrow space; but if they are used the other way around, the narrow space seems less long and narrow, a more pleasantly proportioned area. Consider this principle when paving a terrace which is long and narrow, or one which is too square for good proportions but about which nothing can be done because of space factors beyond your control.

MIXING THE CONCRETE

Accurate measurement of ingredients, plus thorough mixing of them, always makes good, strong, durable concrete. The following proportions will make a good basic mixture:

> 1 part portland cement
> 3 parts of sand (or sand plus gravel or other aggregates)
> 1 part of water (varied for a good workable mixture)

To get some idea of what quantities of materials will be needed, fill your locked-up form with sand, heaping it up a little. Then open the form and pour this sand into the pail or other receptacle you will be using for measuring the ingredients. Three times the quantity of this measure of sand with 1 measure of cement and 1 measure of water will make *three* paving blocks. Although a total of *five* measures have been used, the water carries the fine particles of cement between the particles of sand, so there will be no appreciable increase of quantity over the measure of sand.

A metal wheelbarrow or a wooden mixing box may be used; or you may wish to consider buying one of the small 5-gallon-pail mixers that will save a lot of hand mixing. If you are using the wheelbarrow or mixing box, put in all the dry ingredients first, then thoroughly mix them with the hoe until there are no more streaks of pure cement or pure sand, but just a mass of even grey-brown color. Then make a "saucer" or "bowl" hole in the middle of the heap and pour in a half-measure of water, mixing it in gently and thoroughly until it is absorbed; then add a little more water and mix, repeating until the full measure is used. Be sure to use *only enough water* so that the concrete is loose enough to be easily mixed and worked, but is not watery nor too stiff to spread easily and permit finishing. It is better to mix a little and try, mix a little and try again, than to get the mixture too watery and be forced to add more sand and cement in the attempt to bring it back to the right texture, thus probably upsetting the balance of the ingredients. If by any chance it should become too wet a mixture, add your ingredients in the same ratio, 3 of sand to 1 of cement, using a cup or small tin can, and mixing the dry ingredients in a bucket or box before adding them to the wet mix.

You can tell when the mixture is right by testing it with the trowel or by pushing the hoe away from you with a backward stroke through the mixture. If it is too stiff, it will be grainy and crumbly; if it is too moist, it will look soupy or liquid with a good bit of water working to the surface. But if it is just right it will mix easily and not be grainy nor so wet that it does not hold its shape.

When you are going to be making a large number of stones or when you want a brushed aggregate finish, gravel can be added to the sand in the ratio of 2 parts of gravel to each 3 parts of sand. But use the *same proportion* of 3 parts of the sand-gravel *mixture* to 1 part of cement and 1 part of water. The gravel will merely add bulk and volume to the sand, which may be an asset when you are mixing a large quantity. You can, of course, mix any amount of concrete, from a bucketful to a bathtubful or more, provided the proportions are accurately measured and thoroughly mixed.

CURING THE BLOCKS

Basically the same methods are employed for curing concrete blocks as those used for curing any concrete. (See Chapter 14.) The exact method used here is to release the pin from the hasp of the form and remove the form from the cast block when it has been well dried for a day or two. Let the blocks cure in place for another day, sprinkling them lightly with water or covering them with straw or a 3-inch layer of vermiculite, or doing both the covering and sprinkling to prevent too-sudden drying out in hot weather. If you are casting indoors in a cool basement or garage, it will be necessary to sprinkle only once in a while, every other day or so, but if you are using a heated basement or garage, sprinkle them lightly each day.

After a week or ten days of curing (you can either leave them in the place where they were cast or put them in a pile with 1-inch wood strips between blocks to permit circulation of air between them) the blocks will be ready for permanent placement. They can be laid in various ways, as detailed in Chapter 17, mortared in place or laid on beds of tamped sand, cinders, gravel, or crushed stone.

Naturally blocks which have been cast in position need not be moved, but will be already in place. Fill in around them with sifted soil and plant grass or whatever you wish between the cast blocks. A pleasant style, borrowed from the Orient but now well established in the Occident for garden use, is that of placing stepping stones among large aggregates of selected color and sizes; or of setting them in beds of moss with a variety of tiny wild ferns, miniature violets, and other wildlings sprouting from alongside the stones or carpeting the areas between them. Cast blocks used in this latter way, as well as those set in selected aggregates where the scheme is informal, might better simulate natural stone both in color (subdued, stonelike hues) and in texture and form (rough, irregular surfaces and informal free-form rocklike shapes).

BUY CAST PAVING BLOCKS

Exposed aggregate paving stones in circular as well as square shapes are becoming more widely available. You may wish to spare yourself the labor of casting and finishing exposed aggregate paving blocks, by buying some of these. They can be placed where you wish, and the spaces between them filled with conventional smooth-finished concrete which you mix and lay yourself, thus getting the best of two worlds. Care must be exercised, of course, not to slobber concrete on the surface of the bought blocks. Perhaps covering them with sheets of polyethylene plastic would be the easiest way to protect them, stripping it away when the concrete has hardened.

Whether you use stepping stones as wear-savers for traffic areas in the lawn, or as a feature of the garden in ways such as those we have suggested, you will find them easy to make, most worthwhile additions to your garden scheme.

19

How to Make
Pebble Mosaic Paving

Although at first glance the patterns of Pebble Mosaic may look rather difficult for amateurs to achieve, they are actually quite simple. There is a long history of paving which utilizes pebbles set into cement in patterns, going back at least as far as Roman times and probably even earlier. In recent excavations abroad, villas much older than those in Pompeii have been uncovered to reveal various types of pebble mosaic. The craft reached its greatest height, perhaps, in the gardens of Spain and Italy in the sixteenth and seventeenth centuries, and since then varying degrees of excellence have been achieved in various parts of the world. The art has been revived abroad and is now finding increased favor among enthusiastic do-it-yourselfers in America.

The ingredients are of the commonest sort—pebbles, cement, and sand. While the latter two must be bought, usually, the pebbles can be had for the gathering on beaches, in the beds of brooks and rivers, or on gravelly slopes and fields. It is also possible to buy them in some places where nature has not favored the terrain with a natural occurrence; but most people will want to collect their own, because part of the fun in this creative activity is in picking up your own pebbles.

There are three major sorts which will be useful in achieving good design results: round, ovoid or egg-shaped, and flat oval-shaped. Also you will find color to be important, for it is by matching sizes, shapes, and colors that effective designs are produced. Black or dark pebbles set into light or medium backgrounds, white pebbles used as focal points or as arabesque patterns, and designs set into dark pebble backgrounds—these are the effects for which to strive, with which to make your designs unusual and effective.

Although it is perfectly possible to pave an entire terrace with pebbles, you should be sure you have the time and the ambition to gather sufficient numbers of pebbles to complete the task. Most amateur craftsmen will find that pebble mosaic is best used as a garnish, providing a bit of excitement or a *pièce de résistance* which will give sparkle to some part of the outdoor picture, lifting it out of the ordinary. Edge a walk with pebbles, make a "welcome mat" at an entrance or a gateway, floor a shallow pool, or make an interesting insert in the pavement of a large terrace, and you will have fun without being overwhelmed by the magnitude of the task.

TRIAL RUN

The first step in planning a pebble mosaic is to assemble a sufficient quantity of pebbles so that you can experiment with them and decide upon the nature of the design which you'll want to make. Also it will help you to determine what quantities of what colors and sizes of pebbles will be needed for the entire project. Try to select the colors and shapes which are available in fair quantities, so that you'll not have to make a lifetime project of finding the proper ones. Draft your family as assistants and have them help you to pick up pebbles of the size and kind you finally decide upon. You'll be surprised how the stock pile grows each time you combine a picnic or a beach outing with the gathering of pebbles. Sort them as to size and color, storing them in large cans or small wooden boxes for future use. The economy-sized fruit-juice cans make good containers which are not too large to be handled conveniently, but which will hold a good supply of stones. When your total supply reaches about a half bushel or more, you are ready to begin a trial run or "sketch" of your design.

Knock together a rough wooden frame or better still a shallow box of the size of the project you have in mind. It should not be deeper than 2 to 3 inches and should be filled with sifted sand or loose soil. Sand is better, because it will not adhere to the pebbles, while soil may get muddy and have to be washed off pebbles before you attempt to use them in cement. Water the sand bed and let it settle a bit; then start pushing your pebbles down into it to make the designs and patterns you want to carry out. It is immediately apparent that working in sand is less irrevocable than working in cement mortar; you can make any changes of pattern you wish merely by lifting out the pebbles and replacing them, wetting down the sand again if you need to.

Some craftsmen work out their designs in such shallow boxes of sand beforehand, lay their cement mortar for part of the final design, and then remove the pebbles from the box design one by one and insert them in the mortar for permanent placement. However, you may not want to go to all that bother, for once you have worked out in sand the way the designs are to be made you will know that you can do it in cement, too. But this trial run will give you some idea of the quantities of pebbles needed for the entire job so that you can assemble them before you begin the final design in cement.

HOW TO SET PEBBLES FOR BEST EFFECTS

Flat, oval-shaped pebbles are usually set on edge, working them in rows or lines, each pebble being pushed down closely to the next one, as seen in the illustrations. Other flat pebbles, also set on edge, may be worked across at right angles or at contrasting angles to the first group to give background texture and direction. Round or egg-shaped pebbles might be used as a background for contrast of texture, as well as for color contrast, if you wish. Another effective pattern is obtained by using the flat patterns in a chevron or braid design, with other parts set differently. (See the illustration among the paving examples, Chapter 17.) Don't neglect a change of pace and texture obtainable by using pebbles of several different sizes as well as differently shaped ones.

Your fancy will dictate what pattern and design to use—modern abstract, non-objective designs, or copies of old arabesques, geometric or regular patterns—whatever you may choose to relate to the style of your house and to give the effect you want to achieve where you will be making the mosaic. Beyond integrating the design with the surroundings, the sky is the limit for your inventiveness and creativity.

Try anything which occurs to you in your trial run. Put fair-sized round pebbles (2 to 2½ inches) at regular intervals and surround them with oval-shaped pebbles set on edge to follow the outlines of the circle of each round pebble. Between these regularly placed circles, fill in the background with flat oval-shaped pebbles also set on edge but running in straight lines, choosing a contrasting color. This is merely an example of how you can begin; you will find dozens of other patterns occurring to you as you work in the trial-run sand box. Try anything you like, secure in the knowledge that it is not irrevocable; and then, when you finally make up your mind about what you are going to do, you are ready to prepare for the final steps.

Count the pebbles in the design if you find you do not have enough to complete it and then find more of the proper ones. Don't start until you have more than enough of every kind to complete the job. Then you can prepare to set them in the cement mortar. Bring your various cans or boxes of pebbles out and place them where they can be reached; count them once more to be sure that you have 5 to 10 percent more than you think will be necessary, thus taking care of any miscalculations.

LAYING THE CONCRETE BASE

Dig out the area to a depth of 8 to 9 inches, roughly leveling it, but keeping a slight pitch away from any buildings nearby to insure good drainage. Fill the space with about 4 inches of coarse rubble or other porous material, such as cinders, gravel, or crushed rock; then roll and tamp the layer until it is firm. Set the forms for the concrete base in place, and mix enough concrete to fill the area to a depth of 3 to 4 inches, using a formula of

> 1 part portland cement
> 2 parts sand
> 3 parts coarse gravel, aggregates, or fine crushed stone

(Consult Chapter 14 for method of mixing and making concrete and installing forms.) Be sure to use good *clean* sand and aggregates, for dirt and dust and other matter may cut down or destroy the effectiveness of the concrete. If your sand and gravel piles seem very dusty and full of dirt, they can be hosed down to wash them. Wash off the drainage section, too, using a fine spray so that when you pour your concrete on top of it there will be adherence with the drainage material; washing just before pouring concrete will also keep moisture from being drawn quickly out of the concrete by the drainage base, because the drainage material will be wet.

After the concrete is poured and leveled, cover it with boards, building paper, or wet sacking held off it by strips of wood to prevent its drying out too rapidly. In cool weather it may be simpler merely to shade it to keep the moisture from drying out too quickly.

Remove the covering and test it each day. When it has set enough to support your weight but is not yet really dry (usually two days in summer, and two to four days or more in the cooler times of the year in most places), you will be ready to begin your mosaic work. Mix enough mortar to cover a section of about 12 to 18 inches—a bucketful will be enough to cover a square of this size. Spread it on the concrete, trying to make the edge of the mortar coincide with the edge of the pattern or design you are using, so that the next unit can be joined without an obvious joint showing. Some craftsmen use a board or a piece of heavy metal bent into shape to contain the mortar; others merely place the mortar roughly, leaving an edge which can be cut cleanly away before it dries so that fresh mortar can be placed to abut with and bond to the finished section. Or, if the work is being finished for a day and is to be covered for the night and resumed the next day, the mortar is cut away to the edge of the finished part of the mosaic before being covered. Next day, before placing fresh mortar next to it, all the edges should be carefully wet down to insure bonding.

MORTAR MIXTURE

A prepared mortar mix, bought by the small bagful, may be used. This allows the hard-pressed weekend craftsman more time to spend on the creative side of the job. Or, if you prefer, you can mix your own mortar. If so, use a formula of 1 part portland cement to 2 parts sand, adding water bit by bit and mixing it in well, then adding a little more and mixing that in, until the proper consistency has been reached. This will be apparent when the mortar is still rather stiff but can be easily spread without running. Some craftsmen like to use a bit of lime putty or fire-clay in the mortar mixture to prevent its setting too quickly and also to make it more workable. Proportions suggested are: 3 parts portland cement, 6 parts of sand, 1 part of fire-clay or lime putty. Check with your building materials dealer for the availability of these materials.

Lay the mortar to a depth of ½ to 1 inch, depending upon the size of the pebbles. Remember that if the mortar is too loose and wet the pebbles will sink into it and be lost. If the mortar is too stiff, the pebbles will be difficult to insert in it deeply enough to engage the mortar and make a secure bond. A little practice will enable you to work out just what stiffness and what depth to make the mortar. As your adeptness increases you will be able to increase the size of the area on which you work, but remember that most cement mortar will set within an hour or so, and by two hours will be getting really hard. It is a good plan to keep the area fairly small, thus avoiding disappointment, wasted mortar, or the risk of its not bonding properly and having to be ripped out and done over.

METHOD OF WORKING

The pebbles will bond better with the mortar if they are inserted slightly moist, but with no free water on them. In other words, do not use *wet* pebbles nor *dry* ones. Some craftsmen wash them and keep them in cans (the cans in which they have been sorted for size and color will do nicely), and thus keep dust and dirt off them so that they will be fresh and clean when they are placed in the mortar. Wet them and then let them drain well. The moisture will aid in creating a strong bond. Lay out the design divisions in advance so that you can plan ahead which areas to work on in each session. In the case of steps, for instance, you can do a step at a time, unless they are very broad or very long. For an entrance "doormat" design, divide the pattern into segments, using the important lines as divisions which would look well as joints, if that is possible. Then place the mortar and do each segment separately. If the design does not permit an easy division, then compromise and do the best you can in dividing it up into easy stages.

Note, too, that, just as in other concrete work, large areas will need joints that will permit the concrete to expand and contract with the changes of weather without cracking or buckling. For any project of less than 5 square feet, however, no expansion joint will be needed.

Make the mortar deep enough so that the pebbles can be inserted at least half-way—two-thirds of their depth if possible. In regions where the weather is likely to have sharp changes of temperature and a good bit of frost, it is a good plan to sink them as deeply as possible, so that they are barely above the surface of the mortar. This will prevent the pooling of snow and ice which, by freezing, may cause the surface of the mosaic to open and crack, causing deterioration.

As you work with the mortar, try spreading it and leveling it with a long board before you begin to insert the pebbles. Keep a tap block handy. This can be made of a board with a block or metal drawer handle on the top side to enable you to grasp it easily. Wherever you have finished a section about 8″ x 12″, take the tap block and gently tap the pebbles to force them into the mortar and level them.

After the areas are finished and the mortar begins to set, use an old whisk broom to brush lightly away any loose sand or mortar and to even up the mortar between pebbles. Be sure to use it *lightly* so that the pebbles are not dislodged. An old soft cloth, such as a piece of winter underwear or an old sponge, which is damp but not too wet may also be used to clean the surface of excess cement and sand. Again, use it *lightly* and do not scrub with it.

FINISHING THE JOB

Cover the section with a piece of building paper or wet sacking when you have finished it; if a heavy rain is imminent, it may be well to cover it with a tarpaulin to prevent the rain from falling on the wet cement. Uncover it the next day and sprinkle it with a hose, the nozzle adjusted to the finest spray; then cover it again or put on a layer of 2 to 3 inches of vermiculite, sand, or some other absorbent material. Sprinkle this with the hose twice a day for a week or so. In summer or when the weather is hot, a few days may suffice to cure the cement; but moisture then is even more necessary to its proper curing, so in very hot weather sprinkle it three times daily.

When the mosaic is completely dried and cured, probably ten days to two weeks after completion, brush on a solution of hydrochloric acid (see "Cleaning the Wall," Chapter 12) to remove smears of cement and blobs of mortar. Be careful not to let it remain on too long, because it may injure the mortar; rinsing with a hose and scrubbing with an old broom will do the trick. Also keep the acid from splashing on your clothes or flesh, for it may burn and irritate.

When the mosaic is cleaned and rinsed the job is complete, and you can have an unveiling party to show off your handiwork to your admiring (and envious) friends before settling down to enjoy your creation in the years to come.

20

Driveways for Today and Tomorrow

Practically every house built these days has a garage, or at least a carport, to house the everpresent automobile. Many houses sport a two- or even a three-car garage to accommodate the several cars of the various members of the family. As many car owners have found out, not only must the garage be remodeled and modernized but the driveways, too, should be brought up to date.

Because the number of cars parked on streets is becoming a nuisance of alarming proportions, the wise homeowner will do his best to provide off-street parking for his guests' cars, or for his own when it is not in the garage. As cars have become wider, covering more of the driveway, the necessity for "landing strips," or some sort of paved area alongside the driveway where people may descend from the vehicle, has made itself increasingly felt. Such an area will save wear and tear on the lawn and will provide for dry passage from the car to the house. Clever homeowners are dispensing with the usual walk across the front lawn, placing it instead beside the driveway, thus making the front yard easier to look after and mow, while adding to the visual dimensions of the lot by the unbroken expanse of the lawn achieved by this manoeuvre.

New homeowners, anxious to lay out a plot plan to achieve the utmost efficiency and beauty, should particularly study this section to become cognizant of the problems before embarking on the project of building the driveway. And those with their plots already developed may well examine these pages and reëvaluate their driveways, too, with an eye to improving them. Aside from the convenience and the improved efficiency, there is a cash value in having a modernized driveway. According to realtors, it will be a distinct asset if you should decide to sell your home. Bring your driveway up to date so that it will more than accommodate this year's cars, and you can count on its being adequate for some years to come.

CHECK YOUR REQUIREMENTS

- Is your driveway clear of tree limbs and shrubbery so that no brush or twigs will scratch the top and sides of the car?
- Must you back your car into the street to turn around, or can you turn around on the lot and head into the dangerous street area?
- How wide is your driveway? If it is not wider than the bare minimum of 8 feet, it cannot be negotiated easily by today's cars and by delivery and other trucks.
- Can you descend easily from the car when it is parked in the driveway, without getting your shoes wet or muddy in inclement weather and without injuring the lawn?
- If your driveway turns at a right angle, is the minimum radius 18 feet (inside edge), and if it is a complete circle is the minimum radius of the inside 19 to 20 feet?
- If the driveway has a circular turnaround, is the width of the traffic surface 10 feet or, better still, 11 feet?
- Is the driveway approach to the garage level? If not, icy weather will make it difficult to negotiate when you try to start again after stopping to open the garage doors. There will be the added danger of skidding into the door posts. The driveway should be level or nearly level for a car's length or more out from the garage doors. If the floor is not paved, the garage can be excavated, and the driveway and floor brought into proper relationship. The doors can be lowered to conform with the new level.
- Even though your driveway may be adequate for *your* car, can garbage trucks, delivery trucks, oil trucks, or the oversized cars of

your friends negotiate it without getting scratched by tree limbs or without having trouble with the turns?

- Is the surface of the driveway adequate, or does it need improvement with a permanent paving, such as concrete, or some fairly durable one, such as blacktop?

Once you have the answers to these questions you are ready to make your plans to increase the safety, as well as the usability, of your driveway. The first thing to do is to take measurements of the space to be covered, whether it is an old driveway being modernized or a new one being laid out from scratch. Measure the distance from garage to street or road, the distance from the lot line to edge of driveway from the house to lot line, and note location of any trees or shrubs. Survey plant material to see if any will have to be cut down or moved to prevent trouble in future on the driveway, or if pruning may be necessary.

CHARTING THE DRIVEWAY

Lay out the measurements on squared graph paper and put some tracing paper over the plan. Start sketching the course of the driveway as you think it should go. On small lots, of course, it is usually best to run the driveway directly to the street, the straight course taking up as little as possible of the lot space. On the other hand, it may be to your advantage to consider some other plan, such as those on the chart pages in this section. Try a number of approaches on the tracing paper. It won't cost much, and you'll probably run onto just the one for you if you persevere.

When you think you have a plan which suits your problem, plot it on the graph paper, and then measure it and stake it out on the ground. You can make doubly sure, before committing yourself to the expense or labor of executing the driveway, by waiting for dry weather; then driving your car along the staked driveway; backing up; turning around and in every way you can think of, giving it a final test. Then when you have reset the stakes (if you decide it needs revision) and tested it again, you can start to dig it out or have it bulldozed, or have the paving contractor lay it, if you are not doing the work yourself. Provide sufficient depth for subsurface drainage, which a bed of gravel or cinders will supply, and crown the center to assist in draining it. Then whatever you do in the way of surfacing will be secure.

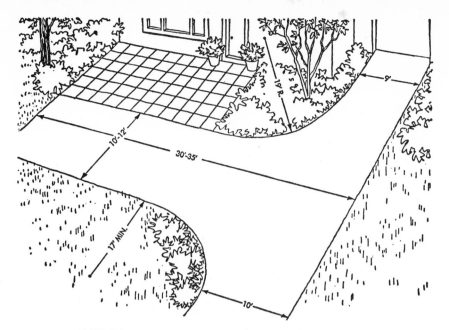

HOW TO LAY OUT THE DRIVEWAY AND TURNAROUND

Avoid the dangers of backing straight into the street by making a back-in turnaround. This can be a means of extending the terrace as shown in the sketch or it can be separate (below) to form extra guest parking space.

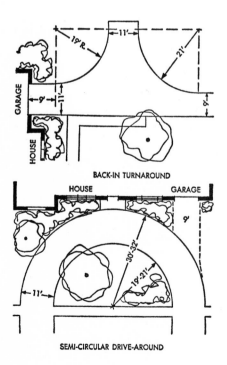

BACK-IN TURNAROUND

SEMI-CIRCULAR DRIVE-AROUND

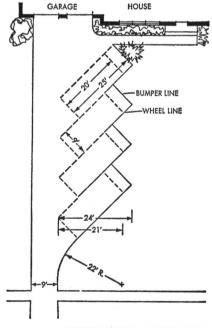

OFF-STREET PARKING SPACE FOR GUESTS

Consider the off-street parking as shown in our charts and see if you can adapt it to your own layout. This is well worth considering now, even though it may not all be done at one time. Bulldozer time spent now will cost less than a second trip later on; and if you are having a contractor figure the job, see how much more it would cost to have the parking area surfaced now, and how much it would be for the same job to be done separately later on. The difference may be more than the cost of the interest on a loan which, if taken out now, would allow you to get the job done and to have the use of it immediately. Payment can be made on a loan over several years. Of course, if you are doing it yourself, you'll probably want to spread the work over a long period of time anyhow, doing the driveway first, and then the parking area in sections. That way you will save the interest on the loan *and* the cost of the labor, too.

PERMANENT PAVING MATERIALS

Concrete laid on a bed of 2 to 3 inches of crushed stone, gravel, or well-tamped cinders, and reinforced (see p. 299), is the best we can recommend. Concrete strips for wheels only may be laid, with wear and permanence provided for by use of flagstone, bricks, concrete blocks, brushed aggregate concrete, or other materials between these tracks. We do not recommend lawn between the strips because of the constant need for repairs due to oil dripping from the cars, or to wheels running off and gouging the lawn in the center as well as on the sides of the driveway. All concrete should have a slightly roughened finish to provide for wet weather traction.

Paving brick when laid on concrete and well mortared together is very satisfactory, giving good wear and providing a certain amount of traction in wet weather. The color of paving brick is usually pleasant, but its cost is commensurate with its permanence and beauty. If you could find an old brick road being ripped up, you might get a good buy on paving bricks. Lay them on 3 inches or more of reinforced concrete base.

Stone paving blocks may also be obtained second-hand in some areas. They may be used on a gravel or sand bed, but are better when

A driveway need be only one car wide, but it should broaden as it approaches a two-car garage or carport, permitting the car to enter the garage squarely on the right, and to back out and reach the narrow driveway with a minimum of complicated maneuvering.

In a sense, the driveway is the welcome mat of the home in these days when guests arrive more often by car than on foot. It behooves homeowners to make sure that the entrance is attractive. An exposed-aggregate paving is handsome, offers traction in winter in cold areas.

Adequate for parking, backout and turning, this driveway offers a dividend of beauty. Flowering curves oppose straight lines of the random-sized oblongs that echo the structural lines of the house. The driveway is paved with light-colored exposed-aggregate concrete.

Note the angles from driveway to street that give vehicles opportunity to enter or leave without damage to the lawn. On a narrower driveway, a curve or quarter-circle from walk to street would be preferable and perhaps a little more graceful in line.

Patterned concrete areas minimize the boring quality inherent in unbroken one-color concrete areas, giving visual interest to the driveway and carport. Colored concrete squares may be bordered with strips of brushed aggregate paving, using varied aggregates.

A semicircular driveway permits easy access and departure and its pleasant curves contrast nicely with the straight lines of the house. In this one, red-brown-colored concrete harmonized well with the red and buff of the house, flattered the green plantings.

laid in concrete. Fitted tightly and set level, they make most interesting patterns, as can be seen in pictures from abroad of driveways, forecourts, and even in streets.

Blacktop, although not really permanent, is definitely cheaper than any of the more permanent paving materials. It will need repair and maintenance in order to be satisfactory, particularly in frost areas, so that the cost over the years may amount to as much as permanent paving. It will heave and crack in zero weather. In summer or hot weather it may also roll under the weight of cars, as it softens in the heat. One other factor should be considered if the driveway is adjacent to the windows of the house. Dark colors absorb heat and, although this may be desirable in snowy weather, in summer the absorbed heat may be unpleasant for hours after sundown. Blacktop provides excellent traction with its rough texture. Should weeds or grass appear in this rough texture, they can be killed by spraying them with weed killers.

NOT PERMANENT, BUT USUALLY SATISFACTORY

Gravel of small size, perhaps pea-size gravel, when well compacted by rolling and use, makes a good driveway surfacing for level or nearly level surfaces. On hillsides or on decided slopes, moderate rains will wash and roll even sizeable gravel considerable distances. On any slope, heavy downpours will strip the driveway to the soil and beyond. But for level surfaces, gravels in any of the various colors are pleasant because they are natural materials.

Crushed rock or stone in large sizes make a good base for gravel, blacktop, concrete, and smaller sizes of crushed stone aggregates. Marble chips, pink granite chips, bluestone, greenstone, and various other sorts and colors of stones may be available in your area, and they make handsome surfacing. The same drawbacks noted above for gravel apply here, perhaps even more so if the grade of crushed stone used contains very small particles. Check local materials and select from them.

NOT RECOMMENDED

Cement and soil mixtures sometimes suggested for use as terrace paving are not satisfactory for surfacing driveways because they will not bear loads or withstand the wear of wheels.

Home-made blacktop is also unlikely to bear heavy loads and resist the punishment of wheel wear. Heavy tamping or rolling (such as is done by contractors) is usually required to make this suitable for heavy duty. You might save money if you arrange with the contractor to let you do the labor of digging out and leveling, leaving the laying of the surfacing to be done by his men. Be sure to work out with him *exactly* what you are to do and what he is to do, so that there will be no misunderstandings later on. Follow his specifications for the under-pinnings so that they will hold up the paving under all use.

Clay or clay mixtures are not satisfactory. Clay is very slippery in wet weather and even when used in a mixture with sand will probably wash out in a short time. Also, it is very messy when tracked indoors.

Oiled soil may be satisfactory for roads but is not recommended for use on home driveways. Around the house, oil splashes on lawns and flower beds adjacent to the driveway, destroying and injuring the plants. It will also track into the house for some time after being laid, ruining rugs, staining floors, and disintegrating some of the synthetic tile which is used so much these days in homes.

Whatever the final choice may be for surfacing, remember that a driveway is only as good as the plan on which it is built. Be sure to give it sufficient advance thought and planning so that it can be used without regret or irritation, or without expensive revisions, later on.

21

Decorative Planters

One of the best ideas in the home garden field in recent years has been the introduction of the decorative plant bed. It may be raised above the level of the terrace so that its front wall becomes a seat wall; it may be a show-place for favorite perennials or annuals; it may be ablaze with bulbs all spring; and in the autumn it may carry the glowing fire of chrysanthemums up to the very gateway of winter. All these things and many more are possible with planters.

On the practical side, too, it has much to offer. It lifts up your plants from ground-level and provides a seat from which to do your weeding, your spraying, your transplanting and maintenance work, taking away the hazards of doing your garden work while kneeling or nearly standing on your head; and it offers an exhibition place for miniature plants where they can be more easily seen, worked with, and studied in the fullest comfort. For those troubled with moles and mice, the planter offers a degree of protection from these pests, particularly from moles. In our own place we once lost all but three tulips of nine dozen planted one year, to moles who were followed by mice. Since we built our

393

planter alongside the house, our tulips have enjoyed security. They like the well-drained bed even better than the other place beside the terrace, which we now devote to narcissus bulbs, having found that rodents leave them alone.

Masonry planters are best built rather low—seat-height alongside a terrace, or even of a lesser height, so that they do not become too obtrusive. Masonry is the most permanent material for a planter, but seat walls of redwood planks or other durable, preservative-coated woods may be adapted, built with the heavy wood frame exposed or concealed on the terrace side, as you wish. Tarring the inside of the wood will help to preserve it, making it resist moisture and bacteria better. Be sure to allow the tar plenty of time to dry and get rid of its noxious gases before filling the planter with soil and planting.

Seats on masonry planters need not be made of masonry. It is possible to embed bolts in the mortar or concrete and fasten wooden seats to them, so that you combine the permanence of masonry with the contrast of wood textures and beauty. And it is possible also to use a wooden frame with corrugated asbestos fastened to it to make the planter; asbestos being impervious to moisture and soil. This makes a very interesting feature in the garden, too; the crisp architectural line of the wood and the regular lines of the corrugations opposing it make a handsome piece.

Masonry planters also give one a chance to use curving lines effectively in the garden—to follow the curving line of a terrace, or to give visual interest to a low retaining wall which would be rather dull if kept straight, but which leads the eye about the garden because of the beauty of its curves.

There are many places where planters may be effectively placed: beside the front or back doors; beneath a picture window so that plantings are lifted up making a foreground to the picture seen through the window; beside or incorporated into the terrace; flanking steps; on the upper side of a retaining wall or on the lower side with the wall as a backdrop for the plants—we could go on detailing the possibilities endlessly, for more come to our notice every month.

The possibilities for plantings to be used in the planters is very wide, too. Roses, perennials, bulbs, annuals, small blossoming shrubs, all-green shrubs, evergreens, which may be clipped into geometric shapes if that

Concrete blocks make it easy to build an over-sized planter. Set on a straight footing at an angle, 4″ solids are placed above and below 8″ hollow-core blocks. Finish off top with 1½″ or 2″ solids, with joints lightly raked to make a handsome permanent plant bed.

An interesting plant box, well-scaled to make a weighty accent at the corner of the terrace, provides plants with the necessary soil and also gives drainage through the pierced bottom into the oversized pebbles around the handsome elevated base. Redwood is used throughout.

Gardens go portable with planters of ¾″ grooved exterior-type plywood. Square box is easiest to build—slant-sided types must have beveled top so mitred 1″ x 2″ finish frame can lie flat. All joints are nailed, waterproof glued; set box on a 2″ x 4″ drainage frame.

Dress up potted plants with an easily made bottomless box made of 14"-long picket-shaped 1" x 4"s spaced ½" apart. One-by-two-inch top and bottom frames are corner-bracked with angle irons to hold the pickets rigidly. Note how corner pickets overlap in rotation.

A planter on wheels with a vine-covered trellis puts color where you want it. Heavily-grooved ¾" exterior plywood makes straight sides, slanted ends; 1" x 2"s form plastic clothesline-laced trellis. Wheels may be conventional type or axles or oversized casters.

A way to extend and widen a terrace with squares, seen here at a flower show, can be adapted to the home. Square plant beds let into the herringbone brick paving make a place for flower color, and if potted plants are used, allow many color changes.

is your hobby, even trees may be placed in a planter and used with low-growing plants around them. For those who are away during the summer we recommend planters with spring bulbs planted under periwinkle, which will give them first a burst of spring blossom and then a neat and orderly green carpet through the summer. Pots of chrysanthemums can be used, either set into the planter in the soil or merely placed on top of it among the periwinkle plants. This will give two bursts of bloom a year.

Others who have small planters keep them gay all summer long by putting a succession of planted pots in them, starting with bulbs forced in pots, and then following with a succession of annuals. Every time a pot of flowers begins to fade, a replacement is brought in from the garden where a battery of pots is started and kept going for just this purpose. In the autumn potted chrysanthemums are brought in as replacements. The trick here is to use pots of the same size throughout, packing peatmoss or vermiculite around them to help to keep them moist and cool and watering them each day; and then putting a replacement pot in the hole left by the one removed. It can be quite a fascinating game, not to mention the interest it will add to your garden. This same idea can be adapted to a planter placed under a picture window so that all season long you will see flowers from the house as you look out.

Modern gardeners will also find other plants and effects which will fit well in the planter. The succulent family—the best known is the "hen-and-chickens"—and the many kinds of stonecrop or *Sedums* make good planter crops, most fascinating effects being obtainable with the varying colors and textures, the various heights, and the unusual blossoms which appear on them at times. Combining them with pebble carpets in which are set stones of various sizes or nestling them among granite or marble chips, using water-washed round stones in large sizes and interesting colors are other ways to make show-pieces of your planters. Some people, taking their cue from Japanese stone gardens, use big rock or irregular stones with only two or three plants and perhaps a piece of driftwood of unusual shape, interesting texture and color to get a very restrained but most artistic effect. A spotlight can be placed to light the planter at night, giving it a function to perform around the clock. Perhaps you will find even more unusual and original ways to fill your planter, ways which will make it a real conversation piece. You might want to make it into a miniature garden, using dwarf evergreens, tiny plants, and little deciduous shrubs. The dwarf roses which grow only a foot or so high would be good subjects, and rocks, tiny waterfalls, and picturesque trees in the Japanese fashion would be good additions.

Don't forget that the parapet or seat wall is also a good place to display pots of plants. Houseplants summering on the terrace can rest on the wall for a few days, and potted plants prepared especially for use around the garden can be placed along the planter wall where they will lend their beauty when it is needed, and be transported to other parts of the garden when they are needed there.

PRACTICAL CONSIDERATIONS

You should be sure to make the footings and foundation for your planter deep enough and strong enough to carry the wall load, so that the planter will not sag or buckle or, if it is being placed against a building, pull away from it. If the inside of the planter is plastered with fine cement mortar it will better resist damage through moisture seepage. Or it may be tarred to prevent damage from moisture. Be sure to wait until the tar is well dried before filling the planter with soil. When the planter is joined to the front of an existing building, a membrane of waterproof material (consult your building materials dealer for what to use) should be placed in front of the existing wall to prevent moisture from soaking or seeping from the planter into the building against which it is placed. It is well to use a membrane *and* a layer of masonry—a single course of bricks, for instance—at the rear of the planter.

Wherever you place your planter, and however you plant it, we are sure you will find it one of the most interesting projects of any you will build in your garden.

CURVES GIVE VERVE
Although only two steps are required between levels, a low planter backed by a low wall will add considerable variety and grace to the edge of a terrace. It may also be used on the level for, being low, it does not interfere with the view or limit a feeling of space, yet traffic is restricted and defined, while the plants contribute color.

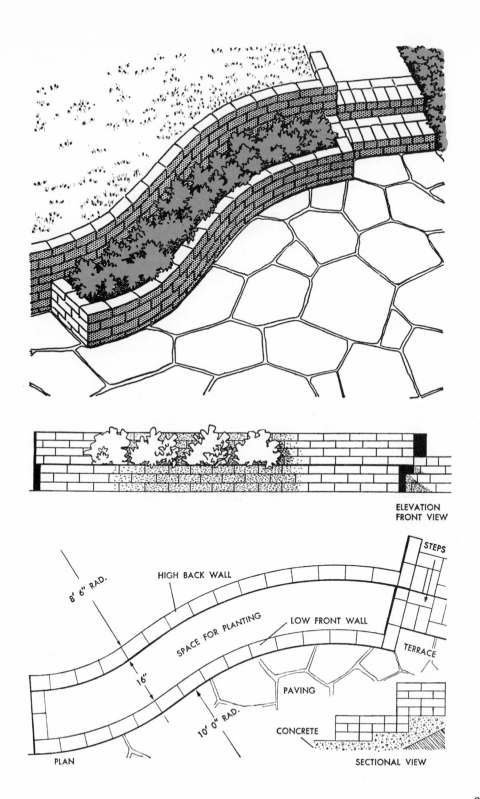

ELEVATION
FRONT VIEW

STEPS

8' 6" RAD.

HIGH BACK WALL

SPACE FOR PLANTING

LOW FRONT WALL

TERRACE

16"

10' 0" RAD.

PAVING

CONCRETE

PLAN

SECTIONAL VIEW

399

ANGLING FOR INTEREST

If your terrace is useful, but seems to lack something, an angular planter will add both visual and architectural distinction. Added seating space for outdoor parties is found on the front wall, backed up by plantings of bright flowers silhouetted against a higher brick wall which may be used to conceal some undesirable aspect or gain privacy.

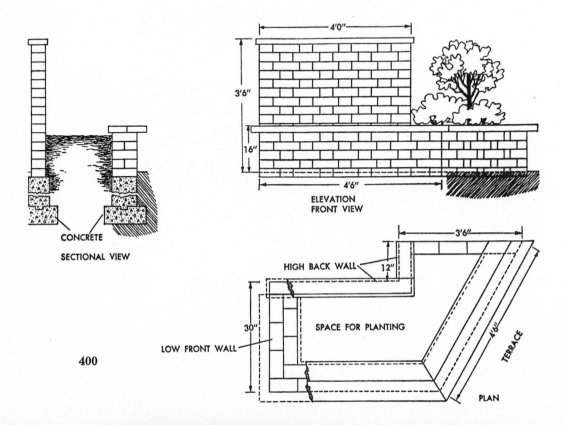

CONCRETE

SECTIONAL VIEW

4'0"

3'6"

16"

4'6"

ELEVATION
FRONT VIEW

HIGH BACK WALL

3'6"

12"

SPACE FOR PLANTING

4'6"

TERRACE

30"

LOW FRONT WALL

PLAN

400

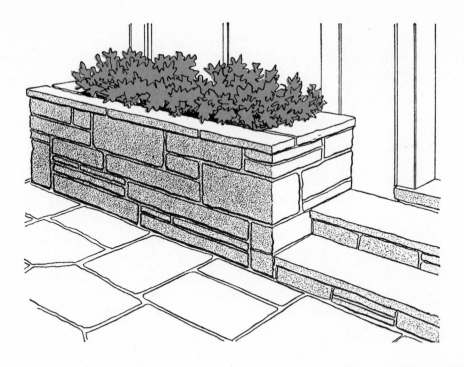

DRESS UP YOUR DOORWAY

The front door, usually so nondescript in a small house, can be featured if a plant box is built beside it. Incorporate it with the steps, using same materials for both—bricks, blocks, stone, or combinations—for best visual value. Note the "membrane." It's a necessary feature with a wood house, a good idea for a masonry one; it gives protection from moisture. Rough-plastering the inside of box with cement conserves water, helps preserve masonry.

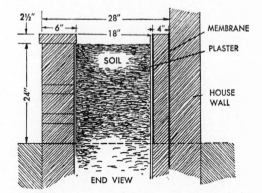

END VIEW

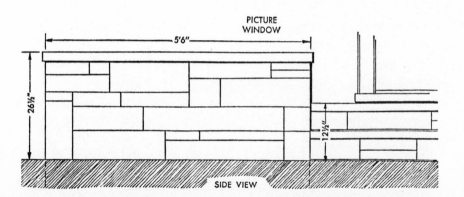

SIDE VIEW

401

22

The Pool in the Garden

Until lately it was only the rich who could afford to have pools in their gardens. Even today, when a garden pool is mentioned, most people envision an elaborate, formal pool with fountains and statuary, requiring a good bit of money and a lot of work to build and maintain it. Happily for most of us, times have changed, and today many a small home garden possesses a charming little pool which adds much pleasure and beauty to our outdoor living. Each year more people discover the possibilities of pools and joyfully add them to their gardens.

Nowadays we take a tip from the Persians and other Far Eastern peoples who know that, however tiny may be the trickle of water running into a small pool, it will give forth a sound which is cooling to our mental climate and conducive, therefore, to physical coolness in the oppressive heat of summer.

No longer is it necessary to have a constant water supply running into the pool and draining out—in a city that can make one's water bills mount up and in the country it can wear out the electric pump. Instead we now use small re-circulating pumps, which take the water that drains from the pool and force it back up through the fountain jet or other inlet. Only occasionally is it necessary to replenish the water lost through evaporation. Once in a while, of course, all pools must be drained, scrubbed out, and refilled, but the cost of this is minor. Recirculators aërate the water constantly as it falls into the pool so that it is kept from stagnating, and many of them have a strainer attachment which prevents débris from fouling the line.

We have learned, too, that pools need not be deep. Even when we want to grow water lilies and other aquatic plants in them, we need only make plant pockets deep enough to contain the roots of the plants at the proper depth and wide enough to hold sufficient soil for their growth. Pockets can be made which are just large enough to hold tubs or boxes of water lily roots. These may be sited outside the main pool in a separate pocket, with a connecting canal to provide water from the main pool. Or the pocket may be placed in the center of the pool. The remainder of the pool may be quite shallow.

Another trick we have learned is to paint shallow pools, 4 to 6 inches deep, a dark color, such as deep green, black, or navy blue. This will give just as much mystery and depth as if the pool had been made 6 feet deep. Many pools today function as mirrors to reflect and to give added dimension to some garden aspect—a picturesque tree, an architectural feature, a doorway—or to give focus to some fine view or garden vista. Being shallow, these pools will save water and at the same time present a safety factor where young children are likely to be playing around the garden.

Another development in late years is the use of the "saucer" pool, a shallow pool which tapers outward and upward from the center. It has many advantages, particularly in regions where winter freezing and consequent damage to the sides of the pool may give pause to the prospective pool builder. In this type of pool, when the ice freezes and expands it must move upward because there are no vertical sides for it to push outward. Hence there is no way the pool can be damaged. These pools are easy to drain, being easily siphoned out, and what water remains in the bottom can be whisked out with a broom. They are easily scrubbed out this same way. No source of water is necessary unless you want one, your garden hose being sufficient when it is needed. Some pools are placed along the course of natural rills or tiny brooks which serve as catch basins here and there, placement under little cascades or small waterfalls being particularly good.

PREFABRICATE YOUR OWN BROOK

It is possible to build your own rill or brook into your hillside, or to build up a little hillock if your place is not naturally hilly. In this can be installed the re-circulating pump. Place rocks in a natural-looking outcropping among and over which the water will flow down in a series of little cascades, to drop into the saucer pools, flow out in a cement cane between stones, and then go over another rock to fall into the next pool.

Saucer pools need not be circular. They can be freeform, somewhat serpentine in shape, even square, so long as the corners are rounded and the pool tapers upward from the center, leaving no angular sides against which ice can press. The edges of the pool may be architectural or natural. A coping may be raised above the level of the terrace 6 inches, 12 inches, or even more, or the pool may be sunk into the terrace with the rim even with the paving. In a natural setting, stones may be set around the edge to overhang it a bit, with soil piled up around them on one side; on the other side the pool may be level with the grass. Ivy or other ground cover can be trained about the edge to mask the masonry or concrete if you wish, making it look as if Nature herself had placed the pool there.

On a terrace it is sometimes interesting to have the pool left in its natural color or painted with a swimming-pool waterproof paint the color of the sky. Beautiful water-washed pebbles gathered from the beach or from along a stream can be arranged in the bottom in patterns to form a pleasant bit of contrast. Not all of them need be the same size, and a range of color will be agreeable, too. Wet stones show their colors to best advantage, so your show-pieces are always at their best in a pool.

Or if you have many pebbles available, you might use them in making a pebble mosaic (see Chapter 19) in some interesting design in the bottom of your pool. This type of pool should have a drain so that it can be kept free of water in winter. Designs can be anything you like— aquatic motives, geometric patterns, nonobjective modern patterns; or you may want to take your inspiration from Japanese gardens where the

A swimming pool that echoes the line of the raised terrace nearby, is graduated from deep at one end to shallow near the circular wading pool where children can play (foreground) and is bordered by a wide walkway, useful for sunning and for keeping debris out of the pool.

A home garden swimming pool of modest size with a raised curb on all four sides can be built by an ambitious amateur with some experience in concrete work—building forms, casting, reinforcing, and curing all surfaces before final finishing with a smooth coat.

A free-form swimming pool adapts well to odd-shaped areas, making a most interesting statement in the garden scheme and adding to the family's pleasure. Poolside areas here are finished in brushed aggregate cement, thus tying in with the large rocks on either side.

sand is raked in swirls to simulate water; you can imitate the swirls across the bottom of your pool. There are many fascinating ways in which a pool can be embellished, even formal mosaic in the manner of the old Romans being a possibility. For this, a shallow pool is ideal because it enables the design to be seen easily.

POOLS FOR WADING, SWIMMING

Another feature of the outdoor life in our country today is the backyard swimming- or wading-pool. Many people build small pools where their children can splash about and wade during the summer. In colder seasons the pool is drained, the drain plugged, and the pool filled with sand to become a sandbox for the youngsters, so that it is used during spring and autumn as well as in the hot seasons. Later on, when the children have grown, the pool can be adapted to growing aquatic plants and made a feature in the garden picture. Thus, for one price, one expenditure of labor, the clever craftsman gets a continuing use of this garden feature for many, many years.

It is now possible to buy prefabricated swimming pools which you can have installed or install yourself. They are not very cheap, however. You can also dig out your own pool and build a perfectly adequate one for yourself, provided that there is a cheap source of water. The number of gallons required for even a minimum-sized pool is formidable, and if you are on a water meter you may want to think twice before getting carried away. Also there should be a good place to drain the pool, either a sewer or some lower spot where it will run away quickly without violating any local laws. Unless your pool has a filter treatment apparatus, pool water should be changed periodically, once every week or ten days, even if you use chlorine or some other disinfectant recommended by health authorities to kill bacteria in the water.

PLACEMENT OF POOLS

This is a problem which deserves serious thought. Overhanging trees will drop their leaves into the water, and may in time shade the pool and make it too cool for comfort on days which are pleasant but not hot. Leaves, grass, débris of all kinds may blow into the pool, too, or be tracked in on the feet of the bathers unless you take measures to prevent it. Leaves and débris can be seined out of the pool before they attract algae and other organisms and begin to disintegrate and decay, and you can always place your pool away from the trees so that they won't shed leaves directly into it. You can pave the verges of the pool so that less trash is tracked in by the bathers; but if you raise the masonry edge to a foot or so above the ground level and install a footbath beside the entrance to the pool's platform for bathers to use on entering it, you will immediately gain some advantages. First and most important, there will be several cubic yards less of soil to be excavated. Then, the coping will prevent surface débris from blowing into the pool; it will serve as a seat wall around the pool; and it will also be a deterrent to people's falling into it at night if they are unfamiliar with the property.

Where the property slopes it is possible to take advantage of this fact in building a swimming pool. The floor of the pool can slope with the land and only enough excavation will be necessary to have proper footings cast and to have enough earth left to fill in around the pool on the sides.

WADING POOLS

These pools, being built for children, should never be very deep. Depending on the age of the child, they may vary from 12 to 27 inches tapering upward to 6 to 9 inches in depth. Small children should never use them without the supervision of an adult; so that it may be well to place the pool where the parent can see it from the house. Five by seven feet is a suggested size, but any size which you find convenient may be used. If you plan to use the pool later on for plants, you may want to check the recommended depth for water lilies and make the deep part of the pool that depth. If it is deeper than you think is safe for your child, fill the pool only part way. A wooden cover may be made to place over the wading pool when it is not in use to prevent children from wandering into your yard and using it without supervision. Small children can have accidents very quickly and it is well to foresee them. The cover will be useful later on as a winter covering for the pool when it is drained.

POOL OVERFLOW AND DRAINAGE

To prevent the pool from overflowing (and even those with no jet or regular source of supply may catch rain water and need drainage facilities when they overflow), some provision must be made in the form of an overflow channel or a removable pipe which screws into the mouth of the regular drain. This pipe should have a strainer or screened cap to prevent the entry of leaves or other débris which would clog the drains. Overflow pipes may be made of brass, also the fittings and piping of the pool. This lasts indefinitely without corroding, and it never will rust, of course. However, today polyethylene piping is often used and has proved quite satisfactory.

The overflow pipe should be just long enough when screwed into the drain to reach the level at which you desire to have the water remain. (See sketches of overflow pipes in the plans.) The recommended mushroom-shaped cover will remain above water level, the excess water flowing freely under the cover and into the pipe. Whenever the pool is to be completely drained it is a simple matter to unscrew the pipe from the drain and allow the water to flow out of the bottom of the pool. First, however, all the débris and leaves which may be in the pool should be removed, or, if this is not possible, reach down to the bottom drain as soon as the overflow pipe is unscrewed and cover the drain with a square of ¼-inch-mesh wire screen, weighting it to hold it in place with a couple of bricks. This will prevent débris from entering the drain and clogging it.

WINTER CARE

In severe climates some provision must be made for the protection of pools in winter—or at least all pools with straight sides, since we have shown that saucer pools need not be protected. For straight-sided pools, some authorities recommend a yearly draining, in autumn. Then a scrubbing out with a coarse scrubbing brush or an old broom, using hot water with a detergent and disinfectant added to it. Then the pool should be covered. For swimming pools and also for straight-sided garden pools, other authorities recommend that they be left filled to assist in resisting inward pressures from the soil as it freezes. Floating a small log or two on the surface of the water will keep ice from becoming too solid on the surface as it freezes. In spring, the pool should be drained, scrubbed out with water with detergent and disinfectant in it, and then refinished with a good waterproofing paint. When it is thoroughly dry, refill and it is ready for the summer season.

For covering the pool, some people use boards covered with a tarpaulin and weight this with more boards to prevent winter winds from whipping it. This is never decorative and not always effective. We advocate making a winter cover of outdoor plywood of a size which will cover the pool completely. A frame of 1″ x 4″s braced as needed across the center if it is a large pool, will make a strong cover and, as it stands up only 4 inches, an unobtrusive one. Plywood can be either nailed or screwed to it. Any water which might seep into the pool will leave through the drain—this should be left open all winter with the wire screen over it—and there should never be enough water gathering at any time to become a problem by freezing. In spring, wash out the pool again as recommended above; remove all leaves and other débris from inside and near it so that spring winds do not blow them into it; install the overflow pipe; and it is ready to be filled so you can enjoy the pleasures of a garden pool again during the green months.

Perhaps you will want to make your winter pool cover into a feature, something which is good to look at and not just a necessary appurtenance. In that case, look over the patterns shown among the fences and trellises in those chapters of the book and choose one to adapt as an appliqué design on the plywood top, using trellis strips or other light wood. Possibly you will want to make a little railing around the edge of the cover, using one of the designs which are shown in other sections of this book for your inspiration.

CASTING POOLS WITHOUT FORMS

The least expensive and in some ways the easiest way to make your pool is to cast it in concrete without forms and preferably with no drain. You merely excavate the soil to the depth of the water you want plus the 6 to 8 inches for the concrete itself, leaving the sides sloping no more steeply than 45 degrees, and preferably 30 degrees or less, so as

A large L-shaped freeform pool with a slide for the kids, a diving board and a handsome retaining wall of split concrete blocks shaped to complement the pool pattern is elaborate and probably should not be attempted by amateurs.

to prevent ice damage in winter. The water line should be accurately determined in advance to make it possible to carry the concrete to a uniform distance above the water line. First decide what the level of the water should be, then outside the concrete area drive a stake and adjust its top 1 or 2 inches above the desired waterline. Drive stakes around the pool outside the concrete area at intervals, using a spirit level—or a mason's level—to insure that they are all at the same height. As you cast the pool, carry the concrete to this point, and then level it to the height established by the stakes. Be careful not to disturb placement of the stakes as you are working about the pool, pouring concrete.

Another way of finding this line will be to fill the pit with water, first placing an old sock or glove over the end of the hose and binding it so that it does not squirt on the sides and cause them to fall in. When the pit is filled to within a couple of inches of the water line you wish, stop the flow and let the water soak into the soil. A heavy mason's cord can be secured three inches above this line by wrapping it around large nails or using large "hairpins" made from wire coathangers. It is to this point (3 inches above water soak line) that you will carry the concrete.

When the water has soaked away, dig out the bottom again to remove any silt which has slid down; and then you are ready to cast the concrete. Use a formula which will make a good stiff mix so it won't run when placed on the sloping sides of the pit. One such is:

> 1 part portland cement
> 2 parts sharp sand
> 3 parts ½" gravel or crushed stone
>   Water enough for stiff mix—about 3-4 gallons.

The pool must be cast all at once, so be sure to have enough material on hand to complete the job. Start in the early morning so that you will have all day for the job and be able to finish it. Light-weight hog wire fencing can be used for reinforcing, or you may use reinforcing rods bent to shape and criss-crossed and wired together, making 6- to 8-inch squares. These reinforcements may be laid on brickbats placed every 18 inches or so to keep them off the ground. The concrete is poured or shoveled carefully over this so that it falls through the mesh until it reaches a depth of 6 inches. Tamp it lightly occasionally to be sure that no air pockets are left in the concrete. Finish the concrete with the trowel and float (see Chapter 14), working from planks laid across the pool if you cannot reach the center from the sides. Level off the concrete at the top even with the line established with the cord, and finish it smoothly. If you wish, you may set stones in the edge of the concrete while it is still wet, or you may mortar them in place afterward. Do not let them overhang below the water level, however, or ice may dislodge them and make the pool leak.

Carefully built, a pool of this sort will fit in any natural landscape and should last for many years, even in climates which are fairly severe. If you want to plant very close to the pool, finish the top edge to about a 3-inch thickness, making a form of plywood or hardboard if necessary to maintain this width. Then the plants can come very close and overhang to obscure the hard edge and make it blend into the landscape. Painting the pool a dark color or using a finish coat of dark-colored concrete will also keep it in its place and obscure the hard edges.

PLANTING THE POOL

Water lilies can be made to grow even in a washtub set into the lawn, given proper care and consideration. Therefore you have every reason to expect them to grow well in a small plant pocket especially constructed for them in a pool, even if it is a shallow one. There are many other plants especially adapted to pool-side or in-pool growing conditions, which will dress up your pool and make it a feature. Once planted, they will need no weeding, no constant cultivation, no watering, but only planting and occasional feeding to make sure that all is well with the plants. In cold climates the lilies and other tender plants must be taken up and stored indoors. In this case the best solution is to plant them in tubs or boxes which can be lifted from the pool when it is drained for the winter and take them indoors.

Water lilies need at least a cubic foot of soil for each root. You can use a butter tub or buy a specially-made lily tub which will hold this amount of soil, using good rich leafmold or rich garden soil with peatmoss added. By making a box with a ½-inch opening between boards and lining it with a burlap bag before filling it, you allow lily roots to emerge to the water if they wish, but the soil is retained in the box. Good roots will enable the lilies to grow large with many leaves spreading in the pool and many flowers. By restricting root area even the larger-growing lilies can be kept to small space and reduced in scale to be appropriate to smaller pools. Cypress or hard pine may be used for the box.

Placing an inch of gravel or sand on top will prevent the soil from washing out of the box and fouling the pool. Check with your plant dealer to see what conditions lilies need, to determine which will be best for your pool. Some need shallower water than others, in which case you can set the box or tub on bricks or blocks to bring it to the proper level for the length of stem the lily will grow.

It is not necessary to change the soil every year. Merely dig out a little hole or two and put a bag of plant food in the soil pockets, covering them with soil and the sand or gravel. Dehydrated cow manure is a good food to use when inserted this way. Don't neglect these yearly feedings, because even though lilies grow in limited amounts of soil they relish rich feedings. Lily tubers may be left in the tub for storage, but you may dig them up if you wish and, after washing them carefully, store them in coarse sand or vermiculite in a box covered with wire screen to keep mice from eating them. Keep them in a dark but airy place, where the temperature is about 50° and the atmosphere is not too dry. In some areas of temperate climate it is possible to keep the lilies in the pool once it is drained. (Check with your plantsman when you buy your lilies on this possibility.) Drain the pool and fill it with leaves to cover the tubs to a foot or more, first covering the tubs with wire or plastic screening to prevent mice from entering and eating the tubers. The pool should be covered with either the kind of cover advocated earlier or two or three layers of tar paper weighted with boards to prevent their blowing off. A tarpaulin over the top of the paper will prevent water from soaking into the pool through openings between sheets of the paper.

OTHER AQUATIC PLANTS

Either actually in the pool or alongside it where they get plenty of moisture, other aquatic plants may also be used. Some prefer shallow water, some need greater depth to grow well, and some need only moist soil and a humid atmosphere to thrive. Siberian and Japanese iris are in this latter category, liking their roots moist but not standing in water; while the yellow iris, *Iris pseudacorus,* will grow either alongside a pool or in the water. There are many other plants with which to dress the edges of pools in natural or formal ways to make them a delightful feature of the home garden. Ferns like moist soil and shade, and so do primroses, Virginia cowslip, forget-me-nots, and many other plants. Most of them are hardy perennials; and they may be supplemented by tender plants, such as caladiums, which have exciting and wonderfully colored leaves; or, if the size of your pool will permit the use of their large relative, elephants ears, you can plant a bulb in the spring and take it up for winter storage each year, too.

If you have a very small pool sunk into your terrace you may want to forget about permanent plantings and enjoy it for itself, or just place a pot or two of houseplants or flowering annuals alongside it to give it interest. Trailing ivy, grape-leaf ivy, succulents, miniature or dwarfed trees, fuchsia, gardenias, flowering begonias of the tuberous- or fibrous-rooted types, coleus, and many other pot plants may be used to give a touch of living green or a flash of color to the pool's edge. Use weathered wood blocks or stack a couple of cinder or cement blocks in an interesting way and set your pots on them. Perhaps a piece of driftwood can be incorporated so that it and the plants are reflected in the pool.

We are sure that whatever the size of pool you build, whether it is just a washtub set in the ground or a large reflecting basin with lilies and other plants in it, you will find that adding water to your garden is a most worth while idea and that you'll enjoy your garden and your terrace all the more because of it.

SPLIT LEVEL POOL FOR THE TERRACE

Does this look familiar to you? It is adapted from a design in the Planter section of the book. Main differences are these: The back wall has the pipe for the inlet inside the block of masonry (a recirculating pump may be housed in it, too, if desired) and the wall of part of the triangular section has been raised to form a pool higher than the seat wall. Also, stone has been used for coping instead of wood, as used for seat in Planter design.

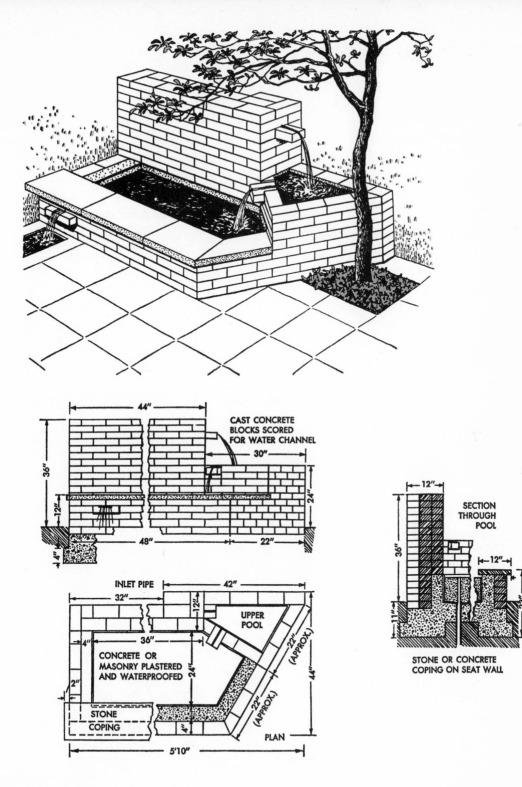

CAST CONCRETE
BLOCKS SCORED
FOR WATER CHANNEL

44"

30"

36"

12"

24"

48"

22"

4"

INLET PIPE

42"

32"

12"

UPPER
POOL

4"

36"

24"

22" (APPROX.)

22" (APPROX.)

44"

CONCRETE OR
MASONRY PLASTERED
AND WATERPROOFED

2"

STONE
COPING

4"

PLAN

5'10"

SECTION
THROUGH
POOL

12"

36"

12"

11"

23"

STONE OR CONCRETE
COPING ON SEAT WALL

415

TWO WAYS TO USE SAUCER POOLS

The various ways in which simple round pools can be used seem to be end-less. Raise them up a few inches above the soil or the pavement of the ter-race, put them in a lawn and sink them to ground level, install a jet with a removable overflow drainpipe or merely depress a portion of the rim to allow for overflow of rain water. (A concrete ditch 2' long will lead overflow water away and keep it from seeping into soil, causing frost trouble beneath the pool.) Saucer pools can be made of concrete smoothly finished, or in brushed aggregate with stones showing in a rough texture, finished with mosaic patterns of your own devising or merely painted with waterproof paints developed for swimming pools.

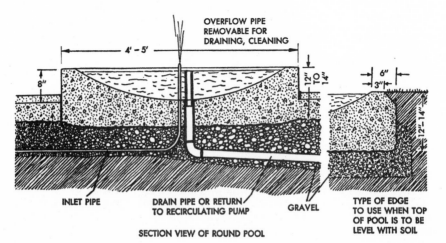

OVERFLOW PIPE
REMOVABLE FOR
DRAINING, CLEANING

4' - 5'

8"

12" TO 14"

6"

3"

12"-14"

INLET PIPE

DRAIN PIPE OR RETURN
TO RECIRCULATING PUMP

GRAVEL

TYPE OF EDGE
TO USE WHEN TOP
OF POOL IS TO BE
LEVEL WITH SOIL

SECTION VIEW OF ROUND POOL

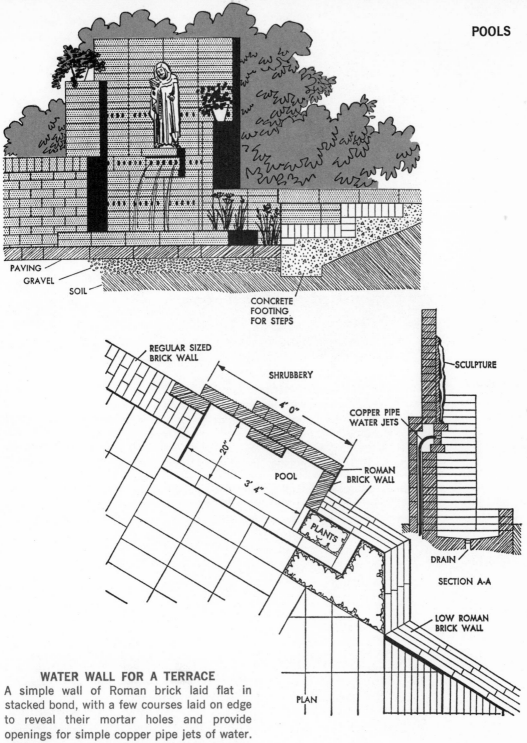

PAVING

GRAVEL

SOIL

CONCRETE
FOOTING
FOR STEPS

REGULAR SIZED
BRICK WALL

SHRUBBERY

4' 0"

20"

3' 4"

POOL

PLANTS

SCULPTURE

COPPER PIPE
WATER JETS

ROMAN
BRICK WALL

DRAIN

SECTION A-A

LOW ROMAN
BRICK WALL

PLAN

WATER WALL FOR A TERRACE

A simple wall of Roman brick laid flat in
stacked bond, with a few courses laid on edge
to reveal their mortar holes and provide
openings for simple copper pipe jets of water.
Although a piece of sculpture is shown here, to
reflect in the pool, a piece of driftwood, painted
ceramic tiles or some other feature might be
used as focal point.

417

POOL FOR NATURAL PLANTINGS

An easy-to-make concrete pool can conform with its natural surroundings if stones are set into the edge of the pool while the concrete is still wet. Paint the pool a dark color and use plants around it to mask even more the fact that it is man-made, and it will fit into its environment. Cattails and other water plants can be grown in an adjacent pool partly filled with soil. Note that both the pools are frostproof saucer shape.

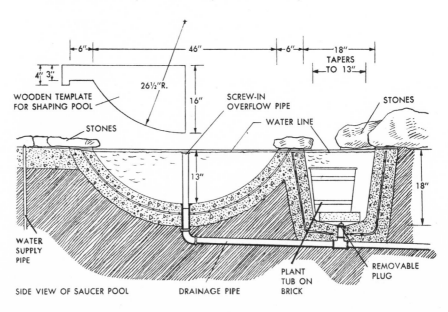

NO-DRAIN SAUCER POOL

Easiest and cheapest to build is the "saucer" pool, since a hose suffices for water supply and when it needs draining for cleaning, the water is swished out with an old broom and it is scrubbed as it is drained. In winter, in frosty regions, its saucer shape prevents any damage from freezing, for there are no sides to break as ice expands. Set rocks into wet edge of cement or place them later and plant between them. The overflow channel takes care of excess water from rains and can be masked by planting or use of rocks.

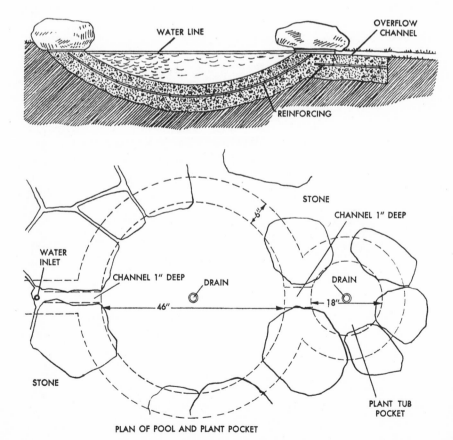

WATER LINE

OVERFLOW CHANNEL

REINFORCING

STONE

CHANNEL 1" DEEP

WATER INLET

CHANNEL 1" DEEP

DRAIN

DRAIN

46"

18"

STONE

PLANT TUB POCKET

PLAN OF POOL AND PLANT POCKET

419

A REFLECTING POOL BESIDE THE HOUSE

To provide reflections, a pool need not be deep, and this one is only 8" in depth with deeper planting pockets provided for the water lilies. Lilies, planted in tubs, may be set on bricks to raise them if they are short-stemmed kinds. Note that drain pipe services both planting pockets, has overflow pipe to draw off excess rain water, although pool is filled with a hose. If a shallow pool is painted a dark color—deep blue, green or black—nobody can tell exactly how deep it is. Also, the reflecting qualities are improved by using a dark color, and where plants are used, soil pockets don't show.

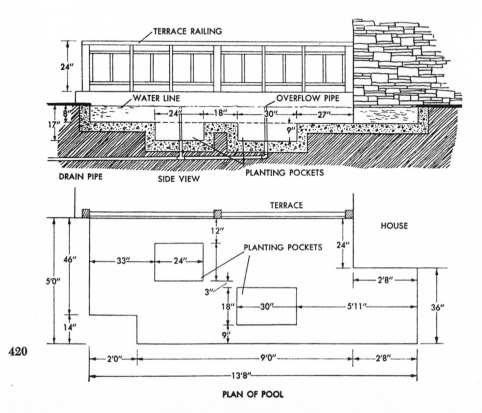

420

PLAN OF POOL

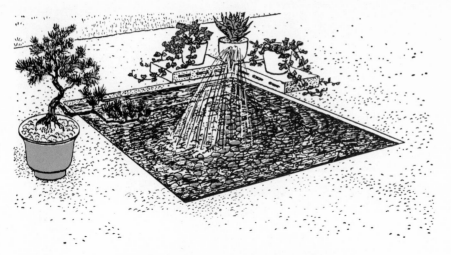

SQUARE POOL SET INTO TERRACE

A pool need not be deep or large to increase the enjoyment of all who behold it. This tiny square pool is adaptable to many locations; as shown here, it is set into a brushed aggregate concrete terrace, a small jet supplying the tinkle of falling water. A recirculating pump permits the use of the same water again and again. Square the pool's edge, or curve it (see detail, below), put loose, water-washed stones in several colors on the floor as indicated in the sketch above. Potted plants put on weathered wood planks or low cinder blocks embellish, as well as protect, the inlet jet from hurrying feet. Dispense with jet—and cut plumbing expenses. Use a hose to fill, siphon pool.

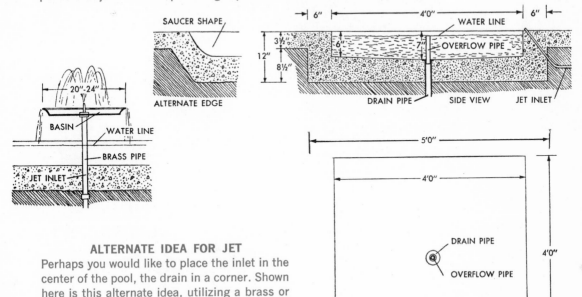

SAUCER SHAPE

ALTERNATE EDGE

WATER LINE

OVERFLOW PIPE

DRAIN PIPE   SIDE VIEW   JET INLET

20"-24"

BASIN   WATER LINE

BRASS PIPE

JET INLET

ALTERNATE IDEA FOR JET

Perhaps you would like to place the inlet in the center of the pool, the drain in a corner. Shown here is this alternate idea, utilizing a brass or aluminum tray with high edge, or a low bowl, for the basin. Drill a hole in the bottom, secure the basin with nuts and washers to the jet pipe. Jet spray falls into the basin, overflows, and in turn falls into the pool, providing sound effects for added delight.

DRAIN PIPE

OVERFLOW PIPE

JET INLET

PLAN OF POOL

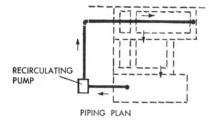

RECIRCULATING PUMP

PIPING PLAN

TRIPLE POOL FOR A TERRACE

Any terrace profits from the sound of running water, for even when it is torrid the sound of splashing water helps one to feel cool. A recirculating pump reuses the water, forcing it from the lowest pool up to the top one where the jet sprays it back into the pool, drains conveying it to the lower pools. Low plantings alongside the lower pools and a rustic fence behind plants enhance the decorative effect.

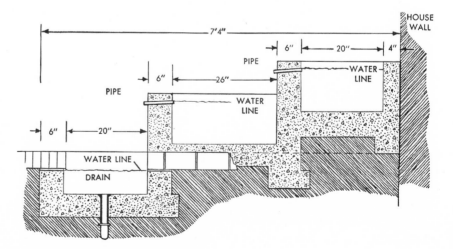

HOUSE WALL

7'4"

6" 20" 4"

PIPE

WATER LINE

6" 26"

PIPE

WATER LINE

6" 20"

WATER LINE

DRAIN

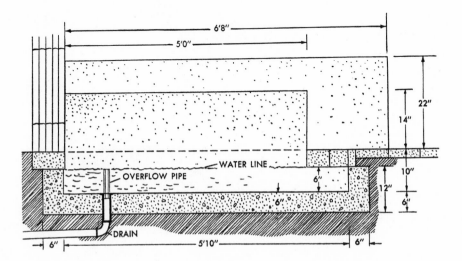

HOUSE WALL

PLANT BOX

4"

20"

PIPE

12"

6"

4'2"

6"

6"

FENCE

6"

12"

6"

30"

6"

26"

PLANT BOX

PIPE

PLANT BOX

6"

7'4"

5'10"

DRAIN

20"

PLAN AND
FRONT VIEW

6'8"

5'0"

22"

14"

WATER LINE

OVERFLOW PIPE

6"

10"

6"

DRAIN

6"

5'10"

6"

12"

6"

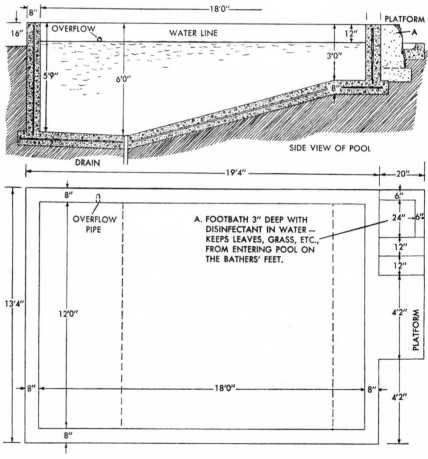

MINIMUM-SIZED SWIMMING POOL

In regions with a cheap water supply, swimming pools are not a luxury. Built partly above ground, this one minimizes excavations, cuts down danger of childrens' stumbling into pool. In warm regions drain may be omitted if siphon or submersible sump pump is used for weekly refilling. Concrete is finished smoothly, painted white or in color each year with a good waterproof paint made for swimming pool use.

8"

18'0"

PLATFORM

16"

OVERFLOW

WATER LINE

12"

A

3'0"

5'9"

6'0"

8"

SIDE VIEW OF POOL

DRAIN

19'4"

20"

8"

6"

OVERFLOW
PIPE

24" — 6"

A. FOOTBATH 3" DEEP WITH
DISINFECTANT IN WATER —
KEEPS LEAVES, GRASS, ETC.,
FROM ENTERING POOL ON
THE BATHERS' FEET.

12"

12"

13'4"

12'0"

4'2"

PLATFORM

8"

18'0"

8"

4'2"

8"

SWIMMING POOL PLAN — DIMENSIONS ADAPTABLE

23

Versatile Walls

When a man's home was his castle—literally—walls had a very definite function. They had to surround the castle, be high, strong, and permanent, able to take the shock of whatever assaults might be made on them from outside. Today, however, their function is less grim and they perform their duties with grace and beauty in a variety of architectural forms. Walls are truly coming into their own.

Today walls may form the boundary line between two properties, or they may be the line of demarcation between two parts of the garden. The wall can separate the beauty of the living portions of a property from the useful and functional, but un-lovely, service areas. It can act as a sound barrier or noise-cushion between a house and a busy street. It may be low enough to become useful as furniture—the seat wall is now an accepted fixture in outdoor planning—or it may rise high enough to screen out the prying eyes of nosy neighbors or curious passersby, or to break the force of winds which might sweep the terrace. Walls may be solid or pierced; smooth or very much textured; may sport any color of the rainbow or be unobtrusively painted to match the house; or be left in the sober and subtle natural hues of the material of which they are built.

A wall may be used to retain and hold back a hillside where it has been necessary to bulldoze out a driveway, walk, or terrace on a sloping site. It may be used to retain the roots of a tree around which the land has to be sliced down and evened as living-space and for which the shade of the tree will be needed. In every case the wall becomes a distinct asset to the property and by its permanence will continue to give satisfaction for many years to come. It has a useful function, but it can also contribute great beauty.

The wall is a year-round joy. In the growing season it will be a good foil for showing off your plants, your trees, and your shrubs. In winter it will give architectural form to the winter landscape, even under the snow, and is a background for bare branches so that their beauty can be appreciated. A garden with walls need never look so forlorn in winter as one which depends entirely on plant material for its effect. The walls are the "bones" of the garden on which the "flesh" of plant material is hung, and, as with humans, good bone structure will make the flesh more charming and beautiful. The wall can be the focal point of the garden, pairing well with steps and pools, with pergolas and shelters, concealing or revealing as the builder chooses, and tying in with the paving material of walks and terraces.

Because of its character, the wall can do wonderful things for the design of your garden. It need not be built always on straight lines, although this may be the easiest and safest course for the amateur builder to follow. It may flow in sweeping curves, be built in geometric arcs or semi-circles, angle off the straight line to make a compelling pointer to the focus of interest of the garden. High walls or low ones that are merely raised edges to define flower beds or terraces—all have their places in the garden picture today.

PRACTICAL WALLS

The easiest walls to build, quite naturally, are low ones. Blocks or bricks are the best materials to use because they are of even size and thus are easy to fit together into a trim structure. But because of their smaller size and lesser weight, bricks are easier to manage than blocks, a factor to remember if you plan to build a high wall. Hoisting heavy

Six-foot-high privacy walls built of diamond-patterned concrete grille blocks set between heavy piers of concrete blocks give distinct character to a garden, help to lift a somewhat usual kind of house into a class of its own, making it more original.

Enclosed by high dark-colored concrete-block privacy walls, a circle of lawn is bordered by a brushed aggregate walk and paving that extends indoors. The fire-pit seat and wall are built of light-colored concrete blocks, softened by the plantings set behind them.

The versatility of concrete is here demonstrated by the curving concrete edge of the flower beds which helps to cut maintenance labor. The split concrete-block retaining wall above the bed echoes curves, and high wall with grilles borders the lot.

Concrete-block grilles give light while maintaining privacy by day and contribute pattern to lighted windows at night. Such grilles keep hot sun from windows, too bright light from rooms. Single-core units set vertically overlap horizontal solid blocks.

Compelling strong curves give skeletal design and direction to the garden scheme. Making a wall of the sort shown here adds dimension as well as form. Stacked bond concrete blocks with well-fitted joints provide a certain pattern and texture.

blocks to shoulder height consumes energy and is tiring to the muscles. On the other hand, a block wall goes much faster because of the size of the units.

Bricks have many tones and textures, mostly warm, and our preference is for the smooth finishes rather than the wire-cut ones. The former, when laid with a raked joint, make a most beautiful texture in sunlight. Blocks are larger and clumsier looking, and not always so pleasant in color, although both bricks and blocks may be painted. (See Chapter 12.) Where a long, low line is wanted, a wall built of 2- or 3-inch solid concrete or cinder blocks will achieve this effect. Where a fairly smooth texture is desired, the wall of blocks with no emphasis on the joints will be a good answer, the texture of the blocks being rather pleasant, especially when painted. Such a wall will be cheaper to build than a brick one, too.

Where walls must be built on a hillside, step them down gracefully from level to level, with posts or piers forming logical points at which to break and lower them when the grade lowers sufficiently to make it necessary. Similarly, a wall may start out high on level plots where concealment is needed, drop by degrees to a seat wall, and dip to form a 6-inch parapet around a pool. It may rise again to make a seat or a privacy wall, to frame a good view, or to blot out an undesirable one.

429

WALLS CAN BE FRIENDLY

Not all walls are built for the purposes of concealment or protection of privacy, nor must they be solid and therefore somewhat unfriendly. The pierced wall, of which we show many examples in the sketches, is a favorite of gardeners today because it encloses an area, yet gives it ventilation (important for humans as well as plants) while maintaining privacy. It will break the force of hot or cold blasts of winds which sweep your land, taming them so that outdoor living can be made not only possible but even pleasant during normally severe weather. The decorative effect of bricks and blocks can hardly be praised enough, for they may be assembled in myriad ways to give practically any textural effect desired. No wall need be a blank enclosure when texture and shadow effects are present, even if no plantings are used with the wall.

Speaking of blank walls reminds us that we must examine the case of the high retaining wall. Often this may be a thing of necessity rather than one of beauty, yet this need not be the case. For the amateur

Open block walls not only allow air to flow freely through, ventilating outdoor living spaces such as patios, but also make handsome and effective architectural points. The two versions shown here make ever-interesting patterns of light and shade for visual enjoyment.

Two more open-block walls demonstrate possibilities offered for texture and pattern. Left, alternate rows laid in half-overlap make an informal pattern. Right, open cube blocks are combined with solid cube blocks to relieve and open up an uninteresting wall.

Light and shade always enhance block textures, but, when laid with alternate blocks recessed, a strong diagonal pattern is achieved by shadows. The wall and circular planter are built with slump blocks, which are notably similar in texture to adobe bricks.

Another patterned wall utilizes shaped blocks for its basic pattern, with single-core blocks set one above the other in staggered formation to give three-dimensional texture by their projection. Or, use solid blocks with open-single-core for wall according to taste.

Decorative effects may be achieved with little effort by using shaped blocks such as the diamond pattern here. Other blocks of many kinds are available in some areas, obtainable readily or on special order. Consult local dealers to see what choices are offered.

craftsman any wall of more than 3 or 4 feet in height is a considerable undertaking, particularly if the wall is to be made of concrete. We advise that, if the wall must be of more than 4 feet in height to retain the soil of a hillside, you hire a contractor to build it for you. If you wish to veneer the concrete later on with bricks or blocks, you will be able to do so if you arrange to have the foundation project enough to hold the veneer.

We wish to raise the question of whether or not retaining walls need to be so high. There is a bleakness which is very grim about the large stretches of unrelieved masonry. They also have problems of bracing and of making heavy footings to hold their weight.

We submit that it would be less troublesome to make terraces down the hillside, with two or more retaining walls spaced three to six feet apart. The amateur could build these himself. Footings would be less expensive and the weight of each wall less, the strain being divided between the two or more walls rather than concentrated on one. Besides all this, the aesthetic effect would be more pleasant. Planting beds in front of each wall would give a chance for shrubs, evergreens, flowers; and the long lines of the walls would make the hillside seem less dominant, because they would expand the visual width of the property and minimize the vertical measure. Vines could be trained to hang over the edge and soften the lines of the walls, and tall-growing shrubs would also break the lines interestingly.

THINGS TO CHECK ON

The first thing that must be determined is your property line, if the wall is to be built on or near it. You must be sure not to encroach on the neighboring property, or you may face a lawsuit or other unpleasantness and expense. It would be well worth the money to have the property re-surveyed and permanent markers inserted at the corners if there is any question about the exact location of your property lines.

You should also check with the local Building Inspector's office to see if there are any ordinances which limit the size, height, material, or character of the wall you plan to build; whether or not a building permit will be necessary for the construction of the wall. A permit may be required in some cities and towns. The Inspector can give you also any restrictions regarding setbacks and finish, and any other information contained in local codes which you will need to know. Codes in most places are mainly concerned with strength and stability of the construction so that no public hazard will result from its being built. In cold sections there may be some requirement for footings or foundations, because frost might heave the wall, crack it, and make it a potential danger. Where retaining walls must be built, even though they may be within the boundaries of your property and not on the edge, there may also be some restrictions locally, so you should check on them, too.

These things sound more important and complex, probably, than they actually are. The important thing is to check on them before you build, before you commit yourself to an order for building materials or even excavate, so that you will know the restrictions within which you must work. Then you can go ahead with an unclouded mind to the planning and building, secure in the knowledge that you're doing the right thing and that your wall won't have to be pulled down and rebuilt.

THICKNESS OF WALLS

Walls up to a foot high do not generally need to be more than one brick course in thickness—approximately 4 inches—and unless they must withstand weight or pressures from soil, as in a retaining wall, they may even go a few inches higher. Retaining walls and those higher than a foot or so are better made two bricks in thickness—about 8 inches—while those used for seat walls—16 to 18 inches may be two or three bricks wide, the three-brick width of 12 inches being preferred where the seat is all masonry with no wooden seat atop it. Walls may also be made of blocks, as we have detailed elsewhere, 4-inch-wide blocks being the narrowest width recommended, 6-, 8-, 10-, and 12-inch wide blocks being adaptable to various walls for various purposes. Concrete may be cast as a wall, too, 4 to 6 inches being the narrowest it is practical to cast for low walls. Above that, use any width that is practical for your purpose. Remember that all walls should be built on a proper footing. The general rule is to make the footing as deep as the width of the wall. That is, an 8-inch wall would have a footing 8 inches deep. The width of the footing should be double that of the width of the wall—16 inches in the case of an 8-inch wall—with the wall centered on the footing.

Block and concrete walls may be topped with a soldier course of bricks (see Chapter 12), finished with coping blocks, cut stone, or cast concrete to provide an overhang so that moisture does not drip or run down the face of the wall. Bolts may be embedded in the mortar or concrete so that wooden planks or other wooden seats may be fastened to the top of the masonry wall. Providing the proper finish on the top of the wall is important not only to the looks of the wall but also to insure its durability. Walls of blocks or concrete may be veneered on the face with bricks where they are used for low retaining walls.

Walls of waist height or higher which are more than moderate length may require the insertion of pilasters or piers every 10 feet. Intervals of 6 to 8 feet would be safer with higher walls. The exception to this rule is the serpentine wall, whose double curve, if based on a proper footing and foundation, will withstand a great deal of stress and strain merely by exerting opposing pressures of its own. A curved template or form built of hardboard or thin plywood, bent to the proper radius, and fastened to a wooden frame will be of assistance in building a serpentine wall quickly and easily. Lay out the basic curves for the first course on the foundation by using a stake and a cord and running an arc. Where this arc joins the arc made from the opposing radial center, the bricks or blocks are laid to join and then go in the opposite curve. For taller serpentine walls deeper curves are needed than for low ones, but for even shoulder-height walls a single-thick course of bricks is all that will be necessary. If you have ever seen the beautiful serpentine walls at the University of Virginia at Charlottesville, Virginia, that were designed by Thomas Jefferson, you know the truth of this statement.

If you wish to make wider serpentines, it is a good plan to run headers to tie together the double or triple courses every four to five courses.

BALUSTRADES

Rather than making solid walls to edge terraces and thus preventing the unwary guest from falling into flower beds below, use a balustrade to give the proper finish and designation to the edge and to signify the change in level. The flowers or shrubs beyond can be seen through the openwork and it will also allow ventilation and air circulation. As shown in the sketches, bricks may be used to build balusters. The details show some of the ways in which bricks can be laid, some high and some low balusters made this way, but all of them practical and tying in well with brick-paved terraces and brick-faced walls.

Balusters are not the easiest of projects for the amateur mason to attempt for they require a precision and skill in handling if they are to look right and stand up to the weather; but if you take your time in building them, use care and take the trouble to fit them as perfectly as you can, you will be rewarded with an unusual balustrade, one that you can be proud of and one that will be much admired by your guests.

STONE WALLS

Stone, as you will note in the pages detailing stone masonry patterns, (Chapter 11), is a beautiful medium. It is rather more difficult to manage than blocks or bricks, but the final result will more than justify all the fuss and fuming you may have to go through. Stones are probably heavier than most other materials and, being irregular in shape, they are harder to fit together. Stone mortaring is more difficult than mortaring other materials, too, because the stones may squeeze the mortar out with their weight. In cases of this sort you can insert chips of stone between courses to hold up the weight and allow the mortar to harden.

If stone is abundant on your property it may be the cheapest material to use, because it can be had for the mere labor of picking it up and taking it to the site of the wall. Use a rich mortar to be assured of a good strong joint, increasing the cement formula by ¼ to ½ part for each other part, making the mortar a bit stiffer than for use with bricks or blocks. It may be pointed out that more mortar may be required for stone walls than for brick or block ones due to the unevenness of the material and the need for filling holes with mortar.

Whether laid in mortar or laid dry (see next paragraph), the stone wall should be "battered"—sloped back—toward the top in the ratio of 1 to 2 inches for every 2 feet of wall height. Where the wall is to be laid dry and also used for a retaining wall, the batter should increase to 8 to 10 inches for every 5 feet of height. This will enable it to withstand the soil pressures.

DRY STONE WALLS

This is one of the most beautiful of stone walls, and, once the trick is mastered, one of the easiest to construct in many ways. The most suitable stones for it are those which are naturally rather flat, those which come from stratified outcroppings or which occur naturally as flattish blocks. Boulders and other rounded rocks are not so suitable and may cause failure of the wall due to their shape. Large boulders can be cracked and broken to secure flat edges, but the smaller ones are definitely to be discarded.

The dry stone wall need not be built on a foundation, the lower stones merely being set in a trench dug 6 to 18 inches below soil level on a base layer of crushed rock or rubble stone which has been well compacted by tamping. This will assure good drainage (it may be omitted in dry and mild climates) and will prevent frost-heaving which might dislodge or unsettle stones and necessitate repairs. A well-built dry stone wall will fit the stones together so carefully that they will hold themselves together by their own weight and the pull of gravity. (See the sketch in the illustration pages which follow for the detailed method of laying it.)

Dry stone walls can be planted with a variety of rock plants, which will enhance their beauty and attractiveness.

A FEW FINAL CAUTIONS

When you are building retaining walls be sure to provide for drainage behind them, either by laying drain tiles to carry away seepage or by leaving weep holes in the face of the wall to allow it to drain away. Footings and foundations are important to all walls. The stability and permanence of your wall will depend as much on the time and care you expend on what *doesn't* show below ground as it will upon the superstructure and the care with which that is built. Don't neglect the underpinnings! (See Chapter 14 for methods of making footings.)

After you have built your wall be sure to plant its surroundings well to give it the setting it deserves. Softening the sharp architectural lines by beautiful plants in front of or behind the wall, training a tendril of vine over the edge or on the face of a wall to contrast its soft lines with the crisp ones of the masonry will help to knit the garden wall into the fabric of the garden's design, quickly, beautifully, and permanently.

Never allow any walls to become so overgrown with ivy or any vine or shrubbery that they can't be seen, or that they lose their character. Remember that walls are the "bones" of your garden which must never be completely obscured by the fat "flesh" of growing things, or the garden design will become a bit flabby and unable to exhibit its true character. A little plastic surgery or pruning will quickly rejuvenate it and bring it back to its youthful loveliness.

A wall, we may repeat, can be the focal point of the garden, a year-round joy. We are sure you will agree as soon as you build one.

CONTINUED ▶

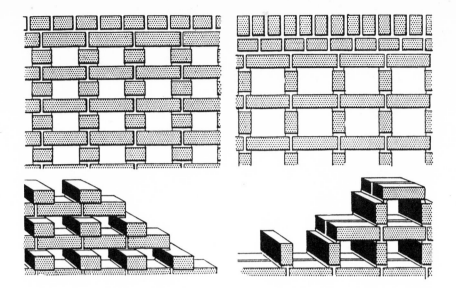

OPEN WALLS AND VENTILATION

All courses are laid flat, every other brick in header courses being omitted and stretcher joints centering on header bricks. The resulting shadows lend great textural interest. In this variation of the Rolok wall, a double stretcher course is laid with the joints centering on the upright headers, spaces between headers being left open for the passage of air.

A SOLID WALL WITH TEXTURAL INTEREST

For a wall beside a driveway, a walk or some other place with no room for plantings, get interest by texture and shadows. Lay a flat stretcher course centered on alternate header courses, one laid flat, the other on edge, both at a 45 degree angle to wall axis. Shawdow effects resulting from this decided texture will make this wall notably interesting architecturally.

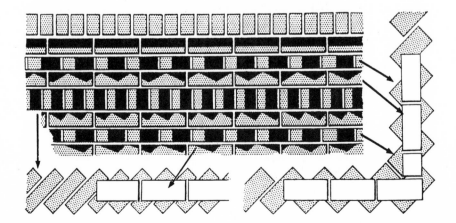

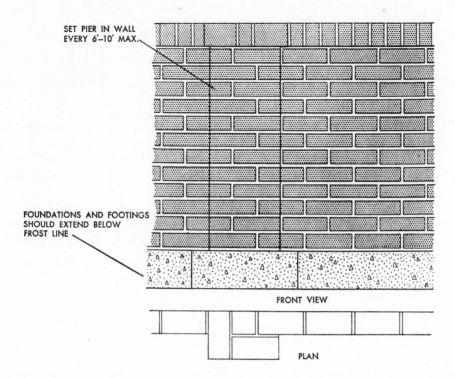

SET PIER IN WALL
EVERY 6'-10' MAX.

FOUNDATIONS AND FOOTINGS
SHOULD EXTEND BELOW
FROST LINE

FRONT VIEW

PLAN

Tall walls that need piers for added strength may project in front the thickness of one brick and may also project on the back of the wall. Headers in alternating courses tie pier into wall (see plan). Retaining walls (below) may be backed up with blocks, faced and capped with bricks. Note walls are centered on footings for stability.

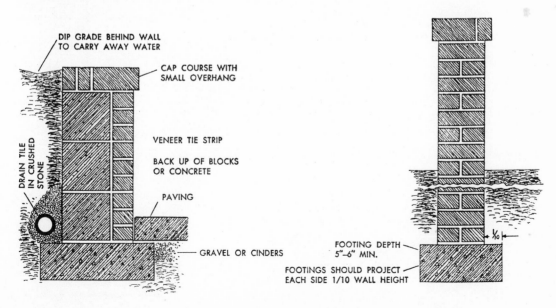

DIP GRADE BEHIND WALL
TO CARRY AWAY WATER

CAP COURSE WITH
SMALL OVERHANG

VENEER TIE STRIP

BACK UP OF BLOCKS
OR CONCRETE

PAVING

DRAIN TILE
IN CRUSHED
STONE

GRAVEL OR CINDERS

FOOTING DEPTH
5"-6" MIN.

FOOTINGS SHOULD PROJECT
EACH SIDE 1/10 WALL HEIGHT

1/10

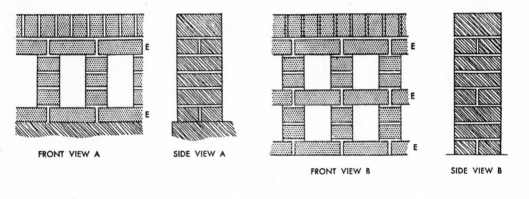

FRONT VIEW A   SIDE VIEW A

FRONT VIEW B   SIDE VIEW B

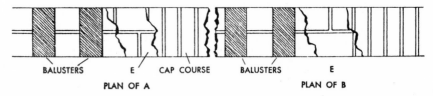

BALUSTERS   E   CAP COURSE   BALUSTERS   E

PLAN OF A   PLAN OF B

Brick balustrades lend a decorative touch, offer craftsmen plenty of scope for ingenuity in creative designs. Openwork brick balusters give strength and support while allowing for ventilation and providing pleasing patterns in the landscape. Balustrades built on the edge of a terrace give extra seating space for outdoor parties.

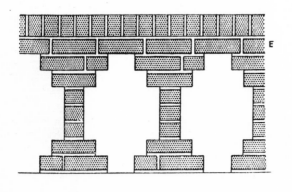

SIDE VIEW OF BALUSTER

BALUSTER   SUPPORTING COURSE E   CAP COURSE   PLAN OF BALUSTER

440

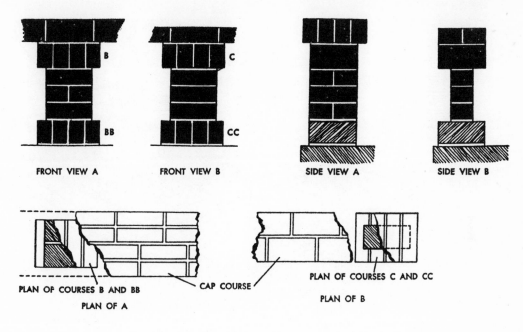

FRONT VIEW A FRONT VIEW B SIDE VIEW A SIDE VIEW B

PLAN OF COURSES B AND BB   CAP COURSE   PLAN OF COURSES C AND CC

PLAN OF A   PLAN OF B

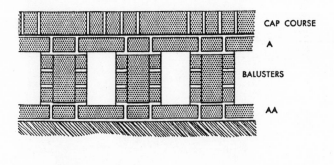

CAP COURSE

A

BALUSTERS

AA

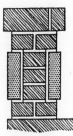

SIDE VIEW

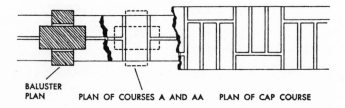

BALUSTER PLAN   PLAN OF COURSES A AND AA   PLAN OF CAP COURSE

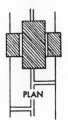

PLAN

441

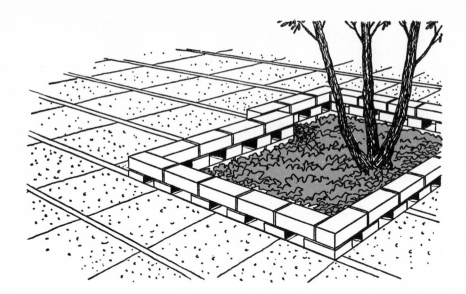

TREE UNDER A TERRACE

When an existing tree is below terrace level, box it in and slope the terrace toward it so that moisture will drain through openings in the bottom course of a low brick wall surrounding it. The deeper the tree hole the higher the wall should be. You may want to make it high and wide enough to use as a seat, employing either wood or masonry.

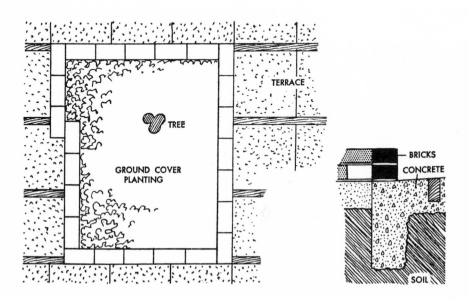

442

" 'ROUND THE MULBERRY BUSH' "

Sometimes a circular wall is the best answer, especially if it ends a terrace. Here a low retaining wall circles the tree, pie-shaped steps rising to winding path. Broken bits of an old cement sidewalk are salvaged for use; their rough sides with gravel showing will give the wall an agreeable ruggedness. Lay with raked mortar joints or lay it dry.

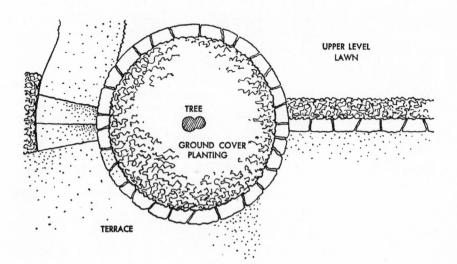

UPPER LEVEL
LAWN

TREE

GROUND COVER
PLANTING

TERRACE

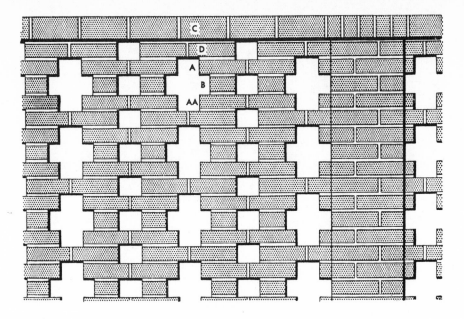

CROSS AND SQUARE PATTERN

Built two bricks thick with three-brick piers inserted every six to nine feet, this open-work pattern is decorative yet strong. Of interest to amateurs is the fact that no bricks need be cut except where it may be necessary for fitting into piers. It may also be adapted to single-brick thickness with two-brick piers at intervals for strengthening.

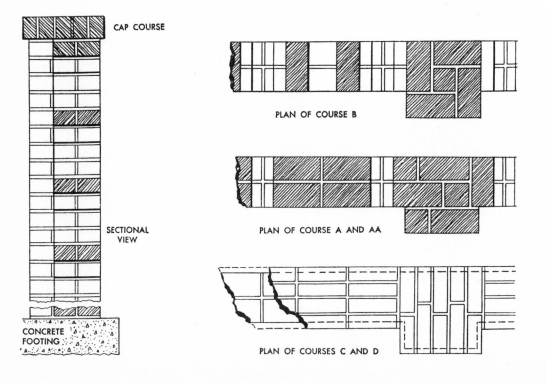

CAP COURSE

SECTIONAL VIEW

CONCRETE FOOTING

PLAN OF COURSE B

PLAN OF COURSE A AND AA

PLAN OF COURSES C AND D

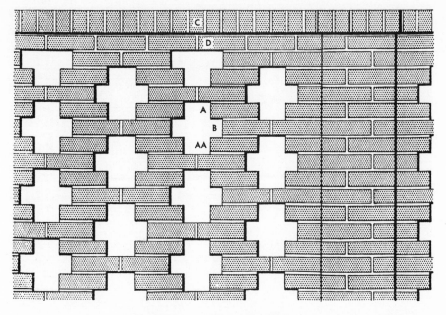

STAGGERED CROSS PATTERN

Another wall design which may be built with a minimum of brick cutting is this single-brick wall, three brick piers set at intervals of six feet, the pleasant cross design set alternately, vertically. Note how the solid top courses provides support for the cap course without materially affecting the regularity of the design or adding undue weight.

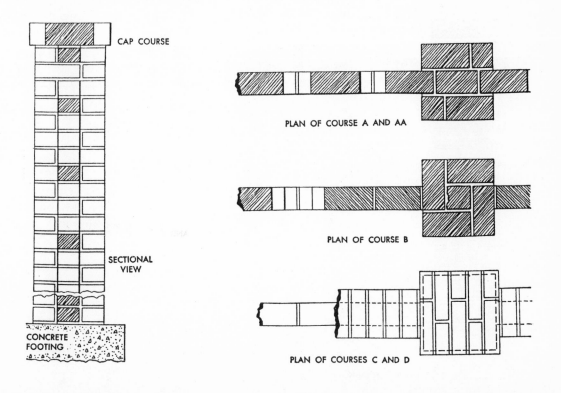

CAP COURSE

SECTIONAL VIEW

CONCRETE FOOTING

PLAN OF COURSE A AND AA

PLAN OF COURSE B

PLAN OF COURSES C AND D

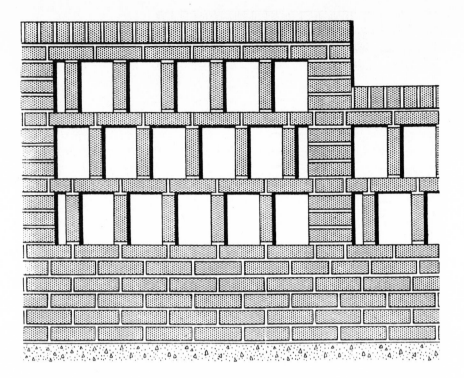

GRAND OPENING

Clean lines and pleasing regularity make a wall compatible with either traditional or modern houses. Two-brick thickness is sued throughout, including the piers which are set no more than six feet apart. Build as shown, above a solid low wall, or keep wall open.

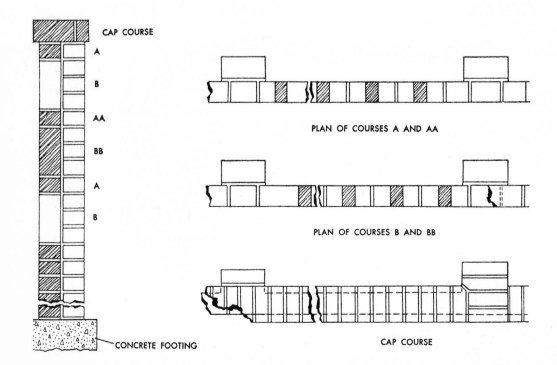

CAP COURSE

A

B

AA

BB

A

B

CONCRETE FOOTING

PLAN OF COURSES A AND AA

PLAN OF COURSES B AND BB

CAP COURSE

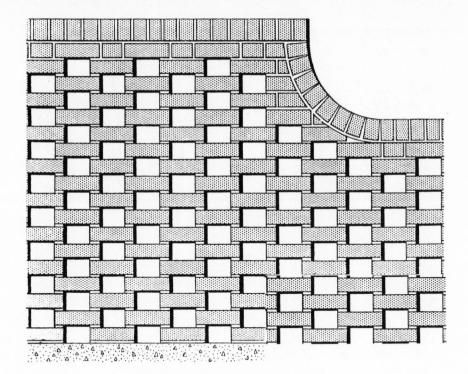

CHECKERBOARD AND CURVES

A single-brick thickness makes a lighter-weight wall, especially when openings are left between bricks equal to one brick-end plus mortar joint. Place piers every six or eight feet; between piers step wall down with a curve as shown, to follow slope.

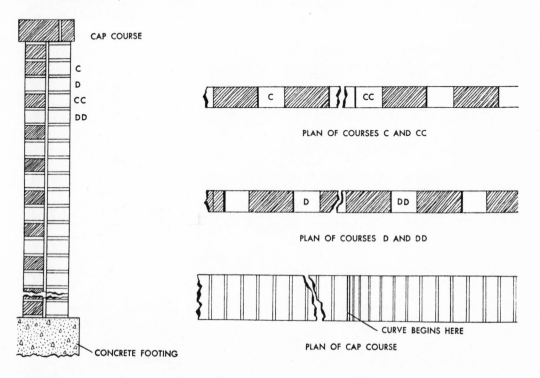

CAP COURSE

C
D
CC
DD

CONCRETE FOOTING

PLAN OF COURSES C AND CC

PLAN OF COURSES D AND DD

CURVE BEGINS HERE

PLAN OF CAP COURSE

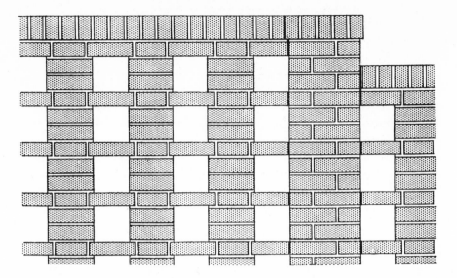

FOURSQUARE AND ON THE LEVEL
Regularity of pattern combined with light weight make a most desirable wall. This one is easy to build and sturdy, too, if a proper foundation is provided and piers are inserted every six to ten feet. Although the wall is only one brick thick, piers should be three bricks square and wall bricks well tied in.

ZIG AND ZAG FOR EXCITEMENT
To dramatize plain surroundings it is a good practice to furnish architectural excitement. It is easy to lay this wall. Leave openings between bricks equal to one brick-end plus a mortar joint; stagger courses one-quarter. Make every 4th course a solid "tie" one.

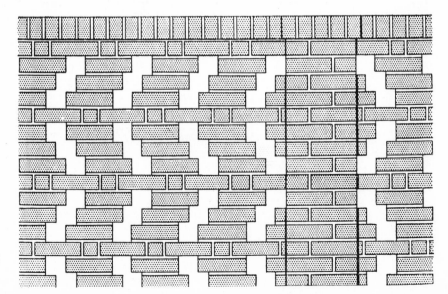

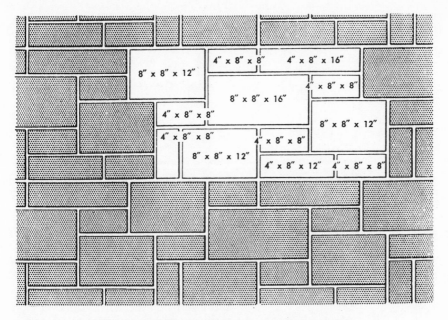

COURSED PATTERNED ASHLAR—TWO VERSIONS

Something of the charm of the ashlar pattern so frequently used in stone masonry may be obtained by grouping various stock sizes of blocks together into a "module" which repeats over and over. At corners and around openings for gates it will be necessary to improvise to fill out the blocks evenly. Don't let too many vertical joints align.

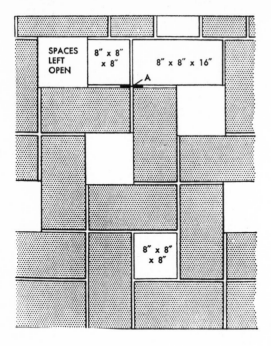

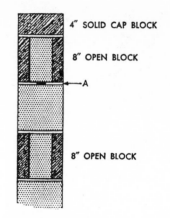

WHIRLING SQUARE PATTERN

By laying every other block vertically as indicated, squares will occur which may be left open as seen in the upper part of the sketch or filled with square blocks as shown below. Scrap iron rods inserted in the mortar (A) will tie the wall together, making it strong and secure however it is constructed.

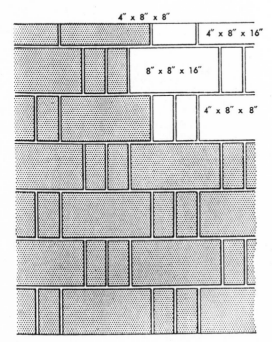

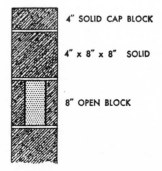

DOUBLE BASKETWEAVE

An interesting variation of basketweave pattern is seen in this wall. Note the 4″ x 8″ x 8″ blocks which are the basis of the linear pattern; or they might be laid horizontally instead, which would give a long-line horizontal effect. If less pattern is desired, 8″ x 8″ blocks may be substituted; also easier to build.

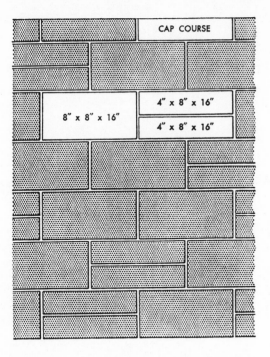

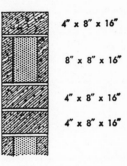

4" x 8" x 16"

8" x 8" x 16"

4" x 8" x 16"

4" x 8" x 16"

REGULAR COURSED ASHLAR

Two 4" blocks laid together space out to be equal, vertically, to an 8" block. Regular designs result from alternating - irregular patterns from laying every second block, as shown here, staggered to give motion to the pattern. Or they may be scattered here and there, to break up the monotony of a plain wall. Raked joints increase effectiveness.

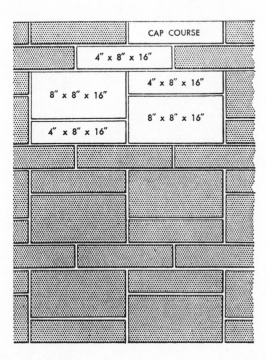

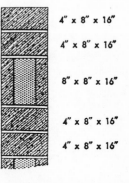

4" x 8" x 16"

4" x 8" x 16"

8" x 8" x 16"

4" x 8" x 16"

4" x 8" x 16"

SPACED COURSE ASHLAR

By laying a 4" block on top of an 8" block and next to it a 4" block under an 8" one, a pleasing variety of pattern is obtained. A 4" block "spacer course" between courses will tie the wall together successfully and give a coherence and unity to the design. Concrete blocks may be used for the main courses, cinder blocks used for "spacers."

451

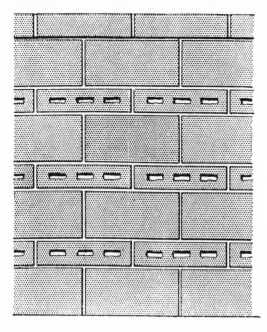

8" x 8" x 16"
OPEN BLOCK

4" OPEN BLOCK LAID
WITH HOLES EXPOSED

TO LENGTHEN, USE LINES

Narrow blocks set with the holes exposed give a strong horizontal line. 4" blocks are used here but 6" or 8" blocks would serve equally well and give greater ventilation.

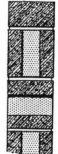

8" OPEN BLOCK

8" OPEN BLOCK LAID
WITH HOLES EXPOSED

PEEKABOO WALL

Ornamental effects are possible when a few blocks are laid to expose their holes in walls otherwise solid. 8" blocks must be used for all units in making this attractive wall.

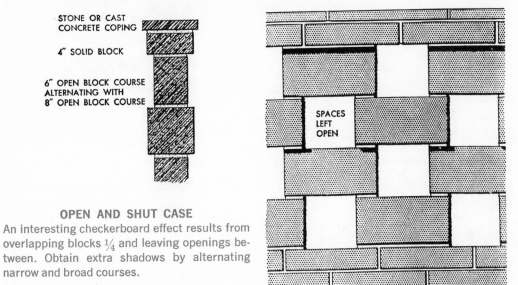

STONE OR CAST
CONCRETE COPING

4″ SOLID BLOCK

6″ OPEN BLOCK COURSE
ALTERNATING WITH
8″ OPEN BLOCK COURSE

SPACES
LEFT
OPEN

OPEN AND SHUT CASE

An interesting checkerboard effect results from
overlapping blocks ¼ and leaving openings between. Obtain extra shadows by alternating
narrow and broad courses.

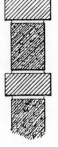

COURSE OF BRICK

6″ BLOCK COURSE
ALTERNATING WITH
COURSE OF BRICK

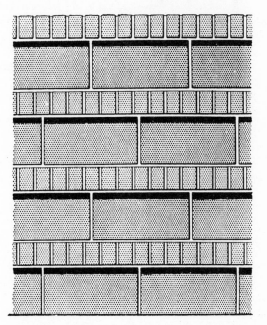

BRICKS AND BLOCKS

Another interpretation of the long line. To give
wall unity, paint it all one color, or paint either
bricks or blocks a color harmonizing with the
natural color of the other.

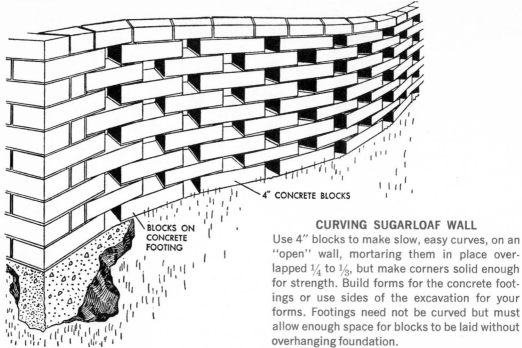

4" CONCRETE BLOCKS

BLOCKS ON CONCRETE FOOTING

CURVING SUGARLOAF WALL

Use 4" blocks to make slow, easy curves, on an "open" wall, mortaring them in place over-lapped 1/4 to 1/3, but make corners solid enough for strength. Build forms for the concrete foot-ings or use sides of the excavation for your forms. Footings need not be curved but must allow enough space for blocks to be laid without overhanging foundation.

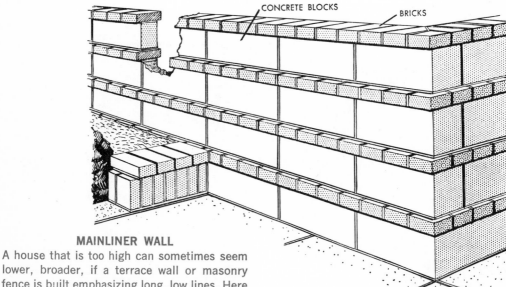

CONCRETE BLOCKS   BRICKS

MAINLINER WALL

A house that is too high can sometimes seem lower, broader, if a terrace wall or masonry fence is built emphasizing long, low lines. Here such a wall seems still lower because a plant-ing box (left) cuts it visually to only three blocks. Brick is used to provide unity, tying in with the bricks used in the wall for alternate courses to make shadow line.

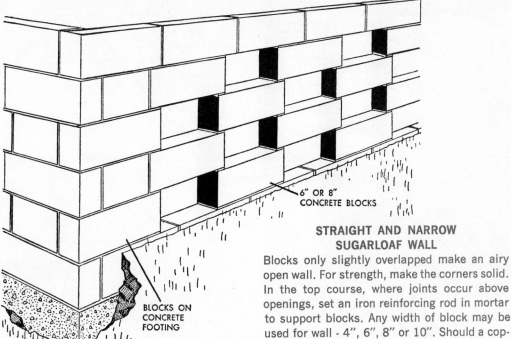

6" OR 8"
CONCRETE BLOCKS

BLOCKS ON
CONCRETE
FOOTING

STRAIGHT AND NARROW
SUGARLOAF WALL

Blocks only slightly overlapped make an airy open wall. For strength, make the corners solid. In the top course, where joints occur above openings, set an iron reinforcing rod in mortar to support blocks. Any width of block may be used for wall - 4", 6", 8" or 10". Should a coping finish be desired, brick, stone or cast concrete may be used.

COPING

MOWING STRIP

UP AND DOWN COMBINATION

This wall combines several features worthy of note: Built entirely of 4" blocks with stone or cast concrete coping, it uses horizontal placement for the higher wall, vertical setting for the low seat wall, thus contrasting visual lines interestingly. Note that mowing strip is cast separately from the footings, with expansion joints to prevent cracking.

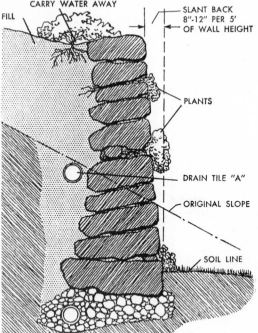

PLANTS IN DRY STONE WALLS

When a sloping plot is being leveled for use as a terrace, a dry stone wall (laid without mortar) often makes a good retaining wall, and offers planting possibilities. Properly constructed, it will last indefinitely, resisting the thrust of frost and gravity's pull. Rock garden plants and succulents are good subjects for planting, adding color and texture.

CONSTRUCTION

DRAINAGE under the wall is mandatory. Use 8"-12" of coarse rubble or crushed rock for a base, excavating so that the base of bottom course is set 6" below soil level, 12" or more in regions of deep frost. In wet climates use a line of drain tile in rubble, with a second line just below the original slope line which will carry away excess water.

SLANT THE WALL back 8"-12" per 5' of wall height (In wet climates use 12"). Tilt up all stones to catch rain, convey it back and downward, thus preventing soil washouts.

MAKE CREVICES for plants as courses are laid. Keep joints narrow at front, broaden toward rear for root space. Plant before laying next course, packing soil firmly and watering well. Mix leafmold or peatmoss together with soil to prevent undue erosion.

OTHER TIPS: a low ditch behind wall keeps mud from washing over wall face; plants in beds along the top soften wall's hard lines. Be sure to chink with stone chips to prop up stones, keep wall firm and secure. Don't space plants too evenly; don't overplant or you'll have a frowzy wall that is never neat.

LOW CHANNEL TO CARRY WATER AWAY

FILL

SLANT BACK 8"-12" PER 5' OF WALL HEIGHT

PLANTS

DRAIN TILE "A"

ORIGINAL SLOPE

SOIL LINE

DRAIN TILE AND/OR CRUSHED STONE, 8" MIN.

456

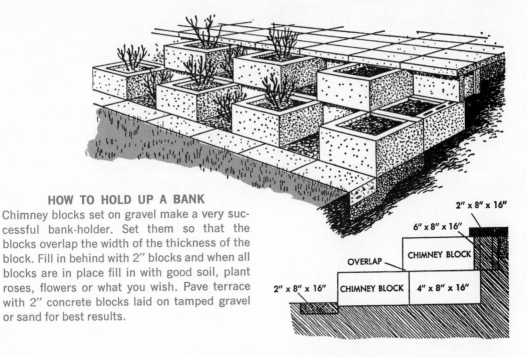

HOW TO HOLD UP A BANK

Chimney blocks set on gravel make a very successful bank-holder. Set them so that the blocks overlap the width of the thickness of the block. Fill in behind with 2″ blocks and when all blocks are in place fill in with good soil, plant roses, flowers or what you wish. Pave terrace with 2″ concrete blocks laid on tamped gravel or sand for best results.

2″ x 8″ x 16″

6″ x 8″ x 16″

OVERLAP   CHIMNEY BLOCK

2″ x 8″ x 16″   CHIMNEY BLOCK   4″ x 8″ x 16″

GENTLE SLOPE HOLDER

A few rows of chimney blocks overlapped as shown and filled with good rich soil will not only hold the bank but when perennials, annuals, herbs or roses are in flower, this bank will be a tremendous garden asset. Wood plank holds back soil while a moving strip of blocks finishes off the bottom of slope.

OVERLAP

CHIMNEY BLOCK

4″ x 8″ x 16″

SPARE THAT TREE—YOU'LL NEED IT

Don't cut down a tree because it interferes with your plans for a terrace. By carefully excavating and providing a masonry box around it you'll keep it to shade your terrace. Roots extend outward to tips of branches; most trees need about ⅔ of these roots to maintain good health. Don't cut root area drastically or tree may die.

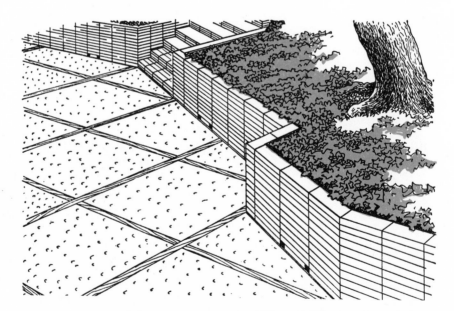

RETAINING WALL RETAINS A TREE

If a slope must be cut back to gain space for a terrace, a retaining wall becomes necessary. Make a jog around an existing tree, gaining architectural interest by slanting the wall. Note that masonry units set in stacked bond in retaining walls require a backing of reinforced concrete or masonry to give walls strength needed to withstand soil pressures.

Index

Aggregates in concrete, 278–80
  coarse aggregates, 279
  exposed aggregates, 296–98
    Finishing Exposed Aggregates, 297 (*fig.*)
    garden steps of, 308 (*fig.*)
  fine aggregates, 278–79
  gravel, 279–80

Balustrades, 435, 440–41 (*figs.*)
Bamboo, 159–60, 172
Beach house, deck for, 344 (*fig.*)
Beginners, project for, *see* Little Projects
Bird feeders, 45
  All Year Bird Shelter-Feeder, 57 (*figs.*)
  foods preferred by garden songbirds or non-game birds, table of, 59
  painting of, 48
  Two Ways to Feed the Birds, 55 (*figs.*)
    Other Ways to Feed Birds, 56 (*figs.*)
Birdhouses, 45, 46–54

building of, adapted to specific birds, 47
cleaning of, 50
  Birdhouse Cleanouts, 53 (*figs.*)
commercial market type of, 48
dimensions, table of, 58
drainage holes in, 50
entrance holes, sizes and placement of, 50–51
for wrens or bluebirds, 52 (*figs.*)
materials for, 47–48, 49–50
Open House for Fresh-Air Birds, 54 (*figs.*)
painting of, 48, 50
where to place, 49
Bird Houses and How to Make Them (U. S. Dept. of Agriculture), 51
Bird shelters, 45
  All Year Bird Shelter-Feeder, 57 (*figs.*)
Bird Study (Boy Scouts of America), 51
Birds, sources of information on attracting and housing of, 51, 52, 58n.
Blades for bench or table saw, 24–26
  carbide-tipped blades, 25

Blades for bench or table saw
  (*continued*)
 dado heads, 25
 molding-cutter head, 25–26
 nail-cutting blades, 25
 plywood cutting blades, 25
Blocks, building, kinds of:
 basic sizes and kinds of, 253 (*fig.*)
  other useful kinds of, 254 (*fig.*)
 colored blocks, aggregates selected, 235
 concrete, 233
 glazed, 234
 grille, 234–35
 ground-face, 235
 shaped-face or sculptured, 235
 slump, 234
 split, 234
Blocks, building, laying of, 241, 251–59 (*figs.*)
Blocks, paving, *see* Paving blocks, casting of
Board foot, definition of, 36
Boxes, old, use of wood of, 40
Breakfront for Plants, 205–6 (*figs.*)
Bricks:
 table of nominal sizes of, 251
 table of quantities needed for 100 sq. ft. of wall area, 272
Bricks, kinds of:
 common, 230
 features found in some bricks, 232–33
 firebrick, 230
 paving, 231, 338, 354
 pressed, 230
 Roman, 231–32
 SCR bricks, 232
 special order types of, 232
 unbaked, 230–31
 used, 231
Bricks, laying of:
 footings and first courses, 240–41
 how to cut bricks, 248
 how to lay bricks, 241–42
 mortar, proper use of, 242–43
 skintled brick work, 271–72
 succeeding courses, 248–49
 tips on, 250–51
 wall, cleaning of, as you go along, 249–50

 wetting of bricks before laying, 239–40
Bricks, patterns for laying of:
 Common Bond, 244 (*fig.*)
 Diaper Pattern, 228, 245 (*fig.*)
 English Bond, 228, 244 (*fig.*)
 Flemish Bond, 228, 244 (*fig.*)
 Flemish Rolok, 245 (*fig.*)
 One-Fourth Bond, 246 (*fig.*)
 One-Third Bond, 246 (*fig.*)
 Rolok Wall, 228, 245 (*fig.*)
 Roman Brick, 246 (*fig.*)
 Stacked Bond, 229, 246 (*fig.*)
Building blocks, *see* Blocks, building headings

Cedar, 32, 33, 41, 159, 288
Cements for patching of concrete work:
 epoxy type, 305
 latex type, 304
 vinyl type, 305
Clay tile, 236
 wetting of, before laying, 239–40
Clay tile, hollow, laying of, 241, 251–59
 typical sizes of hollow tile, 252 (*fig.*)
Coffee tables, outdoor, 202 (*figs.*), 207 (*figs.*), 208 (*figs.*)
Coldframes, 46
 Every Garden Can Use a Coldframe, 67 (*figs.*)
Concrete, and concrete work, 273–305
 aggregates, 278–80
 building of forms, 288–91 (*fig.*)
 coloring concrete, 284
  troweling in color, 285–86
 concrete formulas, 280–81
  two basic formulas, 281
 curing process, 298
 final finishing of the surface, 295–98
  combed rough finish, 295–96
  exposed aggregates, 296–98 (*figs.*)
  medium-rough finish, 295
  smooth finish, 296
 how to lay a concrete slab, 290–93 (*figs.*)
 how to mix concrete, 281–83
 how much concrete to mix for the job, 283

Concrete, and concrete work
    (*continued*)
    table for determining amounts of
        ingredients needed, 283
    making footings, 299–300
    painting, dyeing, staining, 286–87
    patching of, 304–5
    planning for moisture, 302–3
    Portland cement, 277–78
    pouring the concrete, 294
    preliminary finishing, 294–95
    reinforcing concrete slab, 299
    subsurface drainage, allowing for,
        287–88
    tools for concrete work, 226–27
    transit-mix concrete, 275–76
    walls of poured concrete, 300–302
    walls to retain trees, 303–4
    water used in mixing, 280, 281
    "you-haul" concrete, 276–77
    see also Paving blocks, casting of
Copper naphthenate, 42–43
Copper sulfate, 43
Crates, old, use of wood of, 40
Creosote, as wood preservative, 42,43
Cypress, 32, 33, 41, 159, 288

Dead tree, making use of, in garden,
        161–62
    Shade for When a Tree Dies, 175
        (*figs.*)
Decay, as cause of wood deterioration,
        41, 44
Decks, for use as sitting-out spaces,
        343–47 (*figs.*), 356–57
    Use Wood for Decks, 365 (*fig.*)
Do-it-yourself projects, surge of, in re-
        cent years, 2–3
Doors brought up-to-date:
    Ugly Duckling Doors, 46, 68 (*figs.*)
Dooryard gardens, with low fences,
        100, 136–37 (*figs.*)
Douglas fir, 32
Driveways, 383–92
    bringing up-to-date, 383–84
    charting the driveway, 385
    checking requirements, 384–85
    driveway with angles to street, 389
        (*fig.*)
    driveway with dividend of beauty,
        389 (*fig.*)

How to Lay Out the Driveway and
        Turnaround, 386 (*figs.*)
    "landing strips" alongside of, 383
    making entrance attractive, 388
        (*fig.*)
    measurements for, 385
        width of, 388 (*fig.*)
    pattern concrete areas, 390 (*fig.*)
    paving materials for:
        not permanent, but usually satis-
            factory, 391
        not recommended, 391–92
        permanent, 387, 391
    plant material, possible involvement
        of, 385
    semicircular driveway, 390 (*fig.*)

Fence enders, designs for:
    Angle Ender, 128 (*fig.*)
    Fence With a Waggle, 128 (*fig.*)
    Painted Space Division Ender, 128
        (*fig.*)
    Rail Terminals, 129 (*fig.*)
    Tapering Trellis Ender, 129 (*fig.*)
    Trellis Fence Enders, 90 (*figs.*)
    Triple Trellis Terminal, 129 (*fig.*)
Fences, 91–151
    boundary definers, and dooryard
        fences, 100, 136
    children's play-yard fences, 99–100
    concealment fences, 99
    developments of, through man's his-
        tory, 91–92
        fences of yesteryear, 92–93
    dooryard low fences, 100, 136–37
        (*figs.*)
    facing boards and other "skins," 105
        –6
    fence enders, 128–29 (*figs.*)
    fence posts, 103–4 (*figs.*)
    fence rails, 105
    fence shelters, use of variety of ma-
        terials in, 151 (*figs.*)
    fences as walls, 108–9
    frost, and fences, 97–98
    "outdoor room," enclosing of, 100–
        101
    picket fences, 106
    planning the fence in advance, 102–
        3

Fences (*continued*)

privacy fences, 93–97(*figs.*), 204 (*figs.*)

protection, fences for, 97

regulations regarding, 101–2

roll-wire fencing, 108

unusual materials for unusual effects, 107–8

wind deflection, fences for, 98–99

Fences, designs for, 109, 110–51

Adapted from Designs of Thomas Jefferson, 138–39(*figs.*)

Alternating Oblongs, 112(*figs.*)

Animal-Retainer Fence, 113(*figs.*)

Baffle Fence Plant Box, 149(*figs.*)

Band and Baffle, 145(*figs.*)

Basic Frame to Use With Different Coverings, 110–11(*figs.*)

Chevron Design, 131(*figs.*)

Comfort and Privacy, 150(*figs.*)

Completely Baffled, 142(*figs.*)

Concentric Oblongs, 117(*fig.*)

construction of gate, 117(*figs.*)

Cornered and Baffled, 143(*figs.*)

Design With Movement, 133(*figs.*)

construction of gate, 133(*figs.*)

Diagonal Squares, 114(*figs.*)

Diamonds and Oblongs, 140(*figs.*)

Dignified Diamond Pattern Fence, A, 124(*figs.*)

construction details, 124(*figs.*)

Dowel Pickets, Pointed Posts, 135 (*figs.*)

Feature Fences—For Modern, Abstract Effect, 146(*figs.*)

Gate and Fence of Wide Pickets With Cutouts, 122(*figs.*)

how to make pickets, 122 (*fig.*)

Harlequin Pattern, 131(*figs.*)

Interesting Dowel Picket Fence, An, 125(*figs.*)

how to build, 125 (*figs.*)

Interlocking Block Pattern, 144 (*figs.*)

Interlocking Oblongs, 130(*figs.*)

Interlocking Squares, 130(*figs.*)

Interrupted Horizontals, 140(*figs.*)

Lightweight Dooryard Fence, 136 (*figs.*)

Long and Short Dowels, 118(*figs.*)

construction of gate, 118(*figs.*)

Long and Short Oblongs, 115(*figs.*)

Long Lines, Graduated Boards, 137 (*figs.*)

Mainly Horizontal, 114(*figs.*)

Modern Spindle Fence, 116(*figs.*)

construction of gate, 116(*figs.*)

Oriental Peekabo Fence, 144(*figs.*)

Peek-A-Boo Pickets, 120(*figs.*)

construction of gate, 120(*figs.*)

Pointed Posts, Widely Spaced Picket Fence, 123 (*figs.*)

how to assemble pickets, 123 (*figs.*)

Privacy Fence With Let. Down Table, 204 (*figs.*)

Punctuated Regularity, 126 (*figs.*)

construction of gate, 126 (*figs.*)

Ribbons and Bars, 115 (*figs.*)

Round-Tip Pickets, 121 (*figs.*)

Same Fence, Different Treatment, 147 (*figs.*)

Slightly Punctuated, 127 (*figs.*)

construction of gate, 127 (*figs.*)

Square Pickets, 137 (*figs.*)

Square Pickets and Moldings,, 135 (*figs.*)

Squares and Oblongs, 141 (*figs.*)

Squares With Diagonals, 112 (*figs.*)

Strip and Stripe, 145 (*figs.*)

3-Dimensional Oblongs, 113 (*figs.*)

Trellis-Topped View-Breaker Fence, 87–88 (*figs.*)

Unpretentious, Yet Full of Character, 148 (*figs.*)

Up and Down Fence, 111 (*figs.*)

Use Imagination and Make an Interesting Fence (three designs), 134 (*figs.*)

Whirling Oblongs, 141 (*figs.*)

With Diagonal Movement, 119 (*fig.*)

construction of gate, 119 (*figs.*)

Wooden Fence, Brick Style, 132 (*figs.*)

construction of gate, 132(*figs.*)

See also Fence enders, designs for

Fir, 32

Furniture for outdoors, *see* Outdoor furniture

Furniture, table of measurements and heights:

Furniture, table of measurements and
    heights (*continued*)
    for adults, 213
    for children, 213

Garden Shelters, 152–93
    achieving privacy for, 155
    characteristic of today's shelters,
        152–53
    choosing shelter, 155, 157
    displaying and sheltering of summer-
        ing house plants in, 159
    elaborate style not needed in, 154
        (*fig.*)
    flat-roofed exterior-plywood-covered
        storage unit, 191 (*fig.*)
    garden-house-gazebo of unique de-
        sign, 160 (*fig.*)
    garden house with space for every-
        thing for outdoor entertain-
        ing, 192 (*fig.*)
    garden storage house designed like
        a barn, 192 (*fig.*)
    gazebo, revival of, 153
    making use of, 162
    making use of dead tree in, 161–62,
        175 (*figs.*)
    materials for, 159–61
    octagonal garden shelter, 156 (*fig.*),
        157
    placement of posts, 158
    placing of, 155
    playhouse for the kids designed like
        a barn, 192 (*fig.*)
    play-place for children, 159
    screening service yard with, 159
    shelter with off-the-ground storage,
        189 (*fig.*)
    size and height, 158
    tool storage house, 189 (*fig.*)
        in jog in fence, 193 (*figs.*)
    triangular garden house with every-
        thing under one roof, 190
        (*figs.*)
Garden Shelters, designs for:
    Attractive Shelter in a Corner, 154,
        186–88 (*figs.*)
    Eyrie for Eating or Dreaming, An,
        178–79 (*figs.*)
    Garden House With a Modern Fla-
        vor, 182 (*figs.*)

    construction details of, 183 (*figs.*)
    Half-and-Half, Sun and Shade, 174
        (*figs.*)
    Lath House Pavilion, The, 153, 168–
        69 (*figs.*)
    Modern Classic Rose Arbor, A, 153,
        166–67 (*figs.*)
    New Angle on Trellises, A, 176–77
        (*figs.*)
    Shade for When a Tree Dies, 175
        (*figs.*)
    Shadowplay Shelter, A, 180 (*figs.*)
        details on, 181 (*figs.*)
    Sliding Screen for Privacy, 173
        (*figs.*)
    Sliding Screen With Bamboo, 172
        (*figs.*)
    Terrace Sunbreaker, A, 168 (*figs.*)
    Tree House for the Children, A,
        184–85 (*figs.*)
        floor details of, 184 (*figs.*)
    Trellis Room Beside House, 164
        (*figs.*)
        construction details of, 165 (*figs.*)
    Trellis Room Beside Window, 170
        (*figs.*)
        construction details of, 171 (*figs.*)
Garden shrine, 60 (*figs.*)
Garden steps, 306–26
    cast-in-place steps, with planting
        space at ends of, 309 (*fig.*)
    heavy steps leading from wooden
        deck, 309 (*fig.*)
    kinds of, wide choice in, 307
    materials for, 311–13
    planning the steps, 310
    practical considerations, 313
    recommended dimensions for, 311
        (*figs.*)
    relationship of risers to treads, 307
        (*fig.*)
        broad steps with high risers, 311
            (*fig.*)
        high risers and narrow treads, 311
            (*fig.*)
    slope of the land, and choice of steps,
        314 (*fig.*)
    steps of exposed-aggregate concrete,
        boxed by planks of treated
        wood, 308 (*fig.*)

Garden steps, designs for:
 Balanced But Not Bisymmetrical, 322 (*fig.*)
 Boardwalk Steps, 324 (*fig.*)
 Break Straight Lines with Circles, 316 (*figs.*)
 Corner Steps Are Interesting, 317 (*figs.*)
 Low, Broad Steps Offer a Welcome, 325 (*figs.*)
 risers:
  Brick Risers, 324 (*fig.*)
  Concrete Risers, 323 (*fig.*)
  Rough Stone Risers, 324 (*fig.*)
  Staked Wood Risers, 323 (*fig.*)
 Secret Exit from a Terrace, 315 (*figs.*)
 Steps Can Be the Feature of the Garden, 312, 326 (*fig.*)
 Steps With Opposing Circles, 312, 318 (*figs.*)
  More Opposing Circles, 320 (*figs.*)
 Steps With Variety and Unity, 312, 319 (*figs.*)
 Ties That Bind Walls and Steps, 321 (*figs.*)
 Use Ramp Steps for Long Slopes, 323 (*fig.*)
 Wide Steps With Character, 321 (*figs.*)
 With Steps, Easy Does It, 314 (*figs.*)
 Wood and Stone Complement Each Other, 314 (*figs.*)
 Wooden Planks Team With Bricks, 322 (*fig.*)
 Wooden Steps, 324 (*fig.*)
Garden Walks, paving of, *see* Paving garden walks and terraces
Gazebo, revival of, 153

Hardwoods, 34, 356
Hemlock, 32
Homes for Birds (U.S. Dept. of Interior), 51, 58 *n.*
House plants, 45, 398
 Breakfront for Plants, 205–6 (*figs.*)
 outdoor shelves for summering of, 45
 Show-Off for Potted Plants on the Terrace, 61 (*figs.*)
 Summering of, in garden shelters, 159
 Vacation Spot for House Plants, 62 (*figs.*)

Insect damage, as cause of wood deterioration, 41–42, 44

Joints in masonry walls, 266–72
 choosing type to use, 266–67
 various joints and how to make them, 267–72, 269 (*fig.*)
  concave joint, 270
  convex joint, 271
  flush joint, 268
  raked joints, 270
  skintled brick work, 271–72
  struck joint, 268
  tapped joint, 270–71
  tuck pointing, 271
  V-joint, 270
  weather joint, 268–69

Knots and knotholes in lumber, 35–36

Larch, 32
Little Projects, 45–68
 and the beginner, 4–5, 45
 bird feeders, 45, 48, 55–57 (*figs.*)
 birdhouses, 45, 46–54 (*figs.*)
 bird shelters, 45, 57 (*fig.*)
 coldframes, 46, 67 (*figs.*)
 doors brought up-to-date, 46, 68 (*figs.*)
 garden shrine, 60 (*figs.*)
 nesting shelves for birds, 58
 outdoor shelves for summering of house plants, 45, 61 (*figs.*)
 plant boxes, 46, 64–65 (*figs.*), 149 (*figs.*)
 plant shields, 46, 66 (*figs.*)
 plant stands, 45, 61 (*figs.*), 63 (*figs.*)
 see also the specific project
Locust wood, 41
Lumber, choosing of, 31–40
 acutual versus nominal size of lumber, 36–38
  chart of comparison, 37
  and the "manufactured boards," 38

Lumber, choosing of (*continued*)
    grades of wood, 34
        of pine lumber, 33–34
    hardwoods, 34
    how to save money on lumber, 35–36
    measuring and ordering lumber, 36
    plywood, 38–39
    preserving wood, 33
    softwoods, 32–33
    used lumber or salvaged wood, 39–40

Masonry joints, *see* Joints in masonry walls
Masonry, laying of, 237–65
    footings and first courses, 240–41
    how to lay blocks and hollow tile, 241, 251–59 (*figs.*)
    how to lay bricks, 241–51 (*figs.*)
    mortar formula, 237–39
    painting masonry, 255–63
    stone masonry, 263–65 (*figs.*)
    succeeding courses, 248–49
    walls, cleaning of, as you go along, 249–50
    wetting brick or tile, 239–40
    see also Blocks, building, laying of; Bricks, laying of; clay tile, hollow, laying of; stone masonry
Masonry, materials for, 228–36
    blocks of all kinds, 233–35
    brick, kinds of, 230–33
    clay tile, 236
    see also Blocks, building, kinds of; Bricks, kinds of; Clay tile
Masonry, painting of, *see* Painting of masonry
Masonry, tools for, 222–27, 223 (*fig.*)
    care of tools, 227
    level, 222, 224–25
    mason's brick hammer, 222, 224
    tools for concrete work, 226–27
    trowels, 222, 224
    formula for, 237–39
    using mortar properly, 242–43
Mosaic pebble paving, *see* Pebble mosaic paving

Naphthenates, zinc and copper, 42–43
Nesting shelves for birds, table of dimensions for, 58

Outdoor furniture, 194–220
    colors, choice of, for painting, 196
    criteria for choosing of, 195–96
    "demountable" seats and tables, 195
    low walls used as seats, 195
    making your own versus buying of commercial, 194–95
    measurements and heights, 213
    potting benches, 214–20 (*figs.*)
    special features in build-your-own types, 195–96
Outdoor furniture, designs for:
    Bench for Good Companions, 196 (*figs.*)
    Bench, Portable or Permanent, 197 (*figs.*)
        Around the Corner Version, 198 (*fig.*)
        Building the Legs, 198 (*figs.*)
    Breakfront for Plants, 205–6 (*figs.*)
    Coffee Table of Blocks and Slate, 207 (*figs.*)
    Cupboard With Let-Down Table, 203 (*figs.*)
    Demountable Plywood and Block Furniture, 209 (*figs.*)
        Making the Bench, 210 (*figs.*)
    Demountable Square Table, 210 (*figs.*)
    Foldaway Demountable Table, 200 (*figs.*)
    Foldaway Table on Balustrade or Wall, 199 (*figs.*)
    Folding Loveseat or Bench for the Terrace, 201 (*figs.*)
    Large Coffee Table for Outdoors, 202 (*figs.*)
    Place for Everything, A, 217 (*figs.*)
    Privacy Fence With Let-Down Table, 204 (*figs.*)
    Roll-away Service Cart, 211–12 (*figs.*)
    Slate Coffee Table, 208 (*figs.*)
    Two Informal Tables, 208 (*figs.*)
    You Can Move the Table for Lunch, 207 (*fig.*)

Outdoor shelves for summering of house plants, 45
Show-Off for Potted Plants on the Terrace, 61 (*figs.*)

Painting, dyeing, staining of concrete surfaces, 286–87
Painting of Masonry, 259–63
  cleaning of surfaces to be painted, 262–63
  cleanup of rollers, brushes, or sprayers used in painting, 263
  kinds of paint used for:
    cement-base paints, 261
    oil paints, 260–61
    synthetic-base paints, 261–62
  thinning of paints with water, 262
  use of sealants to preserve natural color, 263
Painting of wood to preserve, 41, 44
Patching of concrete work, cements for:
  epoxy type, 305
  latex type, 304
  vinyl type, 305
Patios, 327
Paving blocks, casting of, 367–75
  curing the blocks, 374
  finishing surface, 370
  forms for casting, 368 (*fig.*)
    making the forms, 369
  mixing the concrete, 372–73
  pouring mixed concrete into forms, 370
  preliminary operations, 369–70
  shapes of, 372
  size of blocks, 371–72
  use of bought blocks with home-mixed concrete, 374–75
  using soil for forms, and casting on location, 370–71
Paving garden walks and terraces, 348–66
  basic methods used in laying paving, 350
  choosing paving materials to be used, 348, 350
  paving set in mortar on concrete base, 353–56
    laying the paving, 355–56
    use of bricks in, 354 (*figs.*)

use of edging, 355 (*figs.*)
paving with precast concrete blocks, 349 (*fig.*)
paving with wooden blocks, 356–57
play walk for children, 349
sand base method of laying, 350–53
  laying the paving, 352
  putting on finishing touches, 352–53
Paving garden walks and terraces, patterns for:
  Basketweave, Large Scale, 361 (*fig.*)
  Basketweave, Small Scale, 361 (*fig.*)
  Concentric Squares, 361 (*fig.*)
  Directional Lines, 358 (*fig.*)
  Directional Pattern, 358 (*fig.*)
  Flowing Curves, 357 (*fig.*)
  How To Combine Bricks, 363 (*figs.*)
  Stepping Stones, 364 (*fig.*)
  Traffic Guides, 360 (*fig.*)
  Use Contrasting Materials, 359 (*fig.*)
  Use Directional Contrast, 359 (*fig.*)
  Use Linear Contrast, 359 (*fig.*)
  Use Wood With Brick, 360 (*fig.*)
  Variety Spices the Terrace, 360 (*fig.*)
  Whirling Squares, 358 (*fig.*)
  Why Not Enter on the Bias?, 362 (*figs.*)
Paving garden walks and terraces, use of materials in:
  Blocks of Various Kinds, 365 (*fig.*)
  Crazy Paving, 366 (*fig.*)
  Combination for Walks, 365 (*fig.*)
  Concrete with Wood, 365 (*fig.*)
  Mosaic of Pebbles, 366 (*fig.*)
  Pebbles or Gravel, 365 (*fig.*)
  Random Squares, 366 (*fig.*)
  Small Boulders, 366 (*fig.*)
  Square Wood Blocks, 366 (*fig.*)
  Tile and Concrete, 365 (*fig.*)
  Tree Trunk Blocks, 366 (*fig.*)
  Use Wood for Decks, 365 (*fig.*)
Paving materials for driveways:
  not permanent, but usually satisfactory:
    crushed rock or stone, 391
    gravel, 391
  not recommended:
    cement and soil, 391
    clay, 392

Paving materials for driveways
    (*continued*)
    home-made blacktop, 392
    oiled soil, 392
    permanent:
      blacktop, 391
      concrete, 387
      paving brick, 387
      stone paving blocks, 387, 391
Paving materials for terraces, 335–42
    blacktop, 339
    bricks, 338
    concrete, 339–40
    concrete blocks, 338
    cut flagstone, 341
    pebble mosaic, 341
    slate, 340
    stone, 340
    tile, 340
    wood blocks, 341
Pebble mosaic paving, 376–82
    historical background of, 376
    how to set pebbles for best effects,
      378–79
      Mosaic of Pebbles, 366(*fig.*)
    ingredients of, 376
    laying concrete base, 379
    making effective use of, 341–42, 377
    making trial run, 377–78
    method of working, 381–82
      finishing the job, 382
    mortar mixture, 380–81
    shapes and color of pebbles, 377
    use of, in bottom of pool, 404, 406
Pentachlorophenol (Penta), 42
Pickets for fences:
    how to assemble, 123(*fig.*)
    how to make, 122(*fig.*) 136(*fig.*)
Pine, 32
    grades of pine lumber, 33–34
Plant boxes:
    Baffle Fence Plant Box, 149(*figs.*)
    Ever-Blooming Plant Box, 46, 64–65
      (*figs.*)
Plant shields, 46
    Versatile Plant Shield, 66(*figs.*)
Plant stands, 45
    Block and Trestle Plant Stand, 63
      (*figs.*)
    Show-Off for Potted Plants on the
      Terrace, 61(*figs.*)

Planters, decorative, 393–401
    beauty possibilities of, 393
    dressing up potted plants with bot-
      tomless box, 396(*fig.*)
    effective placing of, 394
    masonry planters, 394
      seats on masonry planters, 394
    oversized planter of concrete blocks,
      395(*fig.*)
    plant box scaled to make weighty
      accent at corner of terrace,
      395(*fig.*)
    planter on wheels, 396(*fig.*)
    plantings for use in, 394, 397
    portable planters of exterior-type
      plywood, 395(*fig.*)
    potted plants used in, 397
    practical aspects of, 393–94
    practical considerations for, 398
    square plant beds in squares used to
      extend terrace, 396(*fig.*)
Planters, decorative, designs for:
    Angling for Interest, 400(*figs.*)
    Curves give Verve, 398–99 (*figs.*)
    Dress Up Your Doorway, 401(*figs.*)
Plywood, 38–39
    interior grades, 39
    outdoor grades, 38
Pool in the garden, 402–24
    backyard swimming- or wading-
      pool, 406
    casting pools without forms, 410–12
    free-form swimming pool,
      405(*fig.*)
    home garden swimming pool with
      raised curb, 406(*fig.*)
    L-shaped freeform pool, 410(*fig.*)
    overflow and drainage of pool, 408
    placement of pools, 407
    planting the pool, 412–13
      use of aquatic plants in or along-
        side pool, 413–14
    prefabricated swimming pools, 406
    prefabricating your own brook, 404,
      406
    saucer pools, 403, 404, 416(*figs.*)
      419(*figs.*)
    swimming pool that echoes line of
      terrace, 405(*fig.*)
    use of pebble mosaic design in bot-
      tom of, 404, 406

Pool in the garden (*continued*)
wading pools, 407–8
winter care of, 409
making winter pool cover into a feature, 409–10
Pool in the garden, designs for:
Alternate Idea for Jet, 421 (*fig.*)
Minimum Sized Swimming Pool, 424 (*figs.*)
No-Drain Saucer Pool, 419 (*figs.*)
Pool for Natural Plantings, 418 (*figs.*)
Reflecting Pool Beside the House, 420 (*figs.*)
Split Level Pool for the Terrace, 414–15 (*figs.*)
Square Pool Set Into Terrace, 421 (*figs.*)
Triple Pool for a Terrace, 422–23 (*figs.*)
Two Ways To Use Saucer Pools, 416 (*figs.*)
Water Wall for a Terrace, 417 (*figs.*)
Portland cement, 277–78
Potting benches, 214–20 (*figs.*)
construction notes, 220 (*figs.*)
Place for Everything, A, 217 (*figs.*)
Potting Bench With Many Other Uses, 219 (*figs.*)
Pusharound Potting Bench, 218 (*figs.*)
Power tools, and woodworking, 19–30
accident prevention, 14 points for, 28–30
factors determining need of, 19
multi-purpose combination tools, 27
portable 20–22
circular saw, 22
drill attachments, 21
electric drill, 20–21
sander, 21–22
stationary, 23–27
bench or table saw, 23–24
blades for bench or table saw, 24–26
drill press, 27
jigsaw, 26
jointer, 26–27
power lathes, 26
word of caution about, 28

Projects for beginners, *see* Little Projects
Redwood, 32, 33, 41, 159, 288, 356
Rules, for woodworking, 15–16
fabric tapes, 15–16
folding, 15
long steel tapes, 15
roll-up steel tapes, 15

Saws, for woodworking, 12–13
crosscut saws, 12–13
keyhole and compass saws, 17
miter box saw (back saw), 17
power tool saws, 22, 23–24, 26
ripsaw, 17
Sealants, use of, to preserve natural color of masonry, 263
Shelters, *see* Bird shelters; garden shelters
Softwoods, 32–33, 356
Songbirds In Your Garden (Terres), 52
Spruce, 32
Stepping stones, 364, 375
casting blocks for, on location, 370–71
size for the stones, 371
suggested patterns for, 364 (*figs.*)
Steps, *see* garden steps
Stone masonry, 263–65 (*figs.*)
errors to avoid:
letting stones project beyond wall face, 265 (*fig.*)
setting stones in slobbery mortar, 265 (*fig.*)
setting stones upright, 265 (*fig.*)
using too many kinds of stones, 265 (*fig.*)
use of, on change in grade, instead of retaining wall, 265 (*fig.*)
Stone masonry, designs of:
Combining Materials on Walls, 264 (*fig.*)
Coursed Ashlar Masonry, 263 (*fig.*)
Fitted Natural Stone Masonry, 264 (*fig.*)
Natural Stone Masonry, 264 (*fig.*)
Random Ashlar Masonry, 264 (*fig.*)
Rubble Stone Masonry, 263 (*fig.*)
Stone walls, 435–36

Stone walls (*continued*)
dry stone walls, 436–37
Storage Garden Houses, *see* garden shelters

Terraces, 327–47
factors for consideration in choice of, 328
factors for consideration in placing of, 329–31
features for enhancing of, 342
how to lay out, 334–35
making use of side yard for terrace, 342 (*fig.*)
materials to use in paving of, 335–42
multiple terraces, 331
on problem sites, 332–34
poured-concrete terraces, 335 (*fig.*)
replacing old porches with, 335
terrace and retaining wall, contrasting textures in, 337 (*fig.*)
traditional definition of, 327
use of, as play-place for children, 329
use of carport-storage project as terrace, 336 (*fig.*)
use of decks as sitting-out spaces, 343–47 (*figs.*), 356–57, 365 (*fig.*)
Ways to Achieve Privacy on Small Properties, 329 (*fig.*)
with concrete "rug" set inside large area of brushed aggregate concrete paving, 336 (*fig.*)
see also Paving garden walks and terraces
Terres, John K., 52
Tools, *see* Concrete and concrete work, tools for; Masonry, tools for; Power tools, and woodworking; Woodworking tools
Tree house for children, 184–85 (*figs.*)
floor details of, 184 (*fig.*)
Trellis Slats, use of, in building garden shelter, 160–61
Trellises, 69–90
choosing a trellis, 69–72 (*fig.*)
color of, 78–79
finishing of, and after-care, 77–78
making of, 76–77

vines for the trellis, 72–74
annual vines, 74–76, 81
what a trellis is not, 79–80
Trellises, designs for:
Doorside Trellis, The, 89 (*figs.*)
Four-Square Trellis, 82 (*figs.*)
French Provincial Trellis, 85 (*figs.*)
From the Spanish, 83 (*fig.*)
New Angle on Trellises, A, 176–77 (*figs.*)
Simple Modern Doorside Trellis, 86 (*figs.*)
Squares on Circles, 83 (*figs.*)
Strictly Vertical, 84 (*figs.*)
Stripes and Rectangles, 84 (*figs.*)
Tapering Trellis, 129 (*figs.*)
Trellis Fence Enders, 90 (*figs.*)
Trellis-Topped View-Breaker Fence, 87–88 (*figs.*)
Tuck pointing, 271

Vines for the trellis, 72–74
annual vines, 74–76, 81

Walks, paving of, *see* Paving garden walks and terraces
Walls, 425–58
balustrades, 435, 440–41 (*figs.*)
concrete-block grilles, 428 (*fig.*)
friendly walls, 430, 432–33
function of, when a man's home was his castle, 425
open block walls, 430–31 (*figs.*)
patterned wall with 3-dimensional texture, 432 (*fig.*)
planting surroundings of, 437
poured concrete, 300–302
capping, 302
expansion joints, 301
reinforcing rods, 301–2
practical walls, 426, 429
privacy walls, 427 (*figs.*)
retaining walls, providing for drainage behind, 437
shadows achieved by diagonal pattern, 431 (*fig.*)
shaped blocks, decorative effects achieved with, 432 (*fig.*)
split concrete-block retaining wall, 428 (*fig.*)
stone walls, 435–36

Walls (*continued*)
dry stone walls, 436–37
strong curves in, 429(*fig.*)
tall walls that need piers for strength, 439(*figs.*)
thickness of, 434–35
things to check on before building, 433
to retain trees, 303–4
today's walls, versatility of, 425–26
Walls, patterns for:
Bricks and Blocks, 453(*figs.*)
Checkerboard and Curves, 447 (*figs.*)
Checkerboard Patterned Ashlar— Two versions, 449(*figs.*)
Cross and Square Pattern, 444 (*figs.*)
Curving Sugarloaf Wall, 454(*fig.*)
Double Basketweave, 450(*figs.*)
Foursquare and on the Level, 448 (*fig.*)
Grand Opening, 446(*figs.*)
How To Hold Up a Bank, 457(*figs.*)
Gentle Slope Holder, 457(*figs.*)
Mainliner Wall, 454(*fig.*)
Open and Shut Case, 453 (*figs.*)
Open Walls and Ventilation, 438 (*fig.*)
Peek-A-Boo Wall, 452 (*figs.*)
Plants in Dry Stone Walls, 456 (*figs.*)
Regular Coursed Ashlar, 451 (*figs.*)
Retaining Wall Retains a Tree, 458 (*fig.*)
Round the Mulberry Bush, 443 (*figs.*)
Solid Wall With Textural Interest, 438 (*fig.*)
Spaced Course Ashlar, 451 (*figs.*)
Spare That Tree—You'll Need It, 458 (*fig.*)
Staggered Cross Pattern, 445 (*figs.*)
Straight and Narrow Sugarloaf Wall, 455 (*fig.*)
To Lengthen, Use Lines, 452 (*figs.*)
Tree Under a Terrace, 442 (*figs.*)
Up and Down Combination, 455 (*fig.*)
Whirling Square Pattern, 450 (*figs.*)
Zig and Zag for Excitement, 448 (*fig.*)
Wood:
deterioration of, from decay and insect damage, 41–42, 44
grades of, 33–34
painting of, to preserve, 41, 44
proper finish of, 41
salvaged, 39
see also Lumber, choosing of
Wood preservatives, 33, 41–44, 159, 161, 288, 356
creosote, 42, 43
some modern preservatives:
copper sulfate, 43
Pentachlorophenol (Penta), 42
zinc and copper naphthenates, 42–43
value of, versus cost of, 44
word of caution on, 43
Woodworking tools, 10–18, 11 (*fig.*)
brace and bit, 16
chisels, 15
drawknife, 17
drills, hand and push, 18
gauges, 15
hammers, claw or nailing, 12
levels, 16
miter box and back saw, 17
planes (smoothing, jack, block), 14–15
pliers, 16
ripsaw, 17
rules, 15–16
saws, 12–13, 17
screwdrivers, 13–14
spoke shave, 17
try-squares, 14
see also Power tools, and woodworking

Zinc naphthenate, 42–43